TOPICS IN CARBON-13 NMR SPECTROSCOPY

TOPICS IN CARBON-13 NMR SPECTROSCOPY

Edited by

GEORGE C. LEVY

Department of Chemistry
The Florida State University
Tallahasee, Florida

Volume 3

A Wiley-Interscience Publication
JOHN WILEY & SONS, New York · Chichester · Brisbane · Toronto

Library of Congress Catalog Card Number 74–10529
ISBN 0-471-02873-8

Printed in the United States of America

10 9 8 7 6 5 4 3 2 1

CONTRIBUTORS

FRANK A. L. ANET
Department of Chemistry, University of California, Los Angeles, California

DAVID E. AXELSON
DuPont of Canada, Ltd., Research Centre, Research Division, Kingston, Ontario, Canada

DAVID L. DALRYMPLE
Nicolet Technology Corporation, Mountain View, California

ERNEST L. ELIEL
W. R. Kenan, Jr., Laboratories, Department of Chemistry, University of North Carolina, Chapel Hill, North Carolina

DAVID M. GRANT
Department of Chemistry, University of Utah, Salt Lake City, Utah

HOWARD HILL
Varian Associates, Palo Alto, California

DAVID I. HOULT
Biomedical Engineering and Instrumentation Branch, National Institutes of Health, Bethesda, Maryland

LeROY F. JOHNSON
Nicolet Technology Corporation, Mountain View, California

GEORGE C. LEVY
Department of Chemistry, The Florida State University, Tallahassee, Florida

W. B. MONIZ
Chemistry Division, Naval Research Laboratory, Washington, D.C.

K. MICHAL PIETRUSIEWICZ
Department of Chemistry, W. R. Kenan, Jr., Laboratories, University of North Carolina, Chapel Hill, North Carolina. On leave of absence from Center of Molecular and Macromolecular Studies, Polish Academy of Sciences, Łódź, Poland

C. F. PORANSKI, JR.
Chemistry Division, Naval Research Laboratory, Washington, D.C.

JACOB SCHAEFER
Corporate Research Laboratories, Monsanto Company, St. Louis, Missouri

JAMES N. SHOOLERY
Varian Associates, Palo Alto, California

S. A. SOJKA
Research Center, Hooker Chemicals and Plastics Corporation, Niagara Falls, New York

E. O. STEJSKAL
Corporate Research Laboratories, Monsanto Company, St. Louis, Missouri

DAN TERPSTRA
Department of Chemistry, The Florida State University, Tallahassee, Florida

D. A. TORCHIA
Laboratory of Biochemistry, National Institute of Dental Research, National Institutes of Health, Bethesda, Maryland

D. L. VANDERHART
Polymer Division, National Bureau of Standards, Washington, D.C.

DAVID A. WRIGHT
Department of Chemistry, The University of Illinois, Champaign-Urbana, Illinois

A. P. ZENS
Department of Chemistry, Tulane University, New Orleans, Louisiana

PREFACE

The phenomenal growth of ^{13}C nmr spectroscopy has become entirely unmanageable for the single author. Thus the multiauthor series *Topics in Carbon-13 NMR Spectroscopy* is increasingly valuable for documentation of current developments. *Topics in Carbon-13 NMR Spectroscopy* is intended to bridge the gap between current research literature and available ^{13}C nmr text and reviews. The third volume of *Topics* chronicles recent advances in ^{13}C nmr theory and methodology. The six chapters of this volume cover a wide range of interests from molecular biology to polymer physics, from molecular dynamics to organic structure analysis.

Chapter 1 represents a unique contribution to the literature: a multi-section, multiauthored discussion of modern experimental techniques used in ^{13}C nmr spectroscopy. The eight sections of Chapter 1 were selected to provide an overview of recent and anticipated advances in experimental methods. Two dimensional Fourier transform ^{13}C nmr and selective pulse excitation methods are discussed, respectively, by Dan Terpstra and Howard Hill. Frank Anet discusses variable temperature methods, while David Hoult, Jim Shoolery, David Grant, and A. P. Zens cover various aspects of sensitivity improvement in ^{13}C spectroscopy. One of the authors of current Nicolet nmr software, David Dalrymple, summarizes FT nmr software characteristics. LeRoy Johnson details several useful ^{13}C nmr spin-decoupling techniques.

Parallel to the development of ^{13}C spectroscopy has been an accelerating interest in varied applications for ^{13}C spin relaxation measurements. In Chapter 2 David Wright, David Axelson, and George Levy discuss physical chemical applications of ^{13}C relaxation measurements, emphasizing studies of liquid molecular dynamics. Chapter 3, by Ernest Eliel and K. Michael Pietrusiewicz, presents for the first time a comprehensive discussion of ^{13}C nmr spectra for saturated heterocyclic compounds.

vii

Chapters 4 and 5 cover two aspects of high resolution ^{13}C nmr in solids. Jacob Schaefer and Ed Stejskal give results for a number of solid polymer samples, showing the wide applicability of the new technology. Dennis Torchia and Dave Vanderhart use high power double resonance ^{13}C techniques to probe the structure and function of important biopolymer systems.

In the final chapter William Moniz, Chester Poranski, and Stanley Sojka summarize ^{13}C CIDNP opportunities for studies of organic radical reactions.

I thank all of the contributors to this volume for producing manuscripts that mirror both the opportunities and challenges of ^{13}C nmr applications. I also thank Mrs. Debby Tyson and the members of my research group for editorial assistance.

George C. Levy

October 1978
Tallahassee, Florida

CONTENTS

6. ^{13}C CIDNP as a Mechanistic and Kinetic Probe
   W. B. MONIZ, C. F. PORANSKI, JR., and S. A. SOJKA

TOPICS IN CARBON-13 NMR SPECTROSCOPY

1 EXPERIMENTAL TECHNIQUES IN ^{13}C SPECTROSCOPY

This chapter may be unique. It is not actually a single chapter; instead it is an assembly of seven short chapters, each describing an experimental facet of ^{13}C nmr spectroscopy. The uniqueness arises from the fact that these sections cover the practical side of ^{13}C spectroscopy, often giving details of methodology that are not in print elsewhere. Each contribution stands as a separate entity, in spite of the fact that there is some overlap in subject areas.

Edited multiauthor chapters on experimental techniques will be regularly incorporated in future volumes of *Topics in Carbon-13 NMR Spectroscopy*.

SECTION

 SPIN-DECOUPLING METHODS
IN ^{13}C NMR STUDIES

LeROY F. JOHNSON

I. INTRODUCTION

The increasingly successful application of ^{13}C nmr spectroscopy to molecular structure determination during the past decade has been largely dependent on development of proton double-resonance methods. An early paper by J. D. Roberts[1] reported tremendous simplification of the spectrum of cholesterol generated with the use of noise-modulated proton spin decoupling and opened the door to determinations of ^{13}C chemical shifts in structurally related compounds.[2]

It is the purpose of this section to review application of straightforward spin-decoupling methods in ^{13}C nmr studies. Particular emphasis is placed

It is the purpose of this section to review application of straightforward spin-decoupling methods in ^{13}C nmr studies. Particular emphasis is placed on detailed use of successive selective decoupling. No discussion of spin decoupling methods in selective excitation sequences or two-dimensional FT studies is presented, since these are covered in a separate section.

II. BROADBAND DECOUPLING

Certainly the most widely used spin-decoupling technique involves simply broadband decoupling of all proton resonances to reduce the ^{13}C spectrum (of most organic compounds) to a set of sharp peaks each directly reflecting a ^{13}C chemical shift. The requirements for broadband decoupling are (1) a sufficiently strong decoupling field strength and (2) method of modulation that will "spread" the decoupling field over the range of proton chemical shifts. Early broadband decoupling results reported by

Ernst[3] were obtained by pseudorandom noise modulation using a multistage binary shift register. This approach was later incorporated in commercial instruments.

An alternate modulation approach described by Grutzner and Santini[4] was found to be more effective when the strength of the decoupling field, $\gamma H_2/2\pi$ is larger than the total spread of proton chemical shifts Δf, which is the typical situation in most ^{13}C instrumentation. In their approach, the decoupling frequency is phase modulated with a 50% duty cycle, 100-Hz square wave. Residual broadening of decoupled off-resonance ^{13}C peaks is significantly reduced using this method in comparison to the pseudorandom noise modulation method. Since the square-wave modulation method is easy to incorporate in systems originally designed for pseudorandom modulation and less expensive to provide as the only broadband decoupling scheme in new systems, it is now being widely used in broadband proton decoupling.

Satisfying the requirement of a sufficiently strong decoupling field strength requires use of an rf power amplifier that is capable of supplying several watts of rf power to the decoupler coil in the probe. Obviously, the limitation here is the ability to remove heat from the probe and the sample with a reasonable air flow. Operating the decoupler at a different level during data acquisition from that used during a delay time can improve the situation.[5] If high-power decoupling, say 25 W, during data acquisition adequately reduces residual broadening from off-resonance decoupling, then lower power decoupling, say 5 W, during a delay period of four times the acquisition period, could serve to maintain the nuclear overhouser enhancement. In this case the average decoupling power is 9 W, which can be much easier to deal with than in the case of a continuous level of 25 W.

III.   OFF-RESONANCE BROADBAND DECOUPLING

When observation of signals from non-proton-bearing carbon groups is desired with minimum interference by signals from protonated carbon groups, the technique of off-resonance broadband decoupling can be useful. Incomplete broadband decoupling generates a residual broadening that is proportioned to the square of the responsible ^{13}C—^{1}H spin-coupling constant.[3] Since one-bond spin-coupling constants between carbon atoms and their directly bonded protons is generally at least 10 times that for two and three bonds, residual broadening induced by off-resonance decoupling is at least 100 times as severe for signals from protonated carbon groups compared to that for any non-proton-bearing carbon groups.

A particularly nice example of off-resonace broadband decoupling was shown by Oldefield, Norton, and Allerhand[6] in their study of protein solutions. In Figure 1.I.1 we see the unsaturated carbon region of the ^{13}C Fourier-transform nmr spectrum of 8.8-mM sperm whale cyanoferrimyoglobin in H_2O at pH 6.8. In spectrum *a*, on-resonance broadband

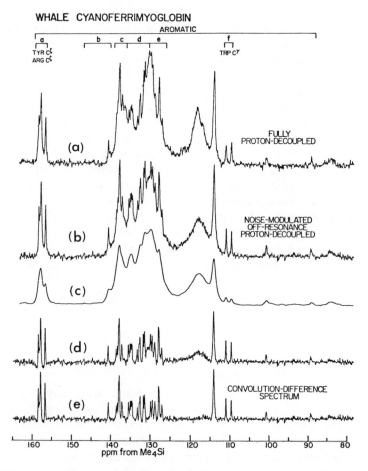

Figure 1.I.1
15.1 MHz ^{13}C spectra of whale cyanoferrimyoglobin. (a) Broadband decoupling; (b) off-resonance broadband decoupling; (c) same data as in (b), but with a stronger exponential multiplication of the fid; (d) difference spectrum equal to (b) −0.9 (c); (e) baseline adjustment of spectrum (d). From Ref. 6.

decoupling was used to generate the narrowest possible signals. In spectrum *b*, the spin decoupler frequency was offset several kilohertz with a resulting large additional broadening of the "methine" aromatic carbons and a negligibly small increase in the nonprotonated aromatic carbons. Spectrum *b* was further treated by using the convolution difference technique.[7,8] Spectrum *c* is generated from the same set of data as used for spectrum *b*, except that an additional 9 Hz of line broadening was produced by an appropriate exponential multiplication of the free induction decay before Fourier transformation. This greatly broadens the signals from the nonprotonated carbons while only slightly modifying the already broad methine aromatic carbon signals. By subtracting an appropriately scaled spectrum *c* from spectrum *b*, the difference spectrum *d* is generated. Here, signals from the broad methine aromatic carbons are greatly reduced, leaving mainly peaks from the nonprotonated carbons. Finally, spectrum *e* was derived from spectrum *d* by means of digital baseline adjustment in which the residual broad components were treated as baseline imperfection. Thus, the combination of off-resonance broadband decoupling and convolution difference data processing has, in this case, resulted in a spectrum that is clearly more suitable for studying the narrow peak components.

IV. OFF-RESONANCE COHERENT DECOUPLING

The application of a coherent rf field of appropriate amplitude near the frequency of the proton nmr spectrum serves to reduce the magnitude of the scalar coupling between the ^{13}C and ^{1}H nuclei.[9,3] The resulting residual splitting is given to a good approximation by

$$J_r = \sqrt{J^2 \left[\frac{(\gamma H_2/2\pi)^2}{(\Delta f)^2} + 1 \right]^{-1}}$$

Here $\gamma H_2/2\pi$ is the strength of the decoupling field expressed in units of hertz, Δf is the difference between the decoupler frequency and the resonance frequency for the proton signal in question, J is the unperturbed ^{13}C—^{1}H spin coupling constant, and J_r is the reduced or residual splitting produced by the off-resonance decoupling.

The technique of off-resonance coherent decoupling has been widely used as an aid in interpretation of ^{13}C spectra. Figure 1.I.2 serves to illustrate the point. Part (*a*) shows the 12 aliphatic carbon signals in the 25.1-MHz broadband decoupled spectrum of estrone methyl ether in CDCl$_3$. The spectrum in part (*b*) was run with off-resonance coherent

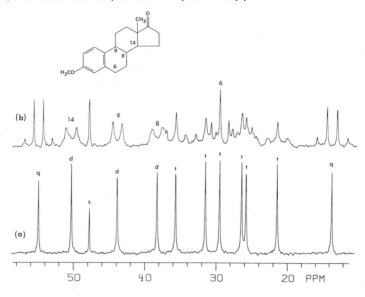

Figure 1.I.2
25.1 MHz ¹³C spectra of estrone methyl ether in CDCl₃. (a) Broadband decoupling; (b) off-resonance coherent decoupling.

decoupling—the decoupler frequency was set 1400 Hz upfrequency from the proton TMS signal. The strength of the decoupler field was about 2700 Hz, and this resulted in residual splittings of approximately 50 Hz. By plotting one spectrum directly over the other, it is easy to pick out the two quartets, six triplets, three doublets, and one singlet that are expected from these aliphatic carbon groups.

Additional information can be obtained from the line widths of the reduced multiplets. Second-order coupling between proton groups can reflect itself in additional splittings, and hence broader lines, in the ¹³C signal from a carbon nucleus that is directly bonded to one of the proton groups.[10] Thus the CH doublet signals in Figure 1.I.2b from C-14 and C-8 are broader than the one from C-9. The proton on C-9 is a benzylic proton that has a chemical shift significantly removed from those on adjacent carbon atoms. Protons on C-8 and C-14 have similar chemical shifts to those on adjacent carbon atoms. Thus, strongly coupled groups in the proton spectra are reflected in patterns observed in the off-resonance decoupled ¹³C spectrum. The same effect is seen in the sharp triplet pattern from the benzylic C-6 methylene compared to the broader patterns from the other methylene groups.

V. SUCCESSIVE SELECTIVE DECOUPLING

Since the magnitude of the residual splitting generated by off-resonance coherent decoupling is dependent on the degree of decoupler frequency offset, it is possible to correlate residual splittings in the ^{13}C spectrum with chemical shifts in the proton spectrum. This can be done by calculation[11-13] or graphical solution.[14,15] A frequently very informative set of ^{13}C spectra can be generated by stepping in the decoupler frequency through the proton spectrum using moderately low-power coherent decoupling. It is possible to acquire such a set of spectra under computer control in which the operator enters a list of decoupler frequencies. Accumulation of data at each decoupler frequency continues until a present number of acquisitions is reached. Each accumulated free-induction decay is stored on disk for subsequent Fourier transformation and plotting. An example of the technique of successive selective decoupling is illustrated in the next series of figures.

Shown in Figure 1.I.3 is the 150-MHz proton spectrum of strychnine in CDCl$_3$ run in the Fourier-transform mode using an NT-150 spectrometer. A few assignments are immediately possible. Because of its position relative to the C-10 carbonyl group, the proton on C-4 produces the signal at highest frequency, 8.1 ppm from TMS. The other three aromatic

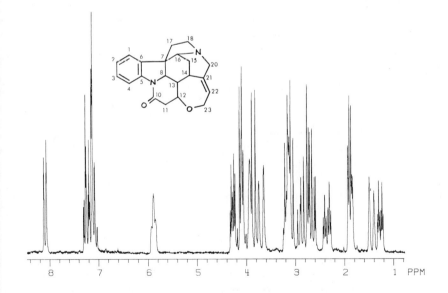

Figure 1 .I.3
150 MHz ^{1}H spectrum of strychnine in CDCl$_3$.

protons produce a complex set of signals between 7.0 and 7.3 ppm. The olefinic proton on C-22 produces the ill-resolved triplet at 5.9 ppm. The proton on C-12, because of oxygen substitution, is probably responsible for the six-line pattern centered at 4.3 ppm. The rest of the spectrum is quite complex and not readily assignable. It is interesting to note that one proton signal, appearing as a doublet of triplets, occurs at quite low frequency, 1.3 ppm from TMS. This could be assigned to a methine proton or one-half of a methylene group.

A series of ^{13}C spectra was accumulated using coherent decoupling with a $\gamma H_2/2\pi$ amplitude of 830 Hz. The decoupler frequency was stepped under computer control in 75 Hz increments beginning at a frequency corresponding to irradiation 75 Hz downfrequency (-0.5 ppm) from TMS and proceeding to 1350 Hz upfrequency (9.0 ppm) from TMS. Four hundred accumulations were made at each decoupler frequency and the total time for accumulation of all 20 sets of data was 2.3 hr. An undecoupled spectrum and a broadband decoupled spectrum were also obtained.

Figure 1.I.4 shows a composite plot of data from the five proton-bearing sp^2 carbons. At the bottom is the undecoupled spectrum, at the top is the broadband decoupled spectrum, and between these is the set of 20 selectively decoupled spectra. Numbers along the side refer to proton irradiation frequencies in parts per million from the proton TMS signal. It is easy to see that the ^{13}C signal near 116 ppm collapses to a singlet when the proton irradiation frequency is approximately 8 ppm from TMS; therefore, this ^{13}C peak is assigned to C-4. Irradiation at 8.5 ppm and 7.5 ppm produces a doublet for C-4, but slightly closer spaced at 8.5 ppm, since the proton chemical shift is at 8.1 ppm. Likewise, the ^{13}C signal near 127 ppm collapses to a singlet when the proton irradiation is at 6 ppm. It is a narrower doublet when irradiation is at 5.5 ppm relative to 6.5 ppm, because the proton chemical shift is at 5.9 ppm. The three remaining ^{13}C peaks all collapse when the proton irradiation is at 7.0 ppm. Irradiation at 7.5 and 6.5 ppm produces ill-defined ^{13}C patterns owing to the strong mixing in the proton spectrum from the associated three protons.[10]

Figure 1.I.5 shows a composite plot of the same type of data for the three substituted sp^2 carbons and the one carbonyl group. Chemical shift reasoning allows assignment of the peak at 169 ppm to the carbonyl group. It is interesting to note that this peak is sharpest when the decoupler frequency is in the 2.5–3.5 ppm range. This is a consequence of selective decoupling of the two- and three-bond couplings between the carbonyl carbon and neighboring aliphatic protons. Note the similar effect in the ^{13}C peak at 140 ppm, which can now be assigned to the C-22 olefinic carbon. The peaks at 132.5 and 140 ppm are sharpest when the

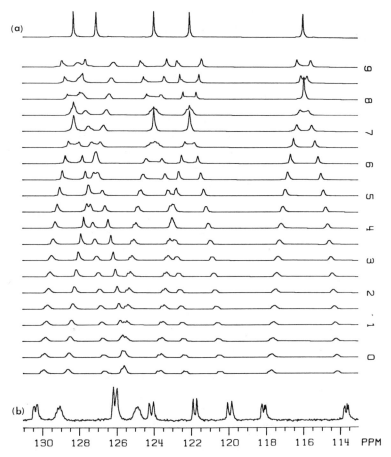

Figure 1.I.4
37.7 MHz ¹³C spectra of strychnine in CDCl₃. (a) Broadband decoupled; (b) undecoupled; (others) selective coherent decoupling at the proton frequency indicated on the right side, in ppm from TMS.

decoupler frequency is in the 6–8 ppm range, a consequence of decoupling long-range spin coupling to other aromatic protons. On chemical shift arguments, the peak at 132.5 ppm is assigned to C-6, and the peak at 141 ppm is assigned to C-5.

Figure 1.I.6 shows data from one ¹³C signal, which is quite near the three signals from the solvent. Reading through the partial solvent interference, it is possible to see that the ¹³C peak is about as well collapsed when the decoupler frequency is at 4 ppm as it is when it is at 4.5 ppm. Thus this peak, which, from its ¹³C chemical shift, can be

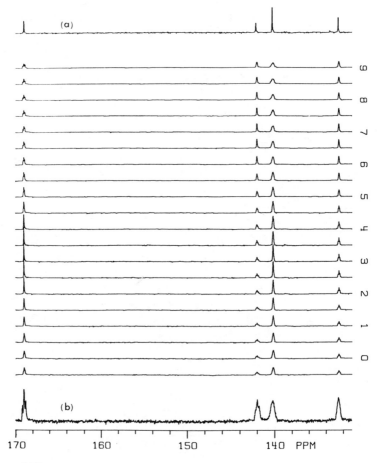

Figure 1.I.5
37.7 MHz ^{13}C spectra of strychnine in CDCl$_3$. (a) Broadband decoupled; (b) undecoupled; (others) selective coherent decoupling at the proton frequency indicated on the right side, in ppm from TMS.

assigned to C-12, provides confirmation of the tentative 4.3 ppm proton assignment made earlier.

Figure 1.I.7 shows the 69–38 ppm region of the ^{13}C spectrum, which contains signals from seven of the aliphatic carbons. From the off-resonance triplet patterns, it is clear that the signal at 64.3 ppm arises from a CH$_2$ group. Best collapse occurs when the decoupler frequency is near (slightly larger than) 4 ppm. Proton chemical shifts in this region would be expected for the two protons on C-23, thus establishing this ^{13}C

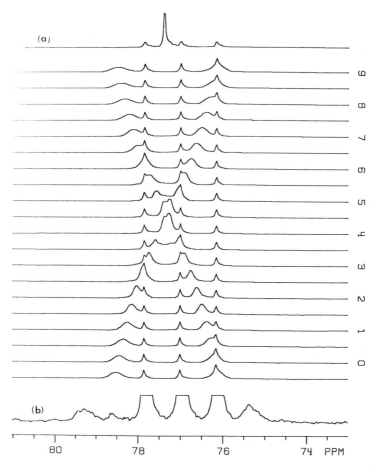

Figure 1.I.6
37.7 MHz ^{13}C spectra of strychnine in CDCl$_3$. (a) Broadband decoupled; (b) undecoupled; (others) selective coherent decoupling at the proton frequency indicated on the right side, in ppm from TMS.

assignment. Two, almost overlapped, ^{13}C signals near 60 ppm arise from CH groups, both of which are associated with proton chemical shifts near 4 ppm. These proton shifts would be expected for the two N—CH groups that are present and suggest corresponding assignments for the ^{13}C spectrum. The ^{13}C signal near 52.5 ppm is not clearly defined in the off-resonance data. Assignment for this signal will be discussed later. The peak at 51.7 ppm shows no multiplicities in the off-resonance data and is therefore assigned to the one quaternary aliphatic carbon, C-7. The signal

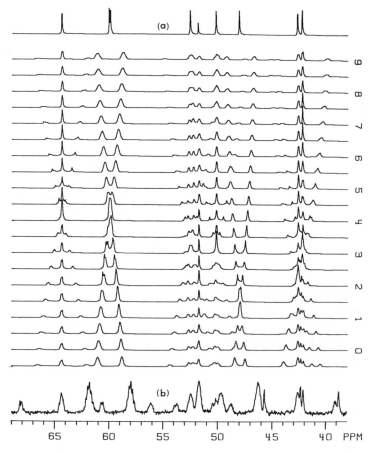

Figure 1.I.7
37.7 MHz ^{13}C spectra of strychnine in CDCl$_3$. (a) Broadband decoupled; (b) undecoupled; (others) selective coherent decoupling at the proton frequency indicated on the right side, in ppm from TMS.

near 50 ppm is obviously from a CH$_2$ group whose proton chemical shifts are near 3 ppm. Candidates for this ^{13}C assignment might be C-11, C-18, or C-20. The signal near 48 ppm clearly arises from a CH group whose proton chemical shift is near 1.25 ppm. The assignment here turns out to be C-13, and the unusual proton chemical shift arises from the C-13 proton being held over the plane of the benzene ring in the conformation of the strychnine molecule. Finally, the two remaining ^{13}C signals between 42 and 43 ppm arise from CH$_2$ groups, but it is difficult to be exact about associated proton chemical shifts.

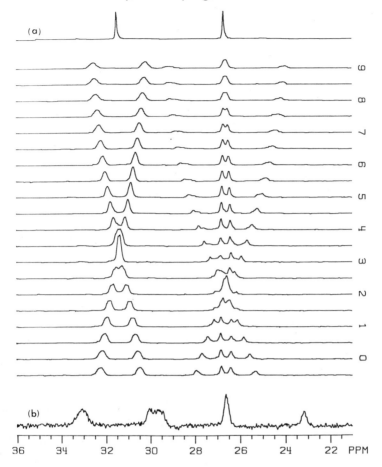

Figure 1.I.8
37.7 MHz ^{13}C spectra of strychnine in CDCl$_3$. (a) Broadband decoupled; (b) undecoupled; (others) selective coherent decoupling at the proton frequency indicated on the right side, in ppm from TMS.

The last remaining section of the ^{13}C spectrum is shown in Figure 1.I.8. Here it is clear that the signal at 31.3 ppm arises from a CH group whose proton chemical shift is about 3.2 ppm. This can be assigned to the only remaining methine carbon, C-14. The signal at 26.6 ppm is assignable to a CH$_2$ group. A lack of clearly defined triplet patterns in the off-resonance spectra is attributable to nonequivalence of the chemical shifts of the two CH$_2$ protons.[10] It is only evident that their average chemical shift is somewhere around 2 ppm.

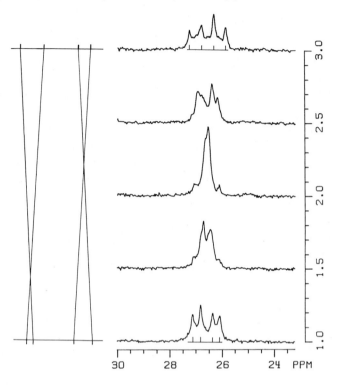

Figure 1.I.9
37.7 MHz ^{13}C data of the C-15 methylene group in the selectively decoupled strychnine spectra. On the leftside-by crossing major and minor splittings, the approximate chemical shifts of the two nonequivalent protons can be determined.

Figure 1.I.9 shows a plot expansion of some of the off-resonance data from Figure 1.I.8. Here, it can be seen that slightly away from the decoupler frequency for best collapse of the CH$_2$ pattern the multiplets appear as pairs of doublets. It is possible to measure major and minor doublet spacings in the patterns generated when the decoupler frequency is at 3 and 1 ppm. These doublet spacings have been indicated at the left side of the figure. By crossing the major splitting at the top with the minor splitting at the bottom, and vice versa it is possible to approximate the chemical shifts of the two nonequivalent protons. The determined values of 1.4 and 2.2 ppm correlate with patterns seen in Figure 1.I.3 at 1.45 and 2.35 ppm. This CH$_2$ group is assigned to C-15 and again the conformation of the molecule is held responsible for an unusual chemical shift of the one of these protons that is more closely over the plane of the benzene ring.

As seen in Figure 1.I.7, the signal at 52.5 ppm shows the same kind of characteristic as the one just discussed. This signal is assignable to the C-20 methylene group. By using the same graphic analysis method, the nonequivalent proton chemical shifts turn out to be about 2.7 and 3.5 ppm. While the 2.7 ppm region in Figure 1.I.3 is very busy, it is possible to see evidence of a doublet pattern showing geminal proton splitting centered at 3.7 ppm.

Thus, the use of successive selective decoupling has not only provided assignments for the ^{13}C spectrum but has been of value in helping to assign features in the proton spectrum as well. The use of a $\gamma H_2/2\pi$ value of approximately 800 Hz was chosen to be large enough to provide relatively well-defined patterns when the decoupler frequency was off-resonance and yet small enough to show a selectivity of about 0.1 ppm in determining proton chemical shifts.

REFERENCES

1. H. J. Reich, M. Jautelat, M. T. Messe, F. J. Weigart, and J. D. Roberts, *J. Am. Chem. Soc.* **91,** 7445 (1969).

2. F. J. Weigert, M. Fautelat, and J. D. Roberts, *Proc. Nat. Acad. Sci. U.S.* **60,** 1152 (1968).

3. R. R. Ernst, *J. Chem. Phys.* **45,** 3845 (1966).

4. J. B. Grutzner and R. E. Santini, *J. Mag. Res.* **19,** 173 (1975).

5. G. C. Levy, I. R. Peat, R. Rosanske, and S. Parks *J. Mag. Res.* **18,** 205 (1975).

6. E. Oldefield, R. S. Norton, and A. Allerhand, *J. Biol. Chem.* **250,** 6381 (1975).

7. I. D. Campbell, C. M. Dobson, R. J. P. Williams, and A. V. Xavier, *Ann. N. Y. Acad. Sci.* **222,** 163 (1973).

8. I. D. Campbell, C. M. Dobson, R. J. P. Williams, and A. V. Xavier, *J. Magn. Res.* **11,** 172 (1973).

9. A. L. Bloom and J. N. Shoolery, *Phys. Rev.* **97,** 1261 (1955).

10. E. W. Hagaman, *Org. Mag. Res.* **8,** 389 (1976).

11. R. A. Archer, R. D. G. Cooper, P. V. Demarco, and L. F. Johnson, *Chem. Soc.* **93,** 273 (1971).

12. M. Tanabe, T. Hamasaki, D. Thomas, and L. F. Johnson, *J. Am. Chem. Soc.* **93,** 273 (1971).
13. K. G. R. Pachler, *J. Mag. Res.* **7,** 442 (1972).
14. R. Freeman and H. D. W. Hill, *J. Chem. Phys.* **54,** 3367 (1971).
15. B. Birdsall, N. J. M. Birdsall, and J. Feeney, *Chem. Commun.* 316 (1972).

SECTION

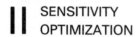 SENSITIVITY
OPTIMIZATION

D. I. HOULT

I. INTRODUCTION

The sensitivity of any nmr experiment is a multiparametered variable where the dominant factor is cost. This determines the strength and homogeneous volume of the main magnetic field $\mathbf{B}_0$ with which one works and often the number of spins that are available. Given that in most cases the would-be experimenter has a magnet, a spectrometer, and a limited number of dollars, the next few pages explain the origins of signal and noise in order that the instrumental factors involved in improving the ratio of the two may be understood.

II. SIGNAL AND NOISE

In any experiment involving electrical induction there is a reciprocal relationship between magnetic field and emf. Thus, if a conductor carrying unit current produces a field $\mathbf{B}_1$, at some point P, a rotating magnetic moment **m** placed at point P will induce in that conductor an emf proportional to $\mathbf{B}_1$. This statement may be quantified[1] as

$$\xi = -\frac{\partial}{\partial t}\{\mathbf{B}_1 \cdot \mathbf{m}\} \tag{1}$$

where ξ is the induced emf. In the case of nmr, the magnetic moment **m** is a nuclear magnetization that precesses about the main field $\mathbf{B}_0$ and for simplicity we consider the results of a 90° pulse that produces a magnetization of $\mathbf{M}_0$ in the xy plane (we assume $\mathbf{B}_0$ is in the z direction). Thus, if we have a sample close to our conductor, the total induced emf is given by

$$\xi = -\int_{\text{sample}} \frac{\partial}{\partial t}\{\mathbf{B}_1 \cdot \mathbf{M}_0\}\, dV_s \tag{2}$$

where dV_s is an elementary sample volume. Clearly, if $\mathbf{M}_0$ is the same over the whole sample (a homogeneous transmitting field), then $\mathbf{B}_1$ should be invariant in direction over the sample *and* in the xy plane, if ξ is to be maximized. This in turn implies that $\mathbf{B}_1$ must be homogeneous over the sample. If this is the case, we may rewrite Eq. 2 as

$$\xi = K\omega_0 M_0 V_s (B_1)_{xy} \cos \omega_0 t \tag{3}$$

where ω_0 is the angular Larmor frequency, $(B_1)_{xy}$ is the component of the B_1 field, perpendicular to $\mathbf{B}_0$, produced by unit current in the conductor, V_s is the sample volume, and K is a correction factor that allows for the B_1 field inhomogeneity. We may note that as M_0 is proportional[2] to B_0 and hence the Larmor frequency, $\xi \propto \omega_0^2$. The calculation of $(B_1)_{xy}$ is feasible for most shapes of coil, and so there is no difficulty in determining the signal induced by the nuclear magnetization in a receiving coil; the problem is to calculate the noise.

Electrical noise is a manifestation of the Brownian motion of the electrons within the conductor and, being a thermal effect, is therefore dependent upon the temperature of the conductor and the strength of the interaction with the lattice. A measure of the latter is the electrical resistance of the conductor, and this in turn is temperature dependent. In a well-designed spectrometer, almost all the noise output should originate from the receiving coil, and a well-known formula for that noise is

$$N = [4kT_c r\, \Delta\nu]^{1/2} \tag{4}$$

where N is the root-mean-square emf, k is the Boltzmann constant, T_c is the temperature of the conductor, r is its resistance, and $\Delta\nu$ is the bandwidth of the receiving system. The unknown in Eq. 4 is the resistance—it cannot be calculated accurately. A full understanding of

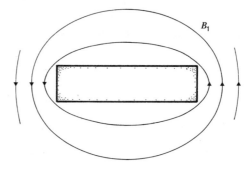

Figure 1.II.1
The current distribution associated with radio frequency skin effect.

why this is so requires considerable insight into the application of Maxwell's equations,[3] but a crude comprehension may be obtained with the realization that for nmr purposes, at frequencies above about 1 MHz, where there is no alternating magnetic field, there is no alternating current. The implication of this statement is demonstrated in Figure 1.II.1. The magnetic field due to alternating current in a conductor of rectangular cross section is shown, along with the interdependent current distribution. Maxwell's equations predict that the current flows in a skin on the surface of the conductor and the rectangular cross section tends to concentrate the current in the corners as shown. A calculation of the effective resistance in even this simple case is difficult: where several conductors are in close proximity the interactions generated render the problem well nigh impossible. In general one can only say that the *skin* and *proximity* effects tend to increase the resistance from the calculated value by a factor of between 1 and 10, typically by a factor of 3. For a very long, straight, cylindrical conductor of length l and circumference p, the resistance at frequency ω_0 is given by

$$r = \frac{l\rho}{p\delta} \tag{5}$$

where ρ is the resistivity of the conductor and δ is the skin depth. If μ is the permeability of the conductor,

$$\delta = \left[\frac{2\rho}{\mu\mu_0\omega_0}\right]^{1/2} \tag{6}$$

For annealed copper at room temperature, $\rho = 1.69 \times 10^{-8}$ Ω/m, and $\delta = 6.5$ μm at 100 MHz. Equations 5 and 6 give an estimate of the resistance of a receiving coil, particularly when multiplied by a "proximity factor" of approximately 3 and a typical value is 0.1 to 0.5 Ω.

From the above, it is clear that the method of construction can influence the resistance of a coil. For example, a solenoid wound of wire will not have the same resistance as one wound of tape, even though the surface areas in the two cases may be the same. A fundamental problem, therefore, is to determine which geometry gives the best sensitivity—the best ratio of $B_1 : \sqrt{r}$. The answer is probably a toroid, but as such a shape is inconvenient for nmr purposes, a single-layer solenoidal winding is probably best.[4] Much attention was given to this problem in the early days of radio communication and the optimum design is shown in Figure 1.II.2.[4,5] The dimensions are not critical and a change of 20% will make little difference.

Unfortunately, the solenoidal configuration is unsuitable in many cases for superconducting systems and in these instances some of the sensitivity gained by the increase in M_0 (Eq. 2) must be sacrificed by the use of saddle-shaped coils.[1] A reasonable design is shown in Figure 1.II.3. The angular width and length shown give the best B_1 homogeneity and a loss of between 2 and 3 in sensitivity may be expected as compared to a solenoid.

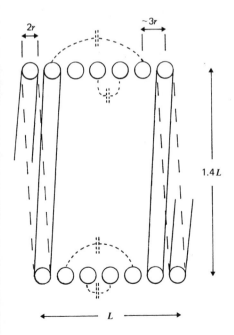

Figure 1.II.2
The optimal design for a single layer solenoid. The number of turns is dependent on the frequency of interest, due to the effect of the self-capacitance (broken lines).

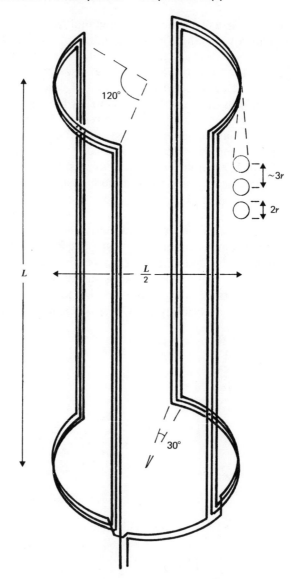

Figure 1.II.3
A low frequency saddle-shaped coil.

III. FREQUENCY RANGE

A factor omitted in Figures 1.II.2 and 1.II.3 is the number of turns on the coil. For a given overall geometry, a small amount of algebra indicates that the sensitivity is largely independent of the number of turns. B_1 and $\sqrt{r}$ have the same dependency on that number. The frequency range is determined by the self-capacitance of the coil. There is distributed capacitance between turns, as shown in Figure 1.II.2, and as the frequency increases the phase shifts introduced by the capacitance spoil the phase coherence of the contribution to B_1 from each element of the coil. The net result is a lowering of the available sensitivity. In addition, excessive self-capacitance provides an electrical (as opposed to magnetic) energy storage mechanism and some of the electric lines of force associated with this energy pass through the sample. The impedance of the coil thereby becomes sample dependent, unless a Faraday screen is incorporated in the probe between the coil and the sample tube. This in turn increases the effective self-capacitance and lowers the sensitivity still more. Clearly, it is essential to avoid the onset of self-resonance (caused by the distributed capacitance) by choosing the correct number of turns for the coil. A Faraday shield is then useful, but harmless, and sensitivity is preserved.

A useful way of checking a coil is to plot its quality factor versus the frequency. Examples are shown in Figure 1.II.4. The correct working

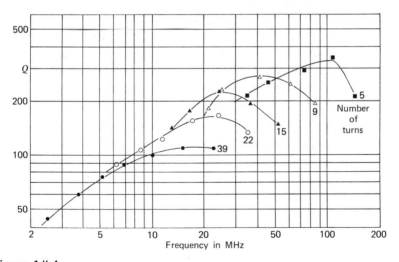

Figure 1.II.4
A plot of Q versus frequency for a set of solenoids 1 cm in diameter and length. The spacing of the turns is three times the wire radius.

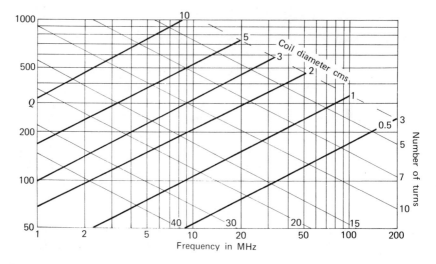

Figure 1.II.5
An indication of the number of turns needed on a solenoid at various frequencies and diameters. Q is approximately given by the formula $Q = 3.5d\sqrt{\nu}$ where d is the coil diameter in meters and ν is the frequency in Hertz. The broken line marks the upper frequency limit above which the number of turns tends to unity and the quality of the coil deteriorates.

range of each coil is on the upward slope of the plot. Figure 1.II.5 provides a rough guide as to the number of turns required on solenoidal coils (designed to the specifications of Figure 1.II.2) at various frequencies and diameters. For a saddle-shaped configuration the number of turns should be approximately halved. A major problem arises if the region of interest is outside that shown in Figure 1.II.5. To minimize self-capacitance, a rather open structure is required (for examples, see Figure 1.II.6) and concomitant with such a structure is the loss in sensitivity already mentioned in connection with saddle-shaped coils. At the time of writing this loss (a factor of approximately two compared with an extrapolation of the lines of Figure 1.II.5) appears to be unavoidable.

IV.   NOISE MATCHING

So far, we have considered only the receiving coil, the signal induced therein, the resistance of the coil, and the associated thermal noise.

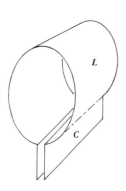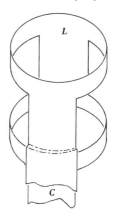

Figure 1.II.6
Single-turn solenoidal and saddle-shaped coils.

However, our next concern must be to amplify the signal and noise as quietly as possible so as to lift them out of the danger area where extra noise can be added by the least error in design or operating procedure. Now electronically, the receiving coil can be considered as an inductive reactance (typically $j50\,\Omega$) in series with a resistance (typically $0.5\,\Omega$) and it is a sad fact that no transistor gives its best performance when operating from such a signal source. A bipolar transistor requires a source impedance, depending on circumstances, of between 50 and $500\,\Omega$ whilst a field effect transistor works best with higher values (200 to $2\,k\Omega$). These impedances are predominantly resistive. They should *not* be confused with the input impedance of the device used, which may or may not be similar. Thus it is necessary to transform the probe impedance, with the minimum of extra resistance, to the optimum source impedance for the particular device used. This is best done with aid of high-quality capacitors as shown in Figure 1.II.7. Many variants are possible including transmission line tuning and an inductive divider, but these two possibilities need especial care in design.

If a single coil probe is used, it is also necessary to power match to the transmitter. Details may be found in Refs. 6, 7, and 8. Clearly two impedance transformations are then required; the probe must be matched to the transmitter (typically $50\,\Omega$) while receiving power, but when receiving signal the $50\,\Omega$ must be transformed to the preamplifying transistor's optimum source impedance. The preamplifier must also be

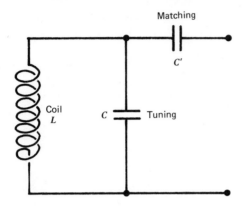

Figure 1.II.7
A probe matching circuit.

protected from the potentially destructive transmitter pulses. A useful measure of the performance of the electronics of a spectrometer, in terms of noise, is the noise figure. This is a measure of the extra noise added to that generated in the receiving coil, and for a single coil system designed to operate at 50 or 75 Ω, the noise figure is especially easy to measure. All that are required, for example, are two 50 Ω sources for the preamplifier input—one at room temperature and the other at 77°K. (A molded half-watt carbon resistor with a room-temperature resistance of approximately 41 Ω has a resistance of 50 Ω at 77°K). Let the noise output from the spectrometer with the room temperature source be α. With the low-temperature source let it be β. Then the noise figure in decibels is given by

$$F = -1.28 - 10\log_{10}\left(1 - \frac{\beta^2}{\alpha^2}\right) \tag{7}$$

A good system has a noise figure of less than 1 dB, and degrades the available signal-to-noise ratio by less than 12%. A mediocre figure of 3 dB degrades the sensitivity by a factor of $\sqrt{2}$. Design details of field-effect transistor (FET) preamplifiers are contained in Refs. 7 and 9.

V.   LEAKAGE

A factor that is often neglected in probe and preamplifier design is the construction of the shield that prevents the intrusion of external radiation. Normally, a completely sealed sample space is impractical as the specimen has to be inserted and removed, and in such instances, a waveguide presents a viable alternative. Below its cutoff frequency ($\nu = 1.69 \times 10^8$/m.diameter) a conducting tube attenuates radiation axially at the rate of 32 dB per diameter penetration, and so a cylinder say six times its diameter in length provides, at its center, excellent shielding. This is ruined if a conductor is inserted into the tube, and any coaxial cable to the probe therefore should be taken up the *side* of the shield. The braiding of the cable should of course be connected to the shield; this is particularly important when the cable reaches the coil and tuning capacitors. Of course if the shield is closed at one end, the receiving coil may be mounted near that end and the inner conductor of the coaxial line brought through the closure. This is shown in Figure 1.II.8. The importance of good "grounding" of the receiving coil to the shield increases at higher frequencies. Lack of grounding can increase substantially the resistance of the receiving coil, the extra term being due to the radiation

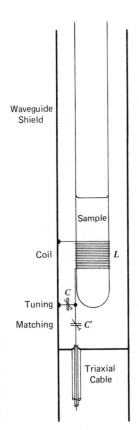

Figure 1.II.8
A long tube acts as a waveguide and the structure shown allows negligible leakage of radiation.

resistance of the outer braiding of the coaxial cable. Also, the diameter of the shield should be at least twice that of the coil to minimize the coupling between the two.

It is not well known that flexible coaxial cable radiates quite strongly, and the use of triaxial cable is strongly recommended. Good rf connectors are also essential and in most chemistry laboratories attention should be paid to the plating on the connectors. Silver alloys corrode remarkably rapidly.

VI. DOUBLE TUNING AND FILTERING

Considerable losses in sensitivity can occur in ^{13}C experiments due to inefficient double tuning and filtering techniques employed when powerful proton decoupling is used. Double tuning with the aid of an extra inductance invariably has losses (at least a factor of 2 in sensitivity) and the use of solid $\lambda/4$ air spaced lines[10] is a far better proposition; one must always beware of the fact that a coil suitable for use at say 270 MHz for protons has little resistance at 68 MHz for ^{13}C, and thus the resistance of any tuning element becomes critical. Figure 1.II.9 shows a design that aims to minimize losses at the lower frequency. The tuning at 68 MHz is performed at the node on the $\lambda/2$ line and therefore does not materially affect the proton tuning. The proton matching capacitor Y' may be neglected at the ^{13}C frequency, and the loss in ^{13}C sensitivity is no more than 10%. The proton losses are a little greater.

Further losses can occur when the ^{13}C preamplifier is overloaded by breakthrough of decoupling power. An efficient, noiseless filter is required to eliminate this effect and, even though efficiency is not too difficult to achieve, this is normally at the expense of sensitivity, for the filter contributes noise, and can degrade the effective noise figure of the

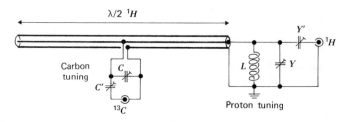

Figure 1.II.9
A double-tuned circuit which minimizes losses at the lower frequency.

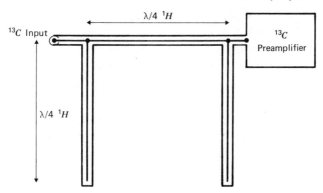

Figure 1.II.10
A low noise filter with 60 dB rejection of decoupling leakage.

spectrometer by as much as 3 dB. Rather than use lumped elements in such a filter, lines may again be usefully employed, and Figure 1.II.10 shows a simple circuit that attenuates the leakage by at least 60 dB while degrading the ^{13}C signal-to-noise ratio by approximately 0.2 dB.

VII. CONCLUSION

By paying attention to the details outlined above, considerable improvements in sensitivity can often be obtained. An exciting future development is the possibility of cooling the receiving coil and preamplifier.[9] The resistivity of copper at 77°K is nearly one-tenth of the room-temperature value, and from Eq. 4, 5, and 6 it is clear that cooling in liquid nitrogen can reduce the noise by a factor of approximately 3.5. Thus, allowing for the insertion of a dewar vessel, we may hope to see improvements of the order of 3 in sensitivity.

REFERENCES

1. D. I. Hoult and R. E. Richards, *J. Mag. Res.* **24,** 71 (1976).

2. A Abragam, *The Principles of Nuclear Resonance*, Clarendon Press, Oxford, 1961, pp. 82–83.

3. B. I. Bleaney and B. Bleaney, *Electricity and Magnetism*, 2nd ed., Clarendon Press, Oxford, 1965, Chap. 10.

4. F. E. Terman, *Radio Engineer's Handbook,* 1st ed., McGraw-Hill, New York, 1943, pp. 77–85.

5. B. B. Austin, *Wireless Eng. Exp. Wireless* **11,** 12 (1934).

6. I. J. Lowe and C. E. Tarr, *J. Phys. E.* Ser. 2. **1,** 320 (1968).

7. D. I. Hoult and R. E. Richards, *Proc. R. Soc. London* **A344,** 311 (1975).

8. D. I. Hoult and R. E. Richards, *J. Mag. Res.* **22,** 561 (1976).

9. D. I. Hoult and R. E. Richards, *Electron. Lett.* **11,** 596 (1975).

10. V. R. Cross, R. K. Hester, and J. S. Waugh, *Rev. Sci. Instrum.* **47,** 1486 (1976).

SECTION

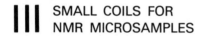 SMALL COILS FOR
NMR MICROSAMPLES

J. N. SHOOLERY

I. INTRODUCTION

The sensitivity of early high-resolution nmr spectrometers was roughly 100 times poorer than modern instruments and led to a widespread belief that a sample of 100 mg or more was needed, even for a proton spectrum. This amount of sample is totally unnecessary today for proton work, and, although it makes ^{13}C spectra obtainable in reasonable data accumulation times, it often places unnecessary demands on the chemist or biochemist who has to provide a reasonably pure sample. The most effective separation techniques, such as gas chromatography, liquid chromatography, and thin-layer chromatography work best on a very small scale where the sample may be micrograms or, at best milligrams. Samples of biological origin may be limited in total amount after purification to a few hundred micrograms. Unless techniques to study such samples are available, the power of high-resolution nmr simply cannot be brought to bear on these problems.

II. DESIGN CONSIDERATIONS AND ASSUMPTIONS

In an effort to overcome this obstacle, the concept of improving the coupling between the nuclear magnetic moments and the receiver coil was explored. For a fixed amount of sample, presumably limited by availability to a few milligrams or micrograms, an approximate expression was derived to estimate the probable improvement in signal-to-noise ratio that would result from varying the diameter of the receiver coil, assuming that as the coil diameter decreased the sample would remain soluble in the volume of solvent needed to fill the smaller sample tube. The derivation is based on the approximation resulting from the assumption that a column of liquid in a sample tube that is long enough so that the magnetic discontinuities due to the end of the tube and the solvent-air interface do not appreciably distort the field is also long enough so that making it longer will not appreciably increase the signal. In other words, high-resolution spectra are generally run with an effectively *infinite column* of liquid, to avoid distortion of the lineshape. This leads to experimentally determined *shortest sample* filling lengths for each coil size.

III. ESTIMATION OF EFFECT OF COIL SIZE ON SIGNAL-TO-NOISE RATIO

The signal from an infinite column containing magnetic nuclei is

$$S(\text{signal}) = QnA_c \frac{d\phi}{dt} \times 10^{-8} \text{ V}$$

where

Q = coil quality factor
n = number of turns
A_c = cross-sectional area
of coil

$d\phi/dt$ is the rate of change of flux in the coil and is equal to $\omega M_0 \eta$, where ω is the nuclear precession frequency, M_0 is the magnetic moment per unit volume of the cylindrical column of sample, and η is the filling factor of the coil, A_s/A_c, which is the ratio of sample cross-sectional area to coil area. Thus $S = QnA_s\omega M_0$. S is proportional to the product of the variables, QnA_sM_0, where M_0 is the mass of sample per unit volume, since the remaining quantities in the expression are constants of the experiment once the nucleus and field strength have been decided upon.

The thermal noise in the coil is given by

$$N(\text{noise}) = (4kT\omega LQ)^{1/2}$$

Again, eliminating the constants of the experiment, we have N proportional to $(LQ)^{1/2}$. For a simple, single-layer coil the inductance is approximated by the formula

$$L = n^2 d^2$$

where d is the diameter of the coil.

We are now in a position to calculate the dependence of the signal-to-noise ratio at the input to the preamplifier as a function of the several variables associated with the coil size.

$$S/N \propto \frac{QnA_sM_0}{(n^2d^2Q)^{1/2}} = \frac{Q^{1/2}\,A_sM_0}{d}$$

M_0 depends upon the volume of solvent used, but M_0V represents the total sample, which is constant in this case. Multiplying top and bottom of the previous expression by V we get

$$S/N \propto \frac{Q^{1/2}A_s}{dV} = \frac{Q^{1/2}}{dl} \tag{1}$$

where l is the depth to which the tube is filled.

A special case of interest is encountered if the spectrometer is designed with interchangeable receiver coils that have the same inductance to minimize tuning when changing from one coil to another. The signal-to-noise expression then becomes

$$S/N \propto \frac{Q^{1/2}n}{l} \tag{2}$$

Equations 1 and 2 are equivalent for a single-layer coil.

It is clear from Eq. 1 that it is quite advantageous to reduce the coil diameter when a limited sample must be studied. For example, if the coil diameter and sample length are both reduced three-fold without loss in

Q, the sensitivity is increased by a factor of 9, and the time required is reduced by 9^2. Even allowing for a drop in Q to half the initial value due to the increased resistive loss in the smaller wire used in coils for capillary samples, the factor is $9 \div \sqrt{2} = 6.4$.

IV. COMPARISON WITH EXPERIMENT

A series of experiments was performed in which the ^{13}C spectrum of a 5-mg sample of cholesterol was obtained with sample tubes of 10, 8, 5, and 1.7 mm o.d., using closely fitting receiver coils that were constructed on cylindric glass forms. A coaxial fitting allowed the coils to be interchanged in the nmr probe. Table 1.III.1 shows the number of transients and total time needed for each sample to give the same signal-to-noise ratio in the spectrum. As an example, Eq. 1 or Eq. 2 can be applied to calculate the relative times needed for the 10-mm and 1.7-mm samples, using the factor 3.5 for the ratio of the number of turns or the ratio of the diameters of the coils, measured from the coils used in the experiment. The depths of the samples were 25 mm for the 10-mm tube and 10 mm for the 1.7-mm tube. Q of the 10 mm coil is approximately twice that of the 1.7-mm coil. Either Eq. 1 or Eq. 2 gives the result that the improvement in signal-to-noise ratio with the 1.7-mm tube should be

$$\frac{l_{10}}{l_{1.7}} \times 3.5 \div \sqrt{2} = 6.2$$

and the ratio of times should be $6.2^2 = 38$, which agrees quite well with the observed ratio of 40.

Table 1.III.1
Constant Amount

Coil (mm)	Sample Weight (mg)	Minimum Volume	Sample Concentration (M)	Number of Transients	Total Time
1.7	5	15 μl	0.87	2,000	33 min
5	5	250 μl	0.052	20,000	5.5 hr
8	5	1.0 ml	0.013	50,000	13.9 hr
10	5	1.5 ml	0.0086	80,000	22.2 hr

Table 1.III.2
Constant Concentration

Coil (mm)	Sample Weight (mg)	Minimum Volume	Sample Concentration (M)	Number of Transients	Total Time
1.7	5	15 μl	0.87	2000	33 min
5	83	250 μl	0.87	72	1.2 min
8	333	1.0 ml	0.87	11	11 sec
10	500	1.5 ml	0.87	8	8 sec

Table 1.III.3
Constant Time

Coil (mm)	Sample Weight (mg)	Minimum Volume	Sample Concentration (M)	Number of Transients	Total Time
1.7	3	15 μl	0.5	5500	1.5 hr
5	9.6	250 μl	0.1	5500	1.5 hr
8	15.2	1.0 ml	0.04	5500	1.5 hr
10	19.2	1.5 ml	0.033	5500	1.5 hr

Table 1.III.2 is calculated to show that if the sample is *not* limited, the larger coils offer a significant saving in time. Interchangeable coils thus offer the advantage of optimizing the experimental time to suit the available sample. Table 1.III.3 shows the amounts of sample required for different sized samples to give the same signal-to-noise in approximately the same amount of time, and can be used to estimate the time required to obtain a usable ^{13}C spectrum from samples of various sizes.

V.   EXAMPLES OF ^{13}C MICROSAMPLE SPECTRA

Figures 1.III.1 and 1.III.2 show the ^{13}C spectra of 1.0-mg gelsemine and 500-μg ethyl vanillin in 1.7-mm capillary sample tubes. These spectra

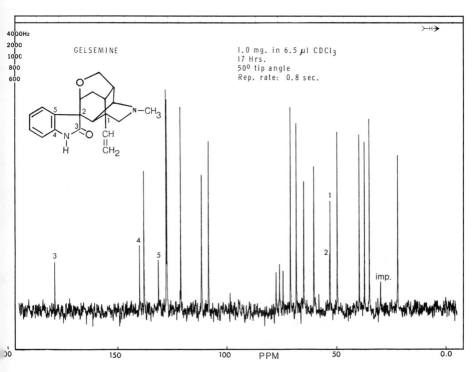

Figure 1.III.1
Gelsemine, ^{13}C spectrum.

illustrate that high-quality ^{13}C spectra from samples of this size can be obtained, even in complex molecules with nonprotonated carbon atoms. Both spectra were run overnight. If only the protonated carbons are of interest, spectra of these samples could be run in 2 to 3 hr.

VI. EXAMPLES OF ^{1}H MICROSAMPLE SPECTRA

Since the sensitivity for protons is approximately 1000 times greater than for ^{13}C, samples of a few micrograms can be studied with small coils optimized for operation at the proton frequency. Samples of 100 μg or larger generally can be run with a few hundred transients, whereas

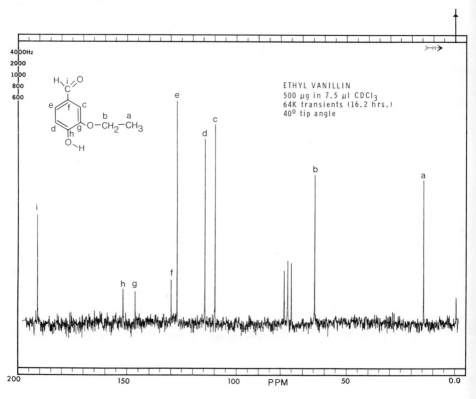

Figure 1.III.2
Ethyl Vanillin, ^{13}C spectrum.

samples as small as one microgram have given good results. Figure 1.III.3 shows the ^{1}H spectrum of 200-μg ethyl vanillin using only 100 transients and illustrates excellent resolution of the spin-spin multiplets typical of small molecules. Figure 1.III.4 is the ^{1}H spectrum of 1.0 μg cortisone acetate, run for 50,000 transients (13.9 hr).

Water and other impurities containing protons (grease, plasticizers, solvent background) can be a serious problem in ^{1}H microsample work and care must be taken to minimize such interfering substances.

The ability to study microgram quantities makes ^{1}H nmr a particularly useful adjunct to gas chromatography (GC) in identifying small amounts of impurities in a wide variety of samples. A 1.7-mm capillary tube, open

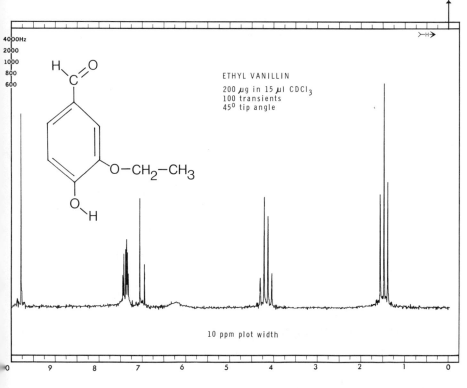

4000Hz
2000
1000
800
600

ETHYL VANILLIN

200 μg in 15 μl CDCl₃
100 transients
45° tip angle

10 ppm plot width

10   9   8   7   6   5   4   3   2   1   0

Figure 1.III.3
Ethyl Vanillin, ¹H spectrum.

at both ends, seems to be a very adequate collector for GC fractions. The entire effluent from a thermal-conductivity detector can be passed through it, and in the case of very small samples 90 to 100% collection efficiencies can be realized by cooling the tube to dry-ice temperature. Sealing the tube at one end, adding solvent, and sealing the other end are trivial procedures that complete the sample preparation. Figure 1.III.5 is the ¹H spectrum of ethyl butyrate collected from a GC separation of pineapple-flavor extract. Larger samples (0.5–1.0 mg) can also be ·collected by heavier loading or larger columns, or by multiple collections of a peak in the same tube, and are sufficient for ¹³C spectra[1] from overnight data accumulations.

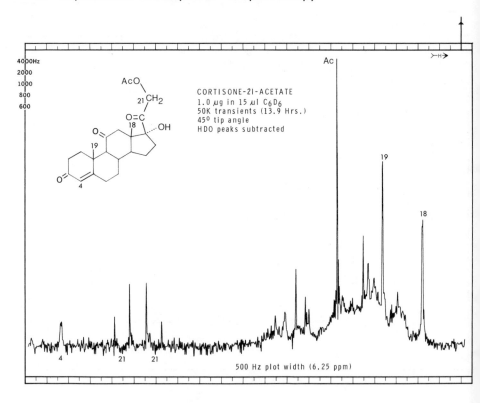

CORTISONE-21-ACETATE
1.0 µg in 15 µl C$_6$D$_6$
50K transients (13.9 Hrs.)
45° tip angle
HDO peaks subtracted

500 Hz plot width (6.25 ppm)

Figure 1.III.4
Cortisone-21-Acetate, ^{1}H spectrum.

VII. EXPERIMENTAL TECHNIQUES

Preparation of the sample involves the largest departure from standard nmr techniques due to the small size of the sample tube and the small volume of solvent involved. In addition to learning to manipulate small quantities of material with micropipettes, it is necessary to learn to avoid procedures that might introduce amounts of impurities that would not normally cause trouble with larger samples. Flames should be kept away from the open mouth of a tube when sealing, and tubes and pipettes should be blown out with dry nitrogen or baked in an oven if very small samples are being prepared for ^{1}H studies. The author has written a brief manual[2] of experimental techniques illustrated with ^{1}H and ^{13}C spectra from a variety of types of samples, which elaborates on some of these points.

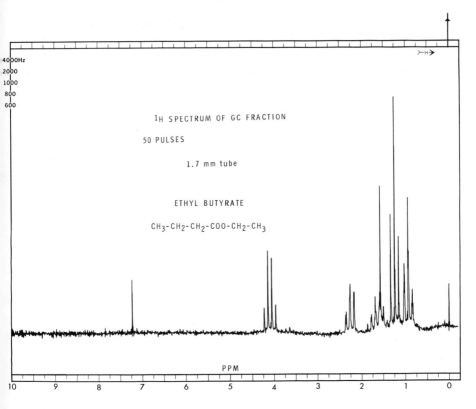

Figure 1.III.5
Ethyl Butyrate, ^{1}H spectrum.

VIII. ADVANTAGES OF SMALL COILS

The use of small coils and the corresponding capillary sample tubes leads to a number of advantages over larger coils and sample tubes.

1. Gain in sensitivity as given by Eq. 1 or Eq. 2
2. Reduced requirement for expensive deuterated solvents
3. Reduced spectral background from the solvent
4. Freedom from spinning sidebands
5. Freedom from problems caused by vortex formation during sample spinning
6. Inexpensive, throw-away cells, typically commercially available 1.7-mm melting point capillaries

Disadvantages include lower lock signal due to the reduced amount of deuterated solvent, need for more care in sample preparation, and slight degradation of the resolution specification of the spectrometer. By comparison with the benefits to be gained, these disadvantages appear to be quite minor and are not difficult to live with. The quality of the spectra obtained from minute samples of quite complex chemical substances allows one to predict confidently that small coils and capillary samples will open up many areas of chemistry to routine nmr studies that previously were totally inaccessible.

REFERENCES

1. J. N. Shoolery and R. E. Majors, *Am. Lab.*, 51 (1977).
2. J. N. Shoolery, *Microsample Techniques in ^1H and ^{13}C NMR Spectroscopy*, Publication No. MAG-2037, Varian Associates, Palo Alto, Calif.

SECTION

IV CONSTRUCTION OF A HIGH-PERFORMANCE LARGE SAMPLE PROBE FOR THE XL-100-15 NMR SPECTROMETER

ALBERT P. ZENS
DAVID M. GRANT

I. INTRODUCTION

The objective of this brief review is to outline some of the principles used in the construction of a high-performance probe for a commercial spectrometer such as the research grade XL-100-15 system. While the XL-100-15 spectrometer is not universally used, most of the principles and techniques discussed should be generally applicable to any high-resolution iron magnet system.

At the present time there are several probe systems employing 18- to 20-mm tubes that have been reported in the literature for iron magnet systems.[1-4] The Allerhand probe, which employs a 20-mm sample tube, was the first large-sample tube probe to allow routine studies on single carbon resonances in proteins. The multinuclide 18-mm probe reported by Byrd and Ellis[5] utilized the multinuclear observe capability of the XL-100 in conjunction with the sensitivity advantages of a large sample tube probe. Commercial probe systems using 18-mm sample tubes have also been reported by Varian Associates and Nicolet Technology Corporation. The sensitivity of all of these probe systems on ^{13}C is nearly equal, although the Allerhand probe uses twice as much sample volume as the 18-mm systems as a means to compensate for the lower field (15 MHz) employed with this probe.

In developing a new probe in this laboratory, an initial goal was set to increase the ^{13}C detection sensitivity by a factor of two or more over that found in iron magnet systems employing 18- to 20-mm sample tubes. It was felt that this increase in sensitivity was a minimal requirement to justify the development effort and to make significant reductions in experimental run times in macromolecular nmr investigations where self-dilution and low solubility are always a problem.

To accomplish the desired goal, the probe had to be optimized with respect to the type of experiments to be routinely performed and with regard to the best choice of receiver and decoupler coil geometry, lock system, and other parameters affecting the performance of the probe. This optimization in most cases had to be empirically determined by carefully varying different parameters while noting the response of the altered probe. Unfortunately the various probe features interact strongly and prevent a simple evaluation of all interacting parameters. The large number of iterations necessary to optimize the selection of all possible parameters probably prevented a perfect adjustment of all factors. Even so, a reasonably well optimized probe system has been achieved that exceeded the initial performance goals by approximately two, giving a probe system having a sensitivity three to four times that presently available for iron core systems and one that slightly exceeds the performance of the very best superconducting large-tube spectrometers.

The customary expression for the signal-to-noise ratio following a 90° pulse[6] is

$$S/N = K\eta M_0 \left(\frac{\mu_0 Q \omega_0 V_c}{4FkT_c\Delta f}\right)^{1/2}. \tag{1}$$

The terms are defined in Ref. 6. For a single-coil Fourier-transform spectrometer operating at a ^{13}C observe frequency of 25.2 MHz and employing a preamplifier that restricts out-of-band conversion of noise in the digital sampling with both the receiver coil and sample at ambient temperature this expression reduces to

$$S/N = ZK\eta \left(\frac{QV_c}{F}\right)^{1/2} \tag{2}$$

where all constant parameters have been collected into Z, K^* is a measure of the effective field produced over the sample volume by a unit flow of current in the receiver coil, η is a measure of the fraction of the effective coil volume occupied by the sample often refered to as "the

---

* This numerical factor is unity for a infinitely long solenoidal receiver coil.

filling factor," Q is the quality factor of the receiver coil, V_c is the volume occupied by the receiver coil, and F is the noise figure of the preamplifier. In Eq. 2, the essential parameters to be optimized are apparent, but unfortunately, as stated previously, these parameters are strongly coupled so that varying one usually results in a significant change in the others. Recently Hoult and Richards have presented an expression for S/N that reduces this interdependence of variables.[7]

$$
S/N = ZKV_s\left(\frac{P}{Fl\zeta}\right)^{1/2}(\rho)^{-1/4} \tag{3}
$$

Where V_s is the volume of sample enclosed by the receiver coil, ζ is the proximity factor, P is the perimeter, l is the length, and ρ is the resistivity of the conductor used to construct the receiver coil. The proximity factor ζ is a measure of the increase in the resistance of the receiver coil due to the nearby presence of other conductors (i.e., decoupler coil, silvered dewars, etc.). In Eq. 3 ζ replaces the traditional parameters Q and η. The major advantage of Eq. 3 over Eq. 2 is that it is applicable to any coil geometry, that is saddle-shaped and solenoidal. Unfortunately, the absence of such traditional terms as Q and η makes it somewhat awkward. The traditional formula (Eq. 2) was retained in this work so that Q and η could be explicitly discussed even though the more extensive expression derived by Hoult and Richards probably provides some important additional features.

II. OPTIMIZATION OF PROBE PARAMETERS

A. Sample Size, Resolution, and the XL-100-15 Magnet

The sample size, the size of the magnet gap, and the resolution obtainable with a given sample size are all related to the resolution, dispersion, and sensitivity required for specific experimental conditions. It was initially determined for large sample tube probes that the resolution necessary for most nmr investigations would be sufficient if one could obtain a half-height resolution of 1 Hz providing reasonably good Lorentzian line shapes could be achieved. Based on this criterion the resolution obtainable for a variety of nmr tubes (18, 20.5, and 22 mm o.d.) was investigated to determine the optimal sample size to be employed in the 34-mm gap of the XL-100-15 magnet. The height of the receiver coil used in these experiments was set at approximately 1.25 times the diameter of the tube size in order to maximize K. The resolution at half-height with all three tube sizes was in every case less than 0.08 Hz on CS_2. Although deviations were noted in the Lorentzian line shapes from that obtained on the standard 12-mm-o.d. sample tubes, these were minimal in all three experiments and were considered to be acceptable. Deviations from the Lorentzian line shape in the base of the resonance peak is particularly destructive to resolution and high sensitivity, and attention was directed to eliminate such non-Lorentzian contributions. Figure 1.IV.1 contains a 2-Hz plot of CS_2 obtained with a 22-mm-o.d. sample tube. The resolution at half-height is 0.04 Hz. From these experiments it was clearly evident that the XL-100-15 magnet can utilize a sample tube of at least 22 mm diameter without serious line shape distortions. The onset of a slight amount of distortion at the base of the peak, however, suggested that further increases may be without profit. Even though a larger sample tube possibly could be employed within the 34-mm magnet gap, these increases place rather severe requirements on the fabrication of the probe body and its component parts. Furthermore, the amount of sample volume necessary to do routine experiments increases dramatically with each modest increase in sample tube size. The quality factor Q of the receiver coil for even larger samples also will deteriorate as the probe side plates are placed in closer proximity to the radio frequency coils employed in the probe. Although we have not explored all aspects of this problem, exploratory work has shown this to be an issue of some consequence.

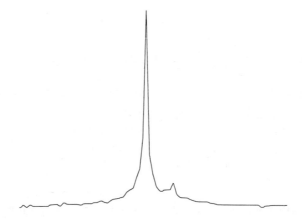

Figure 1.IV.1
The ^{13}C spectrum of 80% CS_2. The spectrum was recorded on a 100-Hz spectral width using 8192 points in the time domain. This spectrum is a 2-Hz portion of the frequency spectrum. The width at half height is 0.04 Hz. This resolution was obtained using a double tuned lock-observe system.

Based upon empirical data of the type referred to above and giving consideration to the physical space limitations, the 22-mm sample tube was finally selected as the best choice for the type of general experiments that have routinely been performed in this laboratory. Some compromise in sensitivity per milliliter of sample volume was made relative to overall sensitivity. It should be evident when smaller sample tubes are required for sample limited studies that the removable 22-mm receiver coil could be readily substituted with a 18-mm or smaller coil thereby reducing the active sample volume by a half or more

B. Receiver Coil Design

The receiver coil geometry was optimized with regard to Q and filling factor within the constraints of an active sample volume of approximately 10 to 11 ml. The receiver coil was constructed from oxygen-free high-conductivity copper ribbon that was 0.055 in. wide and 0.010 in. thick. Regretfully, coils constructed from #15 AWG wire, which were of higher Q, could not be used primarly because of space considerations. The ribbon receiver coil was wound on the inside of a Vespel* coil form with the aid of a threaded mandrill. By placing the coil on the inside of the insert one maximizes the filling factor η. The final dimensions of the 9.5-turn receiver coil were 22.8 mm i.d. and 26.8 mm high. The number of turns and width of the copper ribbon used were empirically determined to give the highest Q at the ^{13}C observe frequency of 25.16 MHz.

The coil form or insert was made to be removable from the probe body to accommodate different size receiver coils and to minimize replacement and maintenance problems. This insert is coupled to the probe through a miniature rf connector (hot lead) that was imbeded in the bottom of the insert. The matching connector is connected to the observe transmitter network. The top lead of the receiver coil is connected to a berylium copper ring at the top of the insert which, when positioned in the probe, grounds the insert through a berylium copper contact ring connected to the probe body. This type of construction provided a receiver coil with a Q just as good as one in which the coil was hard-wired to the probe body. A significant increase in flexibility was realized by using this arrangement.

C. Lock System

The internal ^{2}H lock is derived from a saddle-shaped coil on the coil form (28 mm diam). This type of lock system rather than the traditional

---

* Vespel is the registered trade mark of DuPont polyimide resins.

double-tuned lock-observe system was found to improve the sensitivity of the probe dramatically. The employment of an observe frequency trap in a double-tuned observe-lock system to separate the ^2H lock signal from the observe signal inevitably degrades the Q in the observe circuit and quenches the overall probe performance. Experimental tests using the standard double-tuned ^{13}C observe-^2H-lock system in the XL-100 indicated that the overall performance of the double-tuned circuit had been degraded by nearly a factor of two. The magnitude of this ratio can be explained in part by the high Q of the receiver coil used in the high-performance probe. A smaller receiver coil with a lower Q would undoubtedly not be affected so greatly.

The employment of a separate lock coil rather than the standard double-tuned lock-observe system, unfortunately, did have a noticeable effect on the resolution of the system. Figure 1.IV.2 shows a 2 Hz plot of CS_2 using the separate saddle-shaped lock coil. In Figures 1.IV.1 and 1.IV.2 the resolution at half-height on CS_2 is 0.04 Hz. However, the Lorentzian lineshape of Figure 1.IV.1 is superior to that of Figure 1.IV.2. Various attempts to improve this resolution beyond that shown in Figure 1.IV.2 by varying the initial 29 mm height of the saddle-shaped coil were of no consequence. If we consider the increase in sensitivity, this slight degradation of lineshape will be of little consequence whenever the experiment requires only 1-Hz resolution.

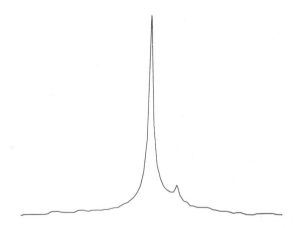

Figure 1.IV.2
The ^{13}C spectra of 80% CS_2. The spectrum was recorded on a 100-Hz spectral width using 8192 points in the time domain. This spectrum is a 2-Hz portion of the frequency spectrum. The width at half height is 0.04 Hz. This resolution was obtained using a separate internal lock coil.

It should be noted that the employment of the separate saddle-shaped lock coil can diminish the ^{2}H-lock channel sensitivity because of a significant loss in filling factor (28 mm i.d.) and also because of the intrinsic inefficiency of saddle-shaped coils.[7] In practice, however, the performance of the Helmholtz-lock coil system was as good as or better than that of the double-tuned observe-lock system. This rather surprising feature is presumably the result of large inefficiencies in the double-tuned observe-lock system similar to those previously mentioned for the observe channel.

D. Decoupler System

The decoupler coil is constructed on the same coil support form as used for the lock coil. The 28 mm diameter of this cylindrical form was selected to minimize degradation of the receiver coil Q resulting from losses directly attributable to the close presence of the lock and decoupler coil. This arrangement maximized the sensitivity of the observe channel, even though the constraints placed on the decoupling system due to the inductive coupling of the lock and decoupler coils and the close proximity of the side plates of the probe were rather difficult to overcome. The best geometry for the decoupler and lock coils was found to be an overlapping rectangle pattern for each of the two-turn saddle-shaped coils. This optimized the γH_2 of the decoupler coil without degrading the resolution attainable from the lock coil. The lock coil was constructed from #24 AWG silvered copper wire and each turn was 131° wide and 29 mm high. The decoupler coil was constructed from #18 AWG copper wire and each turn was 112° wide and 35 mm high. The γH_2 produced by this geometry in the absence of a Faraday screen was 2700 Hz at 100 MHz and 10 W power. A double-tuned lock decoupler coil system utilizing one-fourth wavelength 0.141 in. semirigid coaxial cable was attempted with less success than the two coil system described above.[8]

E. Probe Body, Matching Networks, Preamplifiers, Etc.

A highly modified V4415 probe body was used in this study. The side plates were constructed to maximize the size of the sample cavity and thereby fully utilize the full width of the 34-mm magnet gap. In the immediate vicinity of the receiver coil, the side plates were fabricated from oxygen-free high-conductivity copper in order to minimize losses in the Q of the receiver coil. The spinner housing, constructed in house, utilizes 22-mm-o.d., 20.5-mm-i.d. precision sample tubes (Wilmad Glass Co., Inc., Buena, NJ).

A decoupler-matching network approriate for ^{1}H and ^{19}F was built into the probe body to minimize lead-length losses at 100 MHz. The γH_2 power is generated with a Model 3100 L wideband amplifier (Electronic Navigation Industries, Rochester, NY) and is driven by the low-power output of the XL-100 console. A bandpass filter after the amplifier was used to eliminate harmonics at the observe frequency in the decoupler channel.

The internal ^{2}H-lock matching network consists of a removable assembly to facilitate other lock frequencies and may be used to generate a second decoupler frequency in special experiments. The lock system uses a directional coupler in conjunction with the standard Varian pulse ^{2}H lock and a house-constructed lock preamplifier. The external pulse ^{2}H lock consists of a 3-cm-long, 5-mm nmr tube filled with a doped sample of D_2O. The 6-mm-long pickup coil shielded with a copper can, is wound around the 5-mm tube. Using a coaxial switch the lock system can be switched conveniently from internal to external pulse ^{2}H lock. The single-coil observe-transmitter network is standard in design and uses nonmagnetic high-Q, high-voltage fixed and variable capacitors to minimize loss factors in transforming the impedance of the receiver coil at resonance to 50 Ω resistive. The 2500-V fixed capacitors were obtained from Vitramon Inc. Bridgeport, CN.) while the 4000-V variable capacitors were purchased from Polyflon Corp. (New Rochelle, NY). Both the fixed and variable capacitors have rated Qs that exceeded 5000. At 25 MHz the measured Q of the receiver coil in the probe in the absence of a sample tube is 320, whereas the ^{13}C 90° pulse width is typically 10 μsec with 45 W power.

The observe preamplifier presently in use was purchased from R.H.G. Electronics Laboratory, Inc. (Deer Park, NY). This preamplifier is of single-frequency design and has a tested noise figure of 1.2 dB with 38 dB gain at 25.2 MHz. A particularly notable feature of this commercial unit is the exceptionally good recovery time after the ^{13}C perturbing pulse. Other preamplifiers, both commercial and homebuilt, have not been extensively tested but are referenced as an aid to those wishing to experiment.[9,13] A rf switch constructed on site was placed between the preamplifier and the XL-100 console to provide 50-dB isolation for the observe receiver during the transmitter pulse.

A low-pass filter purchased from Texscan (Indianapolis, IN) was used to isolate the decoupler from the observe channel at 25 MHz. The insertion loss of this unit was measured at 0.2 dB. The isolation provided by this filter at 100 MHz was 85 dB. A Faraday screen was tested as an alternative means of decoupler isolation. The γH_2 of the decoupler with 10 W power at 100 MHz was approximately 2300. Various subtle differences in the performance of these two means of isolation were not tested extensively but the following qualitative trends are noted. Sample heating from the decoupler was less pronounced for ionic solutions with the Faraday screen at the same γH_2. In absence of a Faraday screen the receiver coil was noticeably warm when decoupling with 10 W power. The measured Q of the receiver coil without the Faraday screen was approximately 10% better than with it. From these observations it is clear that a considerable amount of testing is needed to determine the best mode of isolation to use in relation to sensitivity and the type of experiment to be performed. Figure 1.IV.3 shows a block diagram of the probe system in relation to the XL-100 console.

In addition to the construction details of the probe, certain modifications to the XL-100 console were necessary to optimize routine performance. Standard operating gear for the XL-100 not shown in the block diagram of Figure 1.IV.3 is a single-side band crystal filter, a Kron-Hite 48-dB/octive adjustable low-pass audio filter, Kronhite Corporation (Avon, MA), a lab-built 180° alternating pulse phase shifter and an audio oscillator that is used as external modulation for square-wave decoupling.[14] The gate relay driver was modified to extend the off time of the receiver from 50 to approximately 90 μsec to allow the preamplifier to completely recover from the rf pulse.[15] The alpha delay governing the time delay between the receiver on time and analog-to-digital conversion of the observe signal was also extended to reduce audio filter overshoot and pulse breakthrough transients (16).

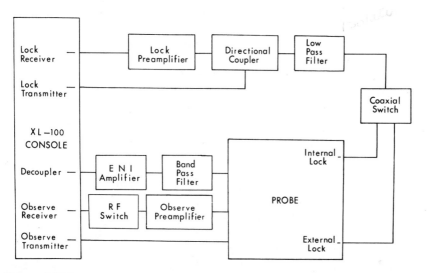

Figure 1.IV.3
The block diagram of the 22-mm probe system interfaced to the XL-100 console.

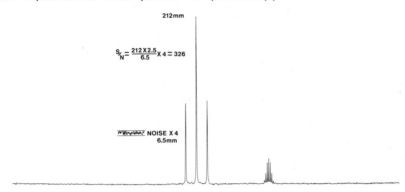

Figure 1.IV.4

^{13}C natural abundance coupled spectrum of 80% dioxane. The signal-to-noise ratio of 326/1 was obtained with one 90° rf pulse and exponential multiplication equal to 1.6 Hz of line broadening.

III. PERFORMANCE

Figure 1.IV.4 shows the ^{13}C sensitivity of the 22-mm probe on 80% dioxane in a coupled mode. The signal-to-noise ratio of 326/1 represents state-of-the-art sensitivity in both overall performance and sensitivity per milliliter of sample volume for large sample tube nmr probes. The minimum volume necessary to obtain maximum sensitivity is 10 to 11 ml sample (33 mm sample in a flat bottom 22-mm-o.d., 20.5-mm-i.d. sample tube). This corresponds to a sensitivity per milliliter of approximately 30 on 80% dioxane. This important criterion is presently not surpassed by any large sample-tube probe. As a comparison, 18-mm nmr probes for the XL-100-15 nmr spectrometer currently use a minimum volume of about 5.5 ml to obtain a maximum signal-to-noise ratio on 80% dioxane of approximately 85/1. This data clearly indicates that the 22-mm probe has nearly a four-fold advantage in sensitivity over these probe systems with a better than a 10-fold reduction in time to obtain the same signal-to-noise ratio at a given sample concentration. In sample limited cases the 22-mm probe retains a substantial three to fourfold time advantage over the smaller 18-mm probe systems. Figure 1.IV.5 shows the performance of this probe on 0.01 M aqueous sucrose. The sensitivity shown is best realized by comparing it with previously published spectra of 10 mM sucrose reported by Allerhand and Byrd.[3,4] We estimate that our signal to noise is four to five times greater than either of these probe systems. Figure 1.IV.6 demonstrates the sensitivity of the probe on 0.004 M HEW lysozyme. The high signal-to-noise ratio was obtained in a few hours and

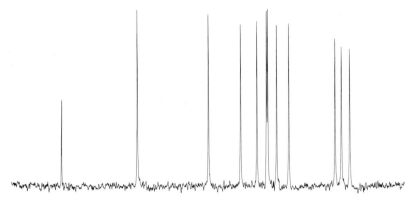

Figure 1.IV.5

Proton decoupled natural abundance ^{13}C spectrum of aqueous sucrose at 32°C. The spectrum represents 4000 accumulations using 90° rf pulses, 4096 points in the time domain, a spectial width of 2048 Hz, and a recycle time of 1.0 sec. An additional 4096 points were added to each time domain signal. Exponential Multiplication equal to 1 Hz was used for this plot which is a 1500 Hz portion of the frequency spectrum.

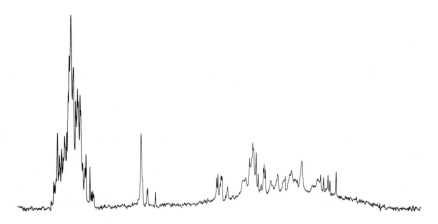

Figure 1.IV.6

The unsaturated carbon region of the 25.2 MHz ^{13}C spectrum of native HEW lysozyme (0.004 M) at 37°C and a pH of 3.1. The spectrum was recorded under conditions of full proton decoupling, with a spectral width of 4000 Hz, 8192 time-domain addresses, 46,000 accumulations and a recycle time of 2.1 seconds, and 1.6 Hz of line broadening.

Figure 1.IV.7
The gas phase ^{13}C natural abundance spectrum of allene at 2 atm pressure. The spectrum represents 20,000 scans under full proton decoupling using a spectral width of 5000 Hz, 4096 time addresses, 0.4 sec recycle time (2.2 hr total time) and 3.18 Hz of line broadening.

is indicative of the utility of the 22-mm probe in native protein studies of this type. Figure 1.IV.7 is the gas phase natural abundance ^{13}C spectrum of allene. Unusual experiments of this type become more practical when the time to do them is reduced from days to a matter of a few hours!

IV.   DISCUSSION AND CONCLUSIONS

The sensitivity reported in this study for our 22-mm proble is comparable to that of state-of-the-art-wide bore superconducting nmr systems. The fact that the probe system is capable of slightly exceeding the performance of the very best large bore superconducting system is explainable from the known greater efficiency of the solenoidal receiver coils of the type used in our 22-mm probe over the Helmholtz or saddle-shaped receiver coil used by these superconducting spectrometers. It is important to recognize that the increases in sensitivity expected from the higher observe frequency of wide bore superconducting systems for ^{13}C is either cancelled or negated by the nearly threefold difference in efficiency between a saddle-shaped coil and a solenoidal coil used in the 22-mm probe. The expected difference in signal to noise as a function of operating frequency for a hypothetical wide bore superconducting system operating at a ^{13}C frequency of 50 MHz assuming that all other factors are equal is expressed by

$$\frac{S/N(50\,\text{MHz})}{S/N(25.16\,\text{MHz})} \propto \left[\frac{\omega(50\,\text{MHz})}{\omega(25.2\,\text{MHz})}\right]^{7/4} = 3.3 \qquad (4)$$

The expected 3.3-fold increase in performance is nearly equal to the theoretical and experimentally determined efficiency difference of three between solenoidal and saddle-shaped receiver coils reported by Hoult and Richard.[6] It is evident therefore that wide bore superconducting systems employing saddle-shaped receiver coils will have to operate at a ^{13}C observe frequency of better than 60 MHz to have a twofold time advantage over the 22-mm probe. Superconducting solenoids capable of utilizing large sample tube probes of this type have not been constructed to date presumably because of problems associated with field homogeneity. Large bore superconducting solenoids utilizing 20- to 30-mm sample tubes operating at ^{13}C frequencies below 50 MHz may equal or exceed the performance of the 22-mm probe; however, the sensitivity per milliliter of sample volume will be considerably less than that reported for our probe.

The disadvantages of the 22-mm probe are related to the lack of dispersion associated with the employment of a low-field high-resolution iron magnet. Although of some concern, this problem in most cases will be of little consequence. For native protein studies of nonprotenated carbons, the increased dispersion of resonances by the use of a wide bore superconducting spectrometer will be offset to some extent by line broadening from the increased relative effectiveness of chemical shift anisotropy as a relaxation mechanism.[17]

As a final aid to those interested in developing similar probe systems, it should be indicated that throughout the period of development many small improvements were made. In many cases these improvements, such as winding the receiver coil on the inside of the coil form, contributed only a small amount ($\approx 5\%$) to the improvement in sensitivity, but it should be emphasized that when the number of these small improvements becomes large then their cumulative contribution is significant and has resulted in the present level of performance.

It is felt by the authors that the performance of the 22-mm probe gives considerable promise for revitalizing the large number of older commercial spectrometers presently in operation. Only the dispersion disadvantages peculiar to low-field systems remains a problem. As these slight dispersion disadvantages are generally of little consequence for nonproton or nondeuterium nuclei, one can also see high promise in expanding the 22-mm probe to include other nuclei.

ACKNOWLEDGMENTS

We acknowledge the helpful comments of the many people who have shown interest in this effort, and in particular those of Vern Burger of

Nicolet Technology Corporation, Bob Codrington of Varian Associates, and Don Alderman in this laboratory. The work was supported by the National Institutes of Health under research grant GM-8421-17.

REFERENCES

1. Varian Associates, The Varian V4418 Probe, 611 Hansen Way, Palo Alto, CA. 94303.
2. Nicolet Technology Corporation, NT 72518 or NT 254018 probe, 145 E, Mountain View, CA. 94041.
3. A. Allerhand, R. F. Childers, and E. Oldfield, *J. Mag. Res.* **11**, 272 (1973).
4. R. A. Byrd and P. D. Ellis, *J. Mag. Res.* **26**, 169 (1977).
5. C. S. Peters, R. Codrington, H. C. Walsh, and P. D. Ellis, *J. Mag. Res.* **11**, 431 (1973).
6. H. E. W. Hill and R. E. Richards, *J. Phys.* **E21**, 977 (1968).
7. D. I. Hoult and R. E. Richards, *J. Mag. Res.* **24**, 711 (1976).
8. V. R. Cross, R. K. Hester, and J. S. Waugh, *Rev. Sci. Instrum,* **47**, 1486 (1976).
9. J. Reisert, *Ham Radio* (Oct. 1975).
10. D. I. Hoult and R. E. Richards, *Electron. Lett.* **11**, 596 (1975).
11. H. D. Kepnick, R. Eschie, and W. Maurer, *J. Mag. Res.* **22**, 161 (1976).
12. M. S. Conradi and C. M. Edwards, *Rev. Sci. Instrum.* **48**, 1219 (1977).
13. Avantek, Inc., Model #UTO-511, 3175 Bowers Ave., Santa Clara, CA. 95051.
14. J. B. Grutzner and R. E. Santini, *J. Mag. Res.* **19**, 173 (1975).
15. I. J. Lowe and C. E. Tarr, *J. Phys.* **E21**, 320 (1968).
16. E. O. Stejkal and J. Schaefer, *J. Mag. Res.* **15**, 173 (1974).
17. A. O. Clouse, R. Addlemand, and A. Allerhand, *J. Am. Chem. Soc.* **99**, 79 (1977).

SECTION

V NMR SOFTWARE:
AN OVERVIEW

DAVID L. DALRYMPLE

The wedding of the minicomputer to the nmr spectrometer has been one of the most productive developments in nmr in the past 10 years. The cost-to-benefit ratio of computers has decreased steadily to the point where they are both economical in routine spectrometers and essential in research systems. This article will attempt to illustrate how nmr software has made the minicomputer a valuable component of the total nmr spectrometer system.

Since software and hardware are so dependent upon each other in determining the characteristics of the system, it is necessary to define certain minimal features of the minicomputer that make it applicable to nmr. Discussion will be limited to on-line systems, in other words, those that interact continuously with the spectrometer and the operator in acquiring, processing, and displaying nmr data. This necessitates, in addition to a central-processing unit and memory for data storage, a variety of analog and digital input and output ports, real-time clocks for precise and accurate timing of operations, and, usually, an external memory device such as a magnetic tape or disk unit for storage of programs and data.

The characteristics of the spectrometer-computer interface determine the type and degree of control that the program can have over the nmr experiment. In what follows, emphasis will of course be on the hardware and software used by Nicolet, but systems from other suppliers have similar capabilities in most cases.

I. DATA ACQUISITION

Prior to actual data acquisition in which the nmr signals are collected by the computer, the operator is required to set the experimental conditions.

Much of what was formerly done by knobs and switches on the spectrometer is now done through software in the computer. This includes the setting of observe and decoupling frequencies via programmable synthesizers and the selection of digitizer resolution, low-pass input filter type and cutoff frequency, and sample temperature. If a pulsed nmr experiment is to be done, additional acquisition parameters include the lengths, phases, and repetition rate of the pulses and the choice of single phase or quadrature phase detection. If a frequency-swept experiment is desired, the frequency range and sweep time must also be specified.

The timing of the actual experiment is controlled by a programmable timer. When data acquisition is called for by the user, the pulse lengths and delay times appropriate to the experiment are loaded into the timer's memory along with specification for the output levels desired during each interval. These levels may be used by the spectrometer to turn on observe and decoupling rf signals, change the phase of either of these signals, alter a magnetic field gradient for homogeneity "spoiling," trigger the start of data acquisition, or generate an interrupt in the computer's operations. An example of an interrupt-directed operation would be to change the observe rf frequency regularly for a rapid-scan CW time-averaging experiment.

The program can have built into it a library of pulse sequences for common nmr experiments to simplify going from one to another. Examples include simple one-pulse experiments with or without gated decoupling, T_1 measurements by inversion-recovery or progressive saturation, T_2 measurements by the Hahn spin-echo method or its Carr-Purcell or Carr-Purcell-Meiboom-Gill modifications, etc. It is also possible to provide the user the ability to define new pulse sequences so that the program can be modified for unusual needs.

If the spectrometer is capable of changing the phase of the observe rf signal under computer control, the user can direct the program to correct for a nonzero dc level in the input signal (automatic baseline correction) by changing the phase of the rf by 180° every other pulse and subtracting the input signal from memory. Likewise, with quadrature phase detection, the program can direct 90° phase shifts of the rf on alternate pulses and reroute the input signals accordingly to correct for nonorthogonality and/or unequal amplitudes of the input channels.[1]

Since the timing of data acquisition is controlled by the independently running programmable timer and the analog to digital converter (ADC) has independent (direct) access to the computer's memory, the data-processing routines in the program can be executed by the user more or less as though no acquisition were in progress. This capability is commonly referred to as "foreground-background" operation.

In many experiments it is necessary to obtain several sets of data under slightly different conditions to extract the desired information. Examples of this would be selective decoupling at several different frequencies, T_1 and T_2 measurements, and two-dimensional Fourier transform (2DFT) experiments. The software can provide the means of pre-setting these variables for a series of acquisitions, and then acquiring each of the sets of data in turn. Since, in general, the minicomputer will not have enough built-in memory for all of the data acquired, the program must store each set on disk after it is acquired before starting the next acquisition under different conditions. This can also be done via interrupt control, allowing data for several spectra to be collected and stored in the background while the operator processes other data in the foreground.

II. DATA PROCESSING

After having acquired the raw experimental data in memory or on an associated disk cartridge, it is necessary to process it to obtain the desired spectrum or other information. The discussion here will be directed mostly at processing data from pulsed nmr experiments. In the case of rapid-scan CW time-averaging, may of the same operations are used plus cross correlation of the experimental data with an appropriate line shape.

If the pulsed nmr free-induction-decay (FID) signal was not acquired with automatic baseline correction or if quadrature FIDs were not acquired by alternating input channels, it is generally necessary to correct the data for a nonzero dc level and/or for phase and amplitude errors in the data. The former operation is easily done by calculating the average level of the signal and subtracting it from each data point. Only the last part of the FID should be used in calculating the average in case there is a sharp line near the carrier frequency. Quadrature FIDs can be corrected for unequal amplitudes and nonorthogonality by a Gram-Schmidt orthonormalization operation.[2]

If the FID was collected for at least four to five T_2^*'s for high digital resolution (several points per peak), one can inprove the signal-to-noise ratio of the final spectrum by attenuating the end of the FID with a decreasing linear, exponential, or Gaussian apodization function. Generally, an exponential function is used since the resulting line broadening preserves the Lorentzian shape of the peaks. Conversely, data with a high signal-to-noise ratio can be scaled up with an increasing function to artificially narrow the lines with an attendant decrease in signal-to-noise ratio. Several examples of this are shown in Figure 1.V.1.

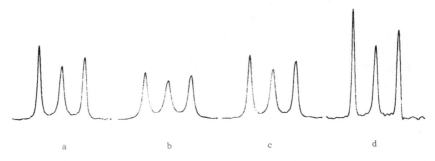

Figure 1.V.1
Effects of apodization on line shape and signal-to-noise. (a) No apodization; line width = 4.5 Hz; S/N = 55. (b) 3 Hz Lorentzian broadening; S/N = 96. (c) 3 Hz Gaussian broadening; S/N = 85. (d) 3 Hz Lorentzian narrowing + 3 Hz Gaussian broadening; S/N = 48.

The Fourier transformation of the time-domain FID into a frequency-domain spectrum is done using variations of the Cooley-Tukey fast Fourier transform algorithm. In the case of quadrature data, a complex FT must be done to convert the phase information into positive and negative offset frequencies. Also, since the spectrometer generally does not provide a correctly phased FID, both the real and imaginary (sine and cosine) parts of the transform are retained in memory for easy phase correction of the resulting spectrum. To retain the frequency resolution present in the FID, it is possible to "zero-fill" it to twice its original length before the transform, thus producing a real spectrum with the same number of points as were present in the FID.

The zero- and first-order (constant and linearly frequency-dependent) phase corrections required to produce a pure absorption-mode spectrum are easily accomplished by correcting two peaks in the spectrum separately. The progam can then calculate the correction required at each point in the spectrum and apply it.

Both the Fourier transformation and the phase correction are speeded up by using a sine look-up table in the program. Rather than calculating (via a series approximation) each sine term, it can be interpolated from the table as needed.

In the case of 2DFT experiments, the 2^m 2^n-word FIDs processed as above must be Fourier transformed in *both* time dimensions. Since the total number of points (2^{m+n}) will generally be larger than the amount of memory available, it is required that the data be reorganized on disk into blocks that can be brought into memory and processed individually. The operation is essentially the transposition of two ($2^m \times 2^{n-1}$) matrices into two ($2^{n-1} \times 2^m$) matrices. The real and imaginary data matrices are kept separate providing $2^{n-1}\,2^{m+1}$-word complex interferograms for Fourier transformation in the second dimension.

III. DATA ANALYSIS AND OUTPUT

For normal spectra, data analysis and output consists of measuring and listing peak positions, intensities, and areas and plotting the spectrum and its integral. Due to the finite resolution of the digital data, the location of the highest data point on a peak will seldom be the location of the center of the peak. If, for example, the digital resolution is 0.5 Hz/point (corresponding to a 2-sec acquisition time), the highest point may be ±0.25 Hz from the center of the peak. It is common practice, therefore, to interpolate the peak position using the coordinates of the highest data point and those of the points on each side of it. Either a Lorentzian or a parabolic interpolation function may be used, generally with essentially identical results. If accurate peak heights are also important, they can also be interpolated using the same formulae.

Given sufficient digital resolution compared to peak widths, simple trapezoidal integration can provide accurate peak areas. The major problem in integrating a spectrum with noise and perhaps a sloping or curved baseline is the discrimination of the beginning and end of a peak. The approach used in Nicolet software is to leave this decision to the user, much as one would measure the height of a plotted integral.

A versatile routine for plotting spectra or integrals is perhaps the most complicated part of the software package. Factors that contribute to this are the digital nature of the data and the need therefore to interpolate between points; plotting at constant pen speed, meaning that horizontal movement slows down when the pen is moving vertically to define a peak; the ability to drive a variety of plotters including analog $x-y$, digital-x analog-y, and fully digital (stepping) $x-y$ recorders; the ability to scale a plot to any user-defined dimensions within the capabilities of the recorder; and finally, doing all plotting in the background allowing the user to process other data while the plot is in progress.

Plot interpolation is generally linear between data points, but if a highly expanded plot is done (relatively few data points per cm) this leads to angular peaks. Also, if the number of data points per peak is small, such a plot will not indicate true peak heights. For these reasons the user is given the option of plotting with cubic interpolation between points using the coordinates of the nearest four data points in determining the actual height of any point in the plot. An example of this is shown in Figure 1.V.2.

Figure 1.V.2
Plot interpolation. The same data are plotted on the left with linear interpolation between data points and on the right with cubic interpolation.

The ability to plot a spectrum or group of spectra in the background implies that the plotting is controlled by interrupt hardware. Since the plotter is a relatively slow device, limited by how fast the pen can move, a clock can be used to interrupt the foreground operation every few milliseconds to allow the background plotting routine to move the pen one step in the x and/or y direction. To make the background plotting operation completely transparent to the user's operations in the foreground, data memory should not be used for the spectrum being plotted. This means essentially plotting data from the disk. When several spectra are to be plotted sequentially with small x- and y-offsets (a stacked plot), as is commonly done with relaxation time studies and 2DFT data, the routine must plot the entire series with offsets and *whitewashing* of hidden lines as specified by the user. An example from a 2DFT experiment is shown in Figure 1.V.3.

Another useful plotting routine is the "registered" plot, where a portion of a previously plotted spectrum is expanded vertically and plotted above the corresponding region of the full spectrum. The same applies to partial integral plots. Both these operations are illustrated in Figure 1.V.4.

For relaxation time studies, additional routines must be provided for extracting peak intensities or areas from a series of spectra and fitting an appropriate function using a least-squares method to determine the independent variables. Typical functions range from $y = A \exp(-t/T_2)$ for T_2 measurements to $y = A(1 - \{1 + P[1 - \exp(-d/T_1)]\} \exp(-t/T_1))$ for fast inversion-recovery T_1 measurements (d is the recovery time after the observe pulse and P is a measure of the rf field homogeneity).[3]

The inclusion of a generalized nonlinear least-squares routine in the progam allows Lorentzian or Gaussian functions to be fit to experimental peaks to measure accurate line widths. Other common features of the nmr software include routines for adding or subtracting two spectra (with adjustable offsets and weighting), generating various projections of 2DFT data, selecting a portion of a spectrum for detailed examination by

Figure 1.V.3
Portion of an offset stacked plot from a proton 2DFT experiment illustrating hidden line whitewashing.

displaying it full-scale on the oscilloscope, and simple but useful operations such as baseline adjustment, signal-to-noise ratio calculations, and spectrum smoothing.

IV. SOFTWARE SYSTEM DESIGN

In developing a software package to implement all of the above features, several hardware characteristics must be considered. Of primary concern is the fact that only a limited amount of computer memory is available for the program. In current Nicolet systems, 8K words are so allocated. The program described (NTCFT) is in fact over 26K words in length, meaning that only a portion of it can be in memory at any one time, the remainder

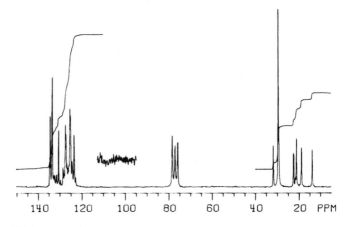

Figure 1.V.4
Carbon nmr spectrum illustrating registered plots of integrals and 10 × expanded region of noise.

being stored on disk until needed for a particular operation. The use of the 8K program area by NTCFT is shown in Figure 1.V.5 to illustrate a typical design.

The first block of memory contains common parameters describing the state of the program, the nature of the data, the experiment's pulse sequence, etc., a subset of which is stored with each spectrum on disk to identify it when it is recalled to memory. The next block of memory contains the utility routines for input and output involving the keyboard/printer, the oscilloscope, and the disk-storage device. Also included in this region are the command interpreter and several commonly used arithmetic routines. The next two blocks contain the data-acquisition and plotting-interrupt service routines for their background operation. All of these sections of the program are resident in memory at all times. The next section of memory is used as the overlay area for the various routines called by the user. When any one of approximately 150 commands is typed by the operator, the command interpreter determines which overlay contains the appropriate subroutine, loads the overlay from disk into memory, and executes the selected operation. Any interrupt

Parameters	I/O routines	Interrupt routines	Overlay area	Scratch /buffer

Figure 1.V.5
Program area allocation in NTCFT-1180.

temporarily transfers program operation to the appropriate interrupt service routine and then back to the foreground operation in the overlay area.

The last section of program memory is used as a scratch or buffer area by various routines. For example, the first half of it is used to store the part of a disk file currently being plotted. It is reloaded from disk with the next section as often as is needed to plot the entire spectrum.

A few other design considerations implicit in an interrupt-driven, real time, disk-based system are related to possible conflicts among the various operations proceeding "simultaneously." Most such situations are avoided by the priorities assigned to the various types of interrupts. Thus an interrupt from the ADC signaling the end of one scan and the need to get ready for the next will automatically take precedence over an interrupt from the clock that is timing the plotting operation. Other more subtle constraints include avoiding the use of common variables and subroutines by both foreground and background operations, protecting data being acquired in the background from inadvertent corruption by the user with a foreground routine, and keeping track of possibly different parameters for data being acquired in one section of memory and data being processed in another.

Hopefully, the software will recognize its limitations and its capabilities in a manner that allows the user to feel totally in control of the operation of the spectrometer and the manipulation of the data it produces.

REFERENCES

1. E. O. Stejskal and J. Schaefer, *J. Mag. Res.* **13,** 249 (1974); also D. I. Hoult and R. E. Richards, *Proc. Roy. Soc.* **A344,** 311 (1975).

2. C. L. Perrin, *Mathematics for Chemists*, Wiley-Interscience, New York, 1970.

3. G. Levy and I. Peat, *J. Mag. Res.* **18,** 500 (1975).

SECTION

VI TWO-DIMENSIONAL FOURIER TRANSFORM ^{13}C NMR

DAN TERPSTRA

I. INTRODUCTION

A perennial problem in nmr studies of complex organic systems is an overabundance of information or, more accurately, too high an information density. An available solution to this problem in the case of ^{13}C nmr has been the removal of ^{13}C—^{1}H coupling information through broadband proton decoupling. A more satisfactory solution, however, would retain all of the information inherent in the experiment and develop alternative methods of presenting this information. The availability of the minicomputer in the nmr laboratory has made this approach feasible. One such technique, that of selective or tailored excitation, is discussed in Chapter 1.VIII. Another experimental approach that has recently been developed is two-dimensional Fourier transform (2DFT) nmr spectroscopy.[1-10] Various applications of the 2DFT technique allow easy access to both scalar[3-8] and dipolar[1,10] heteronuclear coupling constants without interference from other chemically shifted nuclei, provide resolution approaching natural linewidths essentially independent of magnet inhomogeneity,[4-8] and permit homonuclear broadband decoupling.[11,12] The latter technique has particular application for proton nmr but also will be useful for measuring carbon—carbon coupling constants in enriched samples.

II. GENERAL DESCRIPTION

In its most general definition two-dimensional nmr is simply an expression of experimental information as a function of two independent parameters.[13] An example of this is the familiar *stacked plot* of relaxation experiments, where the data is presented as a function both of frequency (field) and of delay time t between pulses. With this definition 2DFT nmr can be considered a subset of two-dimensional nmr. The data in the 2DFT experiment is collected as a function of two independent time parameters and transformed along *both* of these parameters to yield a presentation as a function of two independent frequency parameters.

Since two time parameters are needed to define the data array, the time frame of the experiment is divided into three separate periods as shown in Figure 1.VI.1. These have been called[1] the *preparation, evolution,* and *detection* periods.

During the preparation period, as the name implies, the spins are prepared in a specified state. For ^{13}C this can include cross polarization, as in proton-enhanced experiments[9,10] or, more typically, broadband proton decoupling to allow the establishment of nuclear Overhauser enhancement (NOE).

The evolution period, defined by a time parameter t_1, allows the spins to develop under a defined set of conditions. This can be a field gradient as in FT zeugmatography.[14] During the evolution period the observed ^{13}C nuclei may be coupled to, or decoupled from, another set of spins such as protons; the experiment can also incorporate spin-echo refocusing as in T_2 experiments.[2]

The detection period, at the end of the evolution period, is defined by a data-collection parameter t_2 that is analogous to the running time

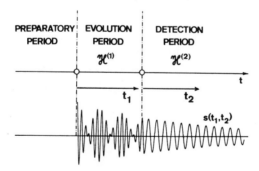

Figure 1.VI.1
Time frame division in a 2-DFT experiment.[1]

parameter of the free induction decay (FID) in conventional FT nmr. The conditions during this period usually differ in some way (for example, the state of the decoupler) from those of the evolution period to permit separation of some of the information in the nmr experiment.

For a given evolution time t_1, the (preparation–evolution–detection) sequence can be repeated and coherently averaged as many times as necessary to provide suitable signal-to-noise levels. The result of such a data collection scheme is a two-dimensional array $S(t_1 t_2)$, stored on a bulk-memory system, such as a cassette or disk. The t_1 axis, defining the columns of this array, is specified by a series of equally spaced delays that define the evolution period. The t_2 axis, corresponding to the rows of the array, is determined by the equally spaced points collected to form the FID. These rows can be treated identically in all respects to a one-dimensional FID regarding exponential weighting, apodization, and Fourier transformation. This can be carried out either before or after each row is stored in bulk memory to yield a time-frequency array $S(t_1 F_2)$ somewhat analogous to a spin-lattice relaxation spectral set.

As in the case of one-dimensional transforms, this first Fourier transformation results in complex frequency domain information in either single-channel or quadrature detection schemes. This is inherent in the Fourier transform itself, since either real or complex time information yields complex frequency information, which can be written in terms of independent sine and cosine components

$$S^S(F) = \int_0^\infty S(t) \sin 2\pi F t\, dt$$

$$S^C(F) = \int_0^\infty S(t) \cos 2\pi F t\, dt$$

By suitable linear combinations of the sine and cosine components, phase errors can be corrected, resulting in pure dispersion (sine) and absorption (cosine) components of the frequency domain spectrum

$$S^S(t_1 F_2)$$
$$S^C(t_1 F_2)$$

Phasing at this point in the data manipulation scheme is nonessential, however, because information along the t_1 axis can be encoded as phase or amplitude modulation of the F_2 axis. The fast Fourier transformation (FFT) algorithm automatically scales data during the course of a transformation to best utilize the digital resolution of the computer word. Therefore it is essential to maintain a record of the scaling factor for each row to allow eventual normalization of the entire array when data manipulation is completed.

Once the transformation of the rows has been carried out, the array must be transposed to allow transformation of the columns. This is easily done on a computer large enough to contain the entire array in active memory, but for minicomputers of the size normally used in nmr laboratories, this would lead to prohibitive restrictions in the size of array. One way to solve this problem[2] involves dividing data memory into two equal sections. For example, if we assume that the array $S(t_1 F_2)$ forms a 64×1024 point matrix and that the minicomputer has 16K of available data memory, each section of memory would contain 8K locations. This allows the first 8 rows to be read into one of the two sections. The first 128 points of each can then be used to begin assembling the first 128 of the necessary 1024 64-point columns in the other half of memory. The remaining 56 rows are then read 8 rows at a time into the first half of the memory to finish assembling the transpose of the first one-eighth of the array. These 128 columns can then be conditioned and transformed in the same manner as the rows, after which they are rewritten into bulk storage, taking care to remember all scaling factors as mentioned previously.

The next 128 columns are then transposed in the same manner and transformed, repeating this procedure until all 1024 columns are acted upon. Thus, for an array of this size, assuming each row to be stored in bulk memory as an independent file (typical for most conventional nmr software), each file must be read for each of $128/1024 = 8$ sets of columns resulting in a total of $64 \times 8 = 512$ disk (or cassette) read operations. If the data acquisition and storage software were rewritten to allow the rows of the array to be stored in 8K blocks in bulk memory, this large number of reads can be reduced by a factor of 8 to 64 operations.

An alternative approach[15] takes advantage of all of data memory rather than just half for performing the second transformation. A variable portion of *program* memory equal in size to one row of the array (1024 words in this example) is set aside as a buffer memory. The rows can then be read into this buffer one at a time and used to assemble transposed columns in the entire data memory, reducing by a factor of two the number of times the sequence must be repeated to transform the entire array. This technique requires 256 read operations and no modification of the file storage structure of the standard nmr program. However, it will be limited by the amount of memory available as data buffer.

In both of these techniques zero-filling[16,17] can be employed to increase the digital resolution along the t_1 axis. After completion of the second transformation and rescaling of the array, the data is stored in bulk memory as a function of two frequency parameters, $S(F_1F_2)$. As with the first transform, the second yields both sine and cosine terms, resulting in an array consisting of four quadrants

$$S^{CC}(F_1F_2): \text{absorption–absorption}$$

$$S^{CS}(F_1F_2): \text{absorption–dispersion}$$

$$S^{SC}(F_1F_2): \text{dispersion–absorption}$$

$$S^{SS}(F_1F_2): \text{dispersion–dispersion}$$

Linear combinations of these four quadrants are necessary to phase the array along both axes.[2]

Obviously, the cosine–cosine quadrant is the most desirable for high-resolution work, but, because of phase anomalies discussed by Bodenhausen et al.[2] and by Bachmann et al.,[18] these arrays have commonly been displayed in an absolute value mode, analogous to power spectra in conventional FT nmr spectroscopy. Recent work has indicated that, by employing a carbon refocusing pulse orthogonal to the sampling pulse just prior to detection, it is possible to sort out absorptive and dispersive components and provide an absorption–absorption display.[18] This technique is applicable in many cases, except in those designed to observe homonuclear coupling or in some experiments involving strong heteronuclear coupling.

Two techniques are currently employed to graphically display the information of a 2DFT experiment. The first approach is that of an intensity contour map.[9,10] Figure 1.VI.5 gives an example of this, showing a separated local field experiment.[7] The second graphical approach is very similar to the *stacked plots* of T_1 relaxation studies. The rows (or columns) of the array are plotted parallel to each other with suitable x and y offsets to give the impression of a three-dimensional surface. Addition of a *whitewash* routine[2] to give the impression of a solid surface is an aesthetic refinement to this technique. The stacked-plot approach is advantageous, not only because it bears close resemblance to familiar nmr spectra, but also because it can be augmented using conventional nmr plotting software.

III.   APPLICATIONS

A.   δ-Sorting

The most straightforward example of ^{13}C 2DFT nmr can be called the δ-sorting experiment.[1,3] As its name implies, the chemical shift (δ) values

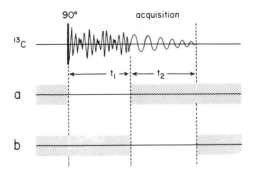

Figure 1.VI.2
Two possible heteronuclear (^{13}C—^{1}H) pulse sequences for the δ-sorting 2DFT experiment. Both sequences provide NOE signal enhancement. In sequence (a) coupled spectra are on the t_1 axis and decoupled spectra are on the t_2 axis. The axes are interchanged in sequence (b).

along the F_1 axis are employed to provide the spread of information into two dimensions. Two heteronuclear pulse sequences for such an experiment are shown in Figure 1.VI.2. Utilization of the *a* sequence will result in a transformed array $S(F_1 F_2)$ in which the F_1 axis (where resolution is limited by the number of different evolution times t_1 that are used) contains proton-decoupled and chemically shifted carbon singlets. The F_2 axis with its greater potential resolution (determined by the number of points in each FID) will contain ^{13}C—H or ^{13}C—X scalar coupling information in addition to chemical shifts. Sequence *b* provides the same information except that now the axes will be interchanged. This poses no disadvantages for cases in which square arrays are employed, but, for those instances in which time rather than bulk memory is the limiting factor, it is advantageous to define the F_1 axis as the informationally less-dense decoupled axis. An example using sequence *b* is shown in Figure 1.VI.3 for *n*-hexane. A square (64×64) array was employed here, but even at this low level of digitization the multiplet patterns are clearly discernible. If the display of Figure 1.VI.3 is viewed as its projection onto the horizontal *x* axis, the high-resolution proton-coupled ^{13}C spectrum can be visualized. Projection onto the vertical *y* axis results in the three singlets of the proton-decoupled spectrum.

2DFT δ-sorting has many potential applications in the ^{13}C nmr analysis of complex organic and biological systems. Until now, most ^{13}C data on such systems has been obtained with broadband proton decoupling to remove the complexity of overlapping multiplets and also to enhance sensitivity through line narrowing and NOE effects. Now, by use of

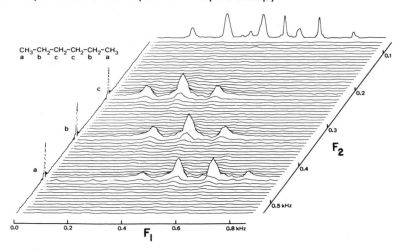

Figure 1.VI.3

δ-Sorting experiment on *n*-hexane using sequence (b). The 64×64 point data array limits resolution but the multiplets are clearly discernable on the F_1 axis. Projection onto the $F2$ axis shows the proton decoupled spectrum. This experiment was performed by Müller et al.[3] on a SXP4-100 nmr spectrometer equipped with a 620/L-200 computer.

decoupler gating techniques and two-dimensional data sorting, the added information afforded by retention of C—H scalar couplings can be readily employed both in spectral assignment as well as in Karplus-type bond angle studies, as demonstrated in polypeptide systems.[19]

B.   Separated Local Field NMR

A two-dimensional experiment that allows access to dipolar coupling constants in solids has been called[9,10] separated-local-field nmr. This technique is described in some detail in Chapter 1.X of this volume; a brief account follows. The pulse sequence required for this technique is shown in Figure 1.VI.4. During the preparatory period, both proton and carbon resonances are irradiated so as to fulfill the Hartmann-Hahn[20] spin-locking condition. The carbon magnetization is allowed to evolve during t_1 under proton coupled conditions; it is then detected during t_2 with proton decoupling. Figure 1.VI.5 demonstrates that this results in a (contour map) display in which dipolar couplings are presented on the F_1 axis and chemical shifts on the F_2 axis.

As discussed in Chapter 1.X, this technique can provide valuable information about the direction of C—H vectors in crystals and oriented polymer samples.

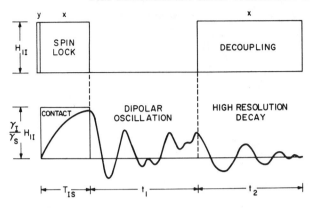

Figure 1.VI.4

Pulse sequence for the SLF experiment.[10] During the preparatory period, T_{IS}, the nuclei are Hartmann-Hahn matched and magnetization is transferred to the carbon S spins. During the evolution period, t_1, the S spins precess in the dipolar field of near-neighbor proton I spins. During the detection period, t_2, the I spins are decoupled and the high resolution S spin FID is recorded. The Fourier-transformed data yields the frequency array $S(F_1F_2)$.

C. *J* Spectroscopy

Spin-echo *J* spectroscopy has been available as an FT nmr technique for a number of years.[21] In spite of the fact that it offers access to natural linewidths (T_2 information), as well as allowing accurate scalar coupling determinations, this technique has never gained widespread application. This results primarily from the fact that in addition to eliminating line broadening effects spin-echo formation also eliminates chemical shift information, generally resulting in serious signal overlap in systems with more than two or three spins. In addition, the method as it was first described was strictly applicable only to first-order spectra. However, with the development of 2DFT techniques, *J* spectroscopy has come to the fore[4-8] as a viable sorting technique for defining one of the axes of the experiment.

The most widely applied pulse sequences for two-dimensional *J* spectroscopy are shown in Figure 1.VI.6 These four sequences can be broken into two sets of two. Sequences *a* and *b*, called *gated decoupling*, apply proton noise decoupling for half of the echo time t_1. This can be the first half of the echo as shown or the last half with similar results. Proton decoupling serves to prevent the complete refocusing of the C—H multiplets, as would occur if no decoupling were employed during this period. Since the coupling is active for only half of the evolution period, the

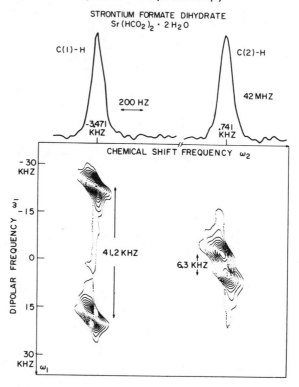

Figure 1.VI.5

Contour map presentation of the results of a SLF experiment on a $Sr(HCO_2) \cdot 2H_2O$ crystal[10] oriented with the C symmetry axis along the magnetic field direction. Chemical shifts are evident on the horizontal (F_2) axis and dipolar couplings can be seen on the vertical (F_1) axis. The C(1)—H and C(2)—H bonds make angles of $\theta_1 = 16.0 \pm 4.5°$ and $\theta_2 = 49.4 \pm 3.1°$ with the external magnetic field, respectively.

resultant J modulation is half of what would be obtained for a conventional coupled experiment. An alternative approach is shown in sequences c and d. Here a π pulse is applied to both carbon and proton spins simultaneously, inverting both spin populations and allowing J modulation for the entire evolution period. This technique, termed *proton flip* results in coupling constants equal to those of the conventional experiment leading to a twofold improvement in resolution over the gated-decoupler technique. As is shown, sequences a and c gate the decoupler off during detection to give coupled spectra along the F_2 axis, whereas

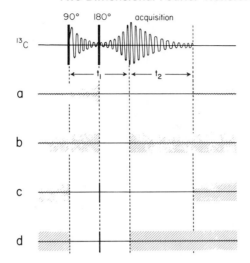

Figure 1.VI.6
The pulse sequences commonly used in two-dimensional J spectroscopy. The top trace shows the carbon irradiation sequence with definition of the evolution (t_1) and detection (t_2) time periods. The bottom 4 traces show various possible proton irradiation sequences. All sequences gate the decoupler on during the preparatory period to achieve NOE signal enhancement. Sequences (a) and (b) demonstrate the "gated decoupler" mode both with and without decoupling during detection. The "proton flip" mode is shown in (c) and (d) with and without decoupling during detection.

sequences b and d gate the decoupler on during this period to provide only chemical shift information along F_2.

Figures 1.VI.7 and 1.VI.8 are examples of sequences a and c for methyl iodide. The difference in apparent coupling constants is clearly visible in these illustrations. Both the gated-decoupler and the proton-flip techniques have been applied to the strongly coupled pyridine system and the results have been discussed in recent publications.[7,8]

The improvement in linewidth as a result of spin-echo formation is clearly illustrated by comparison of Figures 1.VI.8 and 1.VI.9. These figures show the same experimental data for CH_3I plotted in the first case parallel to the F_1 axis and in the second parallel to the conventional F_2 axis.

Also of some interest is the definition of frequencies along the two axes. The F_2 axis has an arbitrary zero value coincident with the carrier frequency as in conventional nmr. The F_1 axis, however, contains no chemical shift frequencies and hence all multiplets are centered around an absolute zero value. This is demonstrated more clearly in the spectrum of

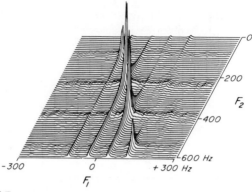

Figure 1.VI.7
An absolute-value mode presentation of the application of the "gated de-coupler" technique to ^{13}C-enriched methyl iodide. Sequence (a) was used in this example resulting in a coupled spectrum along both axes. The frequencies in the F_1 dimension are $\pm 3/4\,J$ and $\pm 1/4\,J$ Hz.

2-methylpentane shown in Figure 1.VI.10. This spectrum was obtained using sequence b described above, and thus it is decoupled along the F_2 axis. Line narrowing is not evident here due to the low level of digitization.

A number of problems must be dealt with in J spectroscopy. First, any deviation from first-order spectra results in asymmetry of the multiplet.

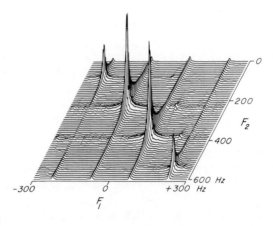

Figure 1.VI.8
An absolute-value mode presentation of the application of the "proton flip" technique to methyl iodide using sequence (c). In this case the frequencies in the F_1 dimension are at $\pm 3/2\,J$ and $\pm 1/2\,J$, providing an increase in resolution by a factor of 2 over the "gated decoupler" technique.

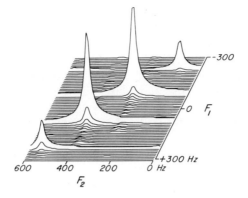

Figure 1.VI.9
The transpose of the array in Figure 1.VI.8. Comparison of these two figures demonstrates the much smaller line widths in the F_1 dimension as a result of spin-echo refocusing. Figures 1.VI.7–9 were taken from Ref. 2.

This makes it essential in non-first-order systems to be able to differentiate positive and negative frequency components. The solution to this problem is inherent in the data manipulation scheme. The Fourier transformation of the first axis (t_2) is always complex. Thus, the input for the second transformation along the t_1 axis is also complex as in the case of quadrature detection. This implies that linear combinations (sums and

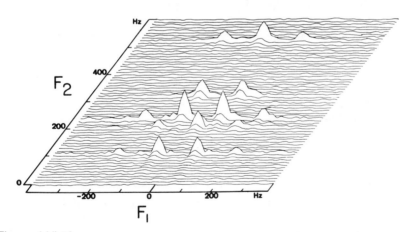

Figure 1.VI.10
A 64×64 point two-dimensional J spectrum of 2-methylpentane[6] collected in the "gated decoupled" mode of sequence (b) in Figure 1.VI.6. All of the multiplets on the F_1 axis are centered around zero frequency as a result of the loss of chemical shift information on this axis.

differences) of the four resulting quadrants will always yield positive and negative frequencies along the second axis. This procedure is described in further detail by Bodenhausen et al.[2]

A second experimental problem in *J* spectroscopy arises from imperfections in the refocusing pulse.[22] This results in two types of extraneous peaks known as *phantoms* and *ghosts*. These spurious signals can be removed (or exorcised) quite effectively by the four-step phase-cycling procedure outlined in Table 1.VI.1. This sequence has been affectionately dubbed "Exorcycle" by its creators.[22] An example of its efficacy, again for methyl iodide, is shown in Figure 1.VI.11, where the π pulse has deliberately been mis-set to 130°.

A third drawback of *J* spectroscopy is not quite as simple to resolve. The axis containing the unique information (i.e., narrow lines and no chemical shifts) is the evolution (F_1) axis. The resolution along this axis is determined by the number of t_1 values and thus by the number of separate data acquisitions performed. Thus, for a high level of frequency resolution to be achieved on this axis, either a large number of experiments must be performed, or a very narrow frequency window must be observed. Analog filtering is not possible along this axis, so care must be taken to avoid signal aliasing difficulties. Fortunately, the absence of chemical shifts along this axis generally limits the necessary frequency window to ±300 Hz.

In spite of the above difficulties, two-dimensional *J* spectroscopy promises to be a valuable technique for sorting complex spin systems. One application where it might prove quite valuable is in nmr spectroscopy using very large sample volumes in superconducting magnets, where magnetic field *stability* is high but field *homogeneity* may be limited.

Table 1.VI.1

A phase-cycling Sequence that Cancels "Phantom" and "Ghost" Responses in Two-Dimensional *J* Spectra

	Phase Angles (deg)		
	90° Carbon Pulse	180° Carbon Pulse	Receiver Reference
(a)	0	0	0
(b)	0	90	180
(c)	0	180	0
(d)	0	270	180

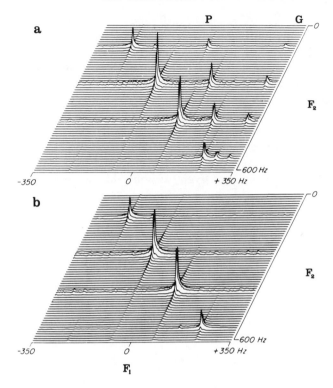

Figure 1.VI.11
Two-dimensional J spectra of ^{13}C in methyl iodide[22] obtained by the "proton flip" technique, but with the ^{13}C refocusing pulse deliberately misset to a flip angle of 135° rather than 180°. In spectrum (a) the phantom (P) and ghost (G) multiplets are clearly visible, whereas in spectrum (b) they have been canceled by using the phase-cycled pulse sequence of Table 1.VI.1.

D. Broadband Homonuclear Decoupling

The final 2DFT technique that will be discussed has as yet only been applied to proton spin systems, but it is equally valid for other nuclei as well. This technique is a form of J spectroscopy applied to coupled homonuclear rather than heteronuclear spin systems.[11,12] The necessary pulse sequence is identical to the one applied to the carbon spins in heteronuclear J spectroscopy. For coupled spin systems such as protons, this results in a two-dimensional frequency array with coupling information along the F_1 axis and both coupling and chemical shift information on the F_2 axis, similar to pulse sequence d of the previous section. Since coupling information is contained on both axes, the spin   multiplet appears to lie on a line bisecting the F_1 and F_2 axes. Perhaps this is most clearly evident in the heteronuclear methyl iodide multiplet of Figures 1.VI.8 and 1.VI.9. This suggests a digital method of producing a singlet from a multiplet presented in such a manner. By shifting each row of the array an appropriate amount and summing it to the next (in other words, by summing the entire array along its bisector), a singlet can be produced from the individual peaks of a multiplet. Chemical shift information will be retained. Of course, this much more readily achieved in the heteronuclear case simply by broadband irradiation of one set of spins. In the homonuclear case, however, it provides a method for indirect homonuclear broadband decoupling. Even without this summation procedure,

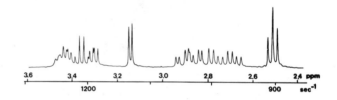

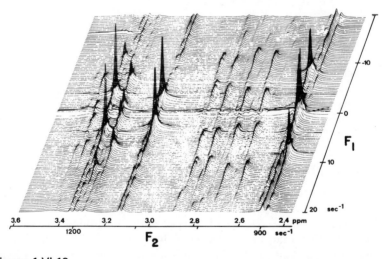

Figure 1.VI.12

Portions of one-(top) and two-(bottom) dimensional ^{1}H nmr spectra at 360 MHz of D$_2$O solution containing 0.1 M of each of the five amino acids Ala, Ile, Met, Tyr, and His, pD = 10.5, T = 25°C. Chemical shifts are relative to internal 2,2-dimethyl-2-silapentane-5-sulfonate (DSS) and are indicated both in hertz and in parts per million.[12]

overlapping multiplets are effectively sorted as is shown in a small portion of the proton spectrum of a mixture of five amino acids, shown in Figure 1.VI.12. Due to the method of presentation of this array, the multiplets do not obviously appear to lie along its bisector.

This technique is not directly applicable to natural abundance ^{13}C studies since the low natural isotopic abundance of ^{13}C leads to an insignificant contribution from ^{13}C—^{13}C coupled multiplets. However, it may prove useful in studies of selectively and especially uniformly ^{13}C-enriched compounds. A presentation such as this would allow straightforward analysis of ^{13}C—^{13}C coupling patterns with no complications introduced by overlapping multiplets. Similar applications can also be envisioned for other spins such as phosphorous or fluorine.

Since this experiment is quite close in nature to heteronuclear J sorting, the problems and limitations are similar as well.

IV. CONCLUSION

Thanks to rapid developments in digital electronics technology, the future of two-dimensional Fourier transform nmr looks promising. Limitations imposed by computational or bulk-memory size and cost and by current computational speeds are temporary at worst. In addition, computation times can be dramatically reduced with present hardware by simply saving only those columns of the data array with high information content following the first transform. These columns are then the only ones that need undergo the second transformation. This greatly reduces the effective size of the array without substantial loss of information. Some of the aesthetic appeal of the display is, however, inevitably lost in such a technique.

The basic unremovable obstacle in 2DFT nmr is spectrometer sensitivity. The amount of time necessary to acquire data for a two-dimensional array is directly proportional to the digital resolution required along the F_2 axis. The severity of this restriction is being reduced with use of larger sample volumes and with continuing improvements in nmr instrumentation.

Regardless of the improvements made in computer or spectrometer technology, the two-dimensional experiment will always require a substantially larger time investment than a conventional nmr experiment, and because of this, care must be exercised to ensure that the increased accessibility of information in the two-dimensional array warrants the additional time investment.

REFERENCES

1. W. P. Aue, E. Bartholdi, and R. R. Ernst, *J. Chem. Phys.* **64**, 2229 (1976).

2. G. Bodenhausen, R. Freeman, R. Niedermeyer, and D. L. Turner, *J. Mag. Res.* **26**, 133 (1977).

3. L. Müller, A. Kumar, and R. R. Ernst, *J. Chem. Phys.* **63**, 5490 (1975).

4. G. Bodenhausen, R. Freeman, and D. L. Turner, *J. Chem. Phys.* **65**, 839 (1976).

5. G. Bodenhausen, R. Freeman, R. Niedermeyer, and D. L. Turner *J. Mag. Res.* **24**, 291 (1976).

6. L. Müller, A. Kumar, and R. R. Ernst, *J. Mag. Res.* **25**, 383 (1977).

7. R. Freeman, G. A. Morris, D. L. Turner, *J. Mag. Res.* **26**, 373 (1977).

8. G. Bodenhausen, R. Freeman, G. A. Morris, and D. L. Turner, *J. Mag. Res.* **28**, 17 (1977).

9. S. J. Opella and J. S. Waugh, *J. Chem. Phys.* **66**, 4919 (1977).

10. E. F. Rybaczewski, B. L. Neff, J. S. Waugh, and J. S. Sherfinski, *J. Chem. Phys.* **67**, 1231 (1977).

11. W. P. Aue, J. Karhan, and R. R. Ernst, *J. Chem. Phys.* **64**, 4226 (1976).

12. K. Nagayama, K. Wüthrich, P. Bachmann, and R. R. Ernst, *Biochem. Biophys. Res. Comm.* **78**, 99 (1977).

13. R. R. Ernst, *Chimia* **29**, 179 (1975).

14. A. Kumar, D. Welti, and R. R. Ernst *J. Mag. Res.* **18**, 69 (1975).

15. D. Terpstra and G. C. Levy, *unpublished results*.

16. R. Freeman and R. C. Jones, *J. Chem. Phys.* **52**, 465 (1970).

17. E. Bartholdi and R. R. Ernst, *J. Mag. Res.* **11**, 9 (1973).

18. P. Bachmann, W. P. Aue, L. Müller, and R. R. Ernst, *J. Mag. Res.* **28**, 29 (1977).

19. V. F. Bystrov, *Prog. NMR Spectrosc.* **10**, 41 (1976).

20. S. R. Hartmann and E. L. Hahn, *Phys. Rev.* **128**, 2042 (1962).

21. R. Freeman and H. D. W. Hill, *J. Chem. Phys.* **54**, 301 (1971).

22. G. Bodenhausen, R. Freeman, and D. L. Turner, *J. Mag. Res.* **27**, 511 (1977).

SECTION

VII VARIABLE TEMPERATURE
^{13}C NMR

F. A. L. ANET

This section is restricted to a review of papers dealing with temperature measurements of samples in spectrometer probes. Problems arise in ^{13}C nmr that do not occur in ^{1}H nmr. For example, ^{13}C nmr spectra are usually obtained with protons decoupled by the application of several watts of radiofrequency power, which can give rise to heating effects and to difficulties in measuring the temperature near the sample by means of thermocouples. The well-established temperature dependence of the OH chemical shifts in methanol or ethylene glycol is very useful as a "proton thermometer,"[1] but it cannot be used with ^{13}C nmr, with the possible exception of dual frequency probes that allow observations to be made of either the ^{1}H or ^{13}C spectrum without removing the sample from the probe.

Most organic compounds show more or less temperature-dependent ^{13}C chemical shifts, but the effects are often small and are not completely understood.[2,3] Commonly used internal references, such as tetramethylsilane, show temperature-dependent ^{13}C chemical shifts[4] when comparisons are made with known temperature-independent nmr chemical shifts.[5] In the case of tetramethylsilane a deshielding has been reported both below approximately 20°C and above 40°C, although the effect above 40°C may be a result of a data extrapolation that was required in the analysis.[4] The small temperature-dependent shifts observed with hydrocarbons and most of their derivatives are not very suitable for use in temperature measurements.

Several systems having relatively large ^{13}C temperature-dependent chemical shifts have been suggested as suitable thermometers for ^{13}C nmr spectroscopy. The equilibria between carbonium and halonium ions can lead to averaged chemical shifts that have a very strong temperature dependence,[6] and this is a desirable feature in an nmr thermometer. However, the compounds are sensitive to water and require special handling, and the temperature range is rather limited.

Methyl iodide and other iodides, such as methylene iodide, have ^{13}C chemical shifts that are appreciably dependent on temperature. Vidrine and Peterson[7] have proposed two ^{13}C nmr thermometers based on these compounds, together with a second component, whose ^{13}C chemical shift is not temperature-dependent. Using a $1:3$ by volume mixture of tetramethylsilane and methyl iodide, these authors established a temperature scale given by

$$\frac{1}{T} = 0.0161165 - 0.000570057 \cdot \Delta\delta \qquad (1)$$

where $\Delta\delta$ is the ^{13}C chemical shift difference in parts per million between the two compounds and T is in degrees Kelvin. For higher temperature work (20–100°C), the same authors propose a $1:5$ by volume mixture of cyclooctane and methylene iodide, in which case

$$\frac{1}{T} = 0.0220027 - 0.000223362 \cdot \Delta\delta \qquad (2)$$

follows.

Another mixture that has been suggested as a ^{13}C nmr thermometer is a $1:1$ by volume mixture of carbon tetrachloride and acetone-d_6 (^{13}C chemical shift of carbonyl group), with the spectrometer locked on the deuterium resonance of the acetone-d_6.[8] The relationship between the absolute temperature (T) and the chemical shift difference $\Delta\delta$ between the two ^{13}C resonances, which is valid between −80°C and 90°C, is given by

$$T = 5802.3 - 50.73 \cdot \Delta\delta \qquad (3)$$

The uncertainty measured by this method at a ^{13}C frequency of 67.9 MHz is about ±1°.

Paramagnetic lanthanide shift reagents should induce strongly temperature-dependent chemical shifts in compounds to which they complex, since the equilibrium constants are generally temperature-dependent, that is, $\Delta H° \neq 0$. A ^{13}C nmr thermometer based on this principle has been proposed[7] and makes use of acetone-d_6 (^{13}C resonance of carbonyl group) in the presence of ytterbium(III) 1,1,1,2,2,3,3-heptafluoro-7,7-dimethyl-4,6-octadionate [Yb(fod)$_3$]. The sample used also contained ordinary acetone and tetramethylsilane, so that it is also a ^{1}H nmr thermometer. The precise sample composition is as follows: 726 mg Yb(fod)$_3$, 738 mg (CD$_3$)$_2$CO, 46 mg (CH$_3$)$_2$CO, 2.646 g CS$_2$, 5.74 g CFCl$_3$, and 1.23 g Si(CH$_3$)$_4$. The absolute temperature (T) is related to the ^{13}C and ^{1}H chemical shifts in parts per million by Eqs. 4 and 5, respectively. The references for the chemical shifts are CS$_2$ (^{13}C) and Si(CH$_3$)$_4$(^{1}H).

$$T = 4505 \cdot \Delta\delta_C^{-1} + 4640 \cdot \Delta\delta_C^{-1} + 70.3 \qquad (4)$$

$$T = 1264 \cdot \Delta\delta_H^{-1} - 745 \cdot \Delta\delta_H^{-2} + 88.6 \qquad (5)$$

The accuracy of both Eqs. 4 and 5 is ±1.5°. The ^{13}C resonance of the carbonyl group of acetone-d_6 is a single line, and thus ^{1}H decoupling is not necessary but can be used. No temperature deviation results from the use of ^{1}H radiofrequency power, at least under the conditions used (Bruker HX90 spectrometer and relatively nonpolar solutions).

Fluorine chemical shifts, e.g., in C_6F_6, show strong temperature dependences and this fact can be used to measure the temperature in a ^{13}C or ^{1}H probe if the spectrometer can be locked to ^{19}F.[9] It is also possible to make use of the temperature dependence of the carbon chemical shift with the magnetic field locked to a deuterium signal.[9] In a mixture of 0.5 ml of perdeuterated toluene, 0.25 ml of hexafluorobenzene, and 5 ml of fluorodichloromethane, the methyl carbon shift varies by about 0.1 part per million per degree over the temperature range −138° to 32°C.

Since most nmr spectrometers make use of a deuterium internal lock for magnetic field stabilization, it is possible to use deuterium chemical shifts in a mixture of, for example, $CD_3OH/CH_3OH/D_2O/H_2O$ (volume ratio of 47:9:30:14).[10] The chemical shift between the CD_3 and OD resonance can be measured by obtaining the chemical shifts of any given nmr resonance, with the spectrometer first locked on the CD_3 line, and then on the OD line. It is also possible to use the CH_3 and OH ^{1}H resonances of the mixture to measure the temperature. There appears to be no isotope effect on the chemical shift differences and thus both the ^{1}H and D data lead to the same temperature. The absolute temperature (T) is related to the ^{1}H and D chemical shift differences ($\Delta\delta_H$ and $\Delta\delta_D$ in parts per million) by Eqs. 6 and 7, respectively, and is valid between room temperature and 90°C.

$$T = 434 - 90 \cdot \Delta\delta_H \tag{6}$$

$$T = 435 - 88 \cdot \Delta\delta_D \tag{7}$$

Any compound showing a dynamic ^{13}C nmr spectrum can in principle be used as an nmr thermometer, but the temperature range is generally limited and peak interference with the spectrum to be observed may be a problem. If the same process is causing both ^{13}C and ^{1}H dynamic nmr effects, then it is possible to compare the temperatures obtained in the two cases, but this has a somewhat limited applicability. The rate constants for Cope rearrangement in bullvalene have been determined over a very wide temperature range by both ^{1}H and ^{13}C nmr,[11] and there is excellent agreement. This of course means that the temperature measurements in the two cases were consistent and therefore probably correct.

The line width and relative peak heights of the ^{13}C nmr spectrum of furfural (1 ml in 2 ml of THF-d_8) have been proposed as an nmr thermometer[12] over the temperature range +10 to −70°C. The process here is restricted rotation, and, since all the resonances are at low fields, there is no interference with saturated carbon resonances. Graphical data and other experimental details are given in the original paper,[12] but unfortunately the frequency of the nmr spectrometer is not given.

A problem with using dynamic nmr as a temperature probe is that the line shape at a given temperature changes with spectrometer frequency, and therefore line widths, etc. are not a linear function of the spectrometer frequency, as are chemical shifts.

The lowest temperatures that have been reached in high-resolution ^{13}C nmr spectroscopy of liquids is around $-180°C$, even when several watts of proton decoupling power is used.[13,14] None of the proposed nmr thermometers reach down to this temperature, although temperatures can still be measured by means of a thermocouple close to the sample tube,[13] or, rather approximately, by making use of the line width of the solvent peaks, and extrapolating from higher temperatures.[14] Deuterium locks become rather unsatisfactory at these temperatures because the high viscosity of the solvents gives rise to very broad deuteron resonances. Fluorine locks, on the other hand, are quite satisfactory.[13]

As mentioned at the beginning of this discussion on measurements of temperatures in ^{13}C nmr spectroscopy, ^{1}H decoupling power can lead to substantial temperature increases in the sample, particularly with conducting solutions. These effects have been examined in detail for aqueous solutions at high frequencies, where they can be quite large.[8] Substantial temperature gradients can be produced in the sample and the temperature calibrations become a function of the decoupling power. The largest effect is observed with conducting samples; hydrocarbons, alcohols, and even pure water are not much of a problem.[8]

Finally, mention may be made of a recent review[15] on dynamic ^{13}C nmr spectroscopy. This review also deals briefly with the temperature dependence of ^{13}C spin-lattice relaxation times. Extensive tables of compounds and activation parameters are included.

REFERENCES

1. a R. Duerst and A. Merbach, *Rev. Sci. Instr.* **36,** 1896 (1965), A. L. Van Geet, *Anal. Chem.* **42,** 679 (1970), M. L. Kaplan, F. A. Bovey, and H. N. Cheng, *Anal. Chem.,* **47,** 1703 (1975).

2. H.-J. Schneider and W. Freitag, *J. Am. Chem. Soc.* **98,** 478 (1976).

3. H. N. Cheng and F. A. Bovey, *Anal. Chem.* **47,** 1703 (1975).

4. H.-J. Schneider, W. Freitag, and M. Schommer, *J. Mag. Res.* **18**, 393 (1975).

5. A. K. Jameson and C. J. Jameson, *J. Am. Chem. Soc.* **95**, 8559 (1973).

6. P. M. Henrichs and P. E. Peterson, *J. Am. Chem. Soc.* **95**, 7449 (1973).

7. D. R. Vidrine and P. E. Peterson, *Anal. Chem.* **48**, 1301 (1976).

8. J. J. Led and S. B. Peterson, *Org. Magn. Reson.*, in press. The author thanks Dr. Led for providing a copy of the manuscript of this paper.

9. S. Brownstein and J. Bornais, *Abstracts of Sixth International Symposium on Magnetic Resonance*, Banff, Canada, May 1977. J. Bornais and S. Brownstein, *J. Mag. Res.*, **29**, 207 (1978).

10. I. Y. Slonim, B. M. Arshava, and V. N. Klyuchnikov, *Zhurnal Fizicheskoi Khimii* **50**, 277 (1976), also *Russ. J. Phys. Chem.* **50**, (1) (1976)].

11. J. F. M. Oth, K. Müllen, J.-M. Gilles, and G. Schröder, *Helv. Chim. Acta* **57**, 1415 (1974).

12. S. Combrission and T. Prange, *J. Mag. Res.*, **19**, 108 (1975).

13. (a) F. A. L. Anet and L. Kozerski, *J. Am. Chem. Soc.* **95**, 3407 (1973). (b) F. A. L. Anet, P. J. Degen, and J. Krane, *J. Am. Chem. Soc.* **98**, 2059 (1976).

14. L. Lunazzi, D. Macciantelli, F. Bernadi, and K. U. Ingold, *J. Am. Chem. Soc.* **99**, 4573 (1977).

15. B. E. Mann, *Prog. NMR Spectrosc.* **11**, 95 (1977).

SECTION

VIII NEW PULSED EXCITATION METHODS

HOWARD HILL

I. INTRODUCTION

Pulsed excitation and Fourier transformation of the subsequent free induction signal is the most commonly used technique for recording ^{13}C spectra. The Fourier transform (FT) relationship between the impulse response and frequency response is well known for linear systems. It was proved to hold for nuclear spin systems by Lowe and Norberg[1] and exploited by Ernst and Anderson[2] to provide a sensitivity advantage over older CW (continuous wave) methods. This improvement in sensitivity arises since the complete spectrum is excited simultaneously by a nonselective pulse and can be re-excited in a time on the order of the relaxation times, whereas the CW sweep method requires that a time on the same order be spent traversing each spectral line.

The use of nonselective pulses in the pulsed FT experiment is very simple and is well suited to the measurement of relaxation times and other transient effects. However, as a method of broadband excitation, it does have some drawbacks. In particular, a high peak rf power is required to excite a wide spectral region; there is a conflict between the conditions for maximum sensitivity and maximum resolution in a repetitively pulsed experiment; and all signal responses are in phase immediately after the pulse, resulting in a high signal-to-noise ratio at the start of the free induction decay. This latter consideration can place stringent requirements on the analog-to-digital converter in a practical spectrometer, particularly when recording proton spectra. In an attempt to overcome these disadvantages, other broadband schemes have been proposed. These include stochastic excitation[3,4] in which the spins are excited by a band of noise, and the use of a rapid frequency sweep over the frequency range of interest in the techniques of correlation spectroscopy[5,6] or chirp spectroscopy.[7]

These schemes, together with the normal pulsed excitation, provide approximately uniform excitation over a wide band of frequencies. However, there are many experiments in which the selective excitation of particular regions of the spectrum is more suitable. For example, excitation of a strong solvent peak may prevent detection of weak signals of interest by exceeding the dynamic range of the spectrometer detection system. Since a solvent peak is not usually of much interest, an excitation scheme that simultaneously excites all other parts of the spectrum while leaving the strong unwanted peak unperturbed would obviously be advantageous. Other experiments, such as selective decoupling and selective relaxation time measurements, call for the irradiation of one or more discrete frequencies. Experiments involving saturation may require the irradiation of a band of frequencies.

An experimental technique that can encompass all these applications is Fourier synthesized, or "tailored," excitation.[8] The frequency spectrum of the excitation is defined to suit a particular application, Fourier transformed and used to modulate the transmitter of the spectrometer. The spectral information from the spin system is calculated by an appropriate analysis of the spectrometer input and output signals. If a computer is used to store the spectrum of the excitation and the modulation signal, complicated excitation patterns may be used. However, for simple excitation spectra, it may be more convenient to generate the modulation signal directly. In an excitation with many frequency components, the form of the modulation signal is strongly affected by the relative phases of those components. For a uniform broadband excitation, the choice of phases that vary with frequency in a linear, quadratic or random manner determine whether the modulation will be a pulse, a frequency sweep or a stochastic signal. Thus these familiar types of excitation may be regarded as special cases of tailored excitation. This section will develop the concepts of this approach to the excitation of nmr spectra and point out some of its applications and limitations.

One particularly important limitation is that it is strictly applicable only for linear systems. A nuclear spin system is not linear and some discrepancies between the anticipated and the observed response are to be expected. Such discrepancies become more obvious as the intensity of the excitation increases, so care must be used in extrapolating the method to experiments involving high excitation levels.

II. TIME AND FREQUENCY DOMAINS

Any signal may be equally well represented by its value as a function of time or by its frequency spectrum, which is characterized by a particular amplitude and phase distribution of frequency components. The two representations of the signal are related by a Fourier transformation. Similarly, a linear system is characterized by its impulse response $h(t)$ or its frequency response $H(\nu)$. For a nuclear spin system, these quantities are the free induction decay and the spectrum, respectively. Since we are usually interested in a nuclear resonance spectrum, or the effect of an excitation on the spectrum, it is convenient to consider the frequency domain representation.

If a system is excited by a signal that has a frequency spectrum $E(\nu)$, the response will be

$$R(\nu) = H(\nu) \cdot E(\nu) \tag{1}$$

The frequency response, or spectrum, of the system may then be calculated directly by dividing the response by the excitation

$$H(\nu) = \frac{R(\nu)}{E(\nu)} \tag{2}$$

Alternately, the spectrum may be calculated from the expression

$$H(\nu) = \frac{R(\nu) \cdot E^*(\nu)}{E(\nu) \cdot E^*(\nu)} \tag{3}$$

where $E^*(\nu)$ is the complex conjugate of the spectrum $E(\nu)$. The term $E(\nu) \cdot E^*(\nu)$ represents the power spectral density of the excitation. If the amplitude of the excitation is independent of the frequency,

$$H(\nu) \propto R(\nu) \cdot E^*(\nu) \tag{4}$$

Many of the excitation schemes used in nmr satisfy this condition, and Eq. 4 provides the basis for the usual detection methods.

While the frequency domain representation gives a convenient picture of the excitation and response, experiments are invariably carried out in the time domain. For example, a spin system is normally excited by modulating an rf carrier with a function $e(t)$, whose Fourier transform is the spectrum $E(\nu)$. The signal $r(t)$ at the spectrometer output is obtained by demodulating the nuclear precession frequencies with a phase sensitive detector using the rf carrier as a reference. (Often several frequency-conversion steps take place in the spectrometer but the net effect is unchanged.) The rf carrier may therefore be ignored in the analysis; the spectrometer output represents the behavior of the spins in the frame rotating at the spectrometer carrier frequency, and the input signal is the modulation function. In the time domain, the response of a linear system is the convolution of the excitation by the system impulse response, i.e.,

$$r(t) = \int_{-\infty}^{\infty} h(\tau) \cdot e(t-\tau) \, d\tau \tag{5}$$

This expression may be easily shown[9] to be the Fourier transform of Eq. 1. The spectral response $H(\nu)$, may therefore be calculated by Fourier transforming both the excitation function and the response and using Eq. 2. When the excitation has an amplitude that is independent of frequency, an alternative approach is to calculate the cross correlation between the input and output signals and to Fourier transform the result. The cross correlation function is

$$g(t) = \int_{-\infty}^{\infty} r(u)e^*(u-t) \, du \tag{6}$$

$$= \int_{-\infty}^{\infty} \left[\int_{-\infty}^{\infty} h(\tau)e(u-\tau) \, d\tau \right] e^*(u-t) \, du \tag{7}$$

$$= \int_{-\infty}^{\infty} h(\tau)[e(u-\tau)e^*(u-t) \, du] \, d\tau \tag{8}$$

Now

$$\int_{-\infty}^{\infty} e(u-\tau)e^*(u-t) \, du \tag{9}$$

is the autocorrelation function of the excitation. It is equal (by the Wiener-Khintchine theorem) to the Fourier transform of the power spectral density $[E(\nu) \cdot E^*(\nu)]$ of the excitation.[10] For an excitation with a uniform power spectral density,

$$\int_{-\infty}^{\infty} e(u-\tau)e^*(u-t) \, du = \delta(t-\tau) \tag{10}$$

and

$$g(t) = h(t) \tag{11}$$

i.e. cross correlation of the excitation and the response gives the transient response of the system, which can be transformed to provide the spectrum. The schemes for analyzing the spectral information are shown in Figure 1.VIII.1.

The relationship between the two domains and the methods of determining a spectral response can be illustrated simply by considering the excitation in the form of a sequence of pulses. Figure 1.VIII.2 shows such

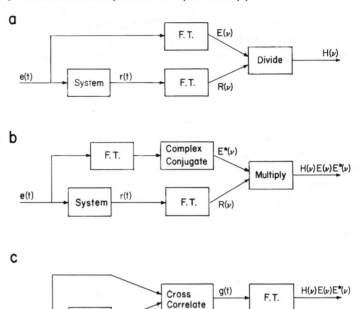

Figure 1.VIII.1
Schemes for calculating the spectrum of a system from the excitation and response functions. Method (a) uses Eq. 2 and is valid for any excitation. Methods (b) and (c) are equivalent and use Eq. 4. The output is proportional to the power spectral density of the excitation so will only give the spectrum $H(\nu)$ for a uniform excitation.

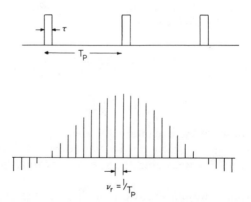

Figure 1.VIII.2
A repetitive pulse sequence and its corresponding frequency spectrum.

a sequence and its corresponding frequency spectrum consisting of discrete sidebands separated by the pulse repetition frequency, $\nu_r = 1/T_P$. If the pulses have a width τ, the envelope of the sidebands is given by

$$A(\nu) = \frac{\sin \pi \nu \tau}{\pi \nu \tau} \tag{12}$$

This expression is the Fourier transform of a single pulse and defines the excitation bandwidth, or "selectivity," of the individual pulses of the sequence.

The interpretation of the frequency spectrum is that the excitation is equivalent to irradiation with a set of *continuous* frequencies with amplitudes and relative phases determined by the corresponding components of the spectrum. For an excitation of finite duration, transient signals are associated with the turn-on and turn-off of that excitation. In a sequence consisting of many pulses, these transient signals can often be ignored, and we may consider that the frequency spectrum is present only for the duration of the excitation signal. This concept is useful for analyzing selective pulse experiments to be considered later. If the intensity of each pulse of the sequence is H_1, the equivalent intensity of the irradiation at a frequency ν_k is

$$H_1' = \frac{H_1 \tau}{T_P} \frac{\sin \pi \nu_k \tau}{\pi \nu_k \tau} \tag{13}$$

For a linear system, the response to the excitation will be the sum of its responses to each individual sideband. In general, since the nmr response is not linear, the sidebands can only be treated independently if their intensity is small compared with their separation, that is if

$$\frac{\gamma H_1'}{2\pi} \ll \frac{1}{T_P} \tag{14}$$

From Eq. 13, this will always be true if

$$\gamma H_1 \tau \ll 1 \tag{15}$$

Now, $\gamma H_1 \tau$ is the flip angle induced by a pulse, so the components of the frequency spectrum may be considered independently if the flip angle of each pulse of the sequence is small.

Pulse sequences of this simple type are used in a number of different ways for exciting nmr spectra. In a pulsed FT experiment the pulse interval is comparable to or long compared with the nuclear relaxation times and the spectral components are closely spaced so the many of them fall within the frequency range of the nmr spectrum, hence the broadband nature of this excitation. The analysis of the spin response in terms of the frequency domain representation of the pulse sequence is simple only for very small levels of excitation when each component may be considered independently. This is often not true even for the normal recording of spectra and is certainly not true for the measurement of relaxation times when 180° perturbing pulses are used. A complete analysis for all levels of excitation is much easier in the time domain representation. For a simple spin system, the use of the Bloch equations is adequate but for a complex spin system, a density matrix approach must be used.

If the pulse interval, T_P, is very short compared with the inverse of the spectral width, only one frequency component of the pulse sequence falls within the nmr spectrum. With certain approximations, the nmr response is as if this were the *only* frequency component present and is therefore equivalent to a normal CW excitation. The effective intensity of the irradiation at each frequency may be large and still satisfy Eq. 14. The individual pulses have only a small effect but they are cumulative for those frequencies for which the pulse interval is equal to an integral number of cycles, that is, when

$$\nu = \pm \frac{k}{T_P} \tag{16}$$

When this type of pulse modulation is combined with simultaneous gating-off of the spectrometer receiver, it is referred to as "time-share modulation."

By modulating the pulse sequence, further sidebands are produced. For example, suppose the train of pulses is amplitude modulated by a cosine function with a period $T/2$ as shown in Figure 1.VIII.3. The spectrum of this sequence consists of components at frequencies $\nu_k \pm 2/T$, that is each component of the unmodulated sequence is decomposed into two components shifted by $\pm(2/T)$. There is an obvious symmetry in the frequency spectrum; the basic frequency range from 0 to $1/2T_P$ has a mirror image in the range 0 to $-1/2T_P$, and the complete pattern from $-1/2T_P$ to $+1/2T_P$ is then repeated in each direction along the frequency axis. If the modulation is derived from a sine function instead of a cosine function, the spectral representation consists of sidebands shifted in phase by 90°. In this case, however, the image frequencies in the range 0 to $-1/2T_P$ are opposite in sign from those in the range 0 to $+1/2T_P$, that is, the sign of the 90° phase shift is different for positive and negative frequencies. For a general form of modulation the sidebands may have any phase angle and are represented by two orthogonal components $A(\nu_k)$ and $B(\nu_k)$. The symmetry about zero frequency may be removed by using a pulse sequence consisting of complex values. When used as a modulation, the

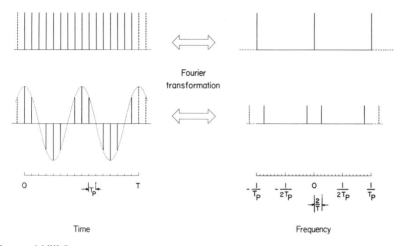

Fourier
transformation

Time   Frequency

Figure 1.VIII.3
The effect of modulating a pulse sequence. The sidebands of the unmodulated
sequence are split by ±(the modulation frequency).

real components correspond to 0° or 180° phase shift of the carrier,
depending on the sign of the modulation, while the imaginary component
corresponds to a 90° or 270° phase shift. In most applications to date, the
modulation function has been real, but in a spectrometer employing
quadrature phase detection, in which the carrier frequency is in the center
of the spectrum, the use of a complex excitation function is essential if an
asymmetric excitation function is required.

A general method for carrying out a steady-state nmr experiment is to
visualize it from the point of view of the frequency domain, to define the
frequencies required in the excitation sequence, and then to synthesize
the modulation sequence by Fourier transformation. This is the concept
of Fourier-synthesized or tailored excitation.[8]

III. IMPLEMENTATION OF TAILORED EXCITATION

Using a computer, the transformation between the two domains is carried
out by means of a *discrete* Fourier transform. The functions in both
domains are represented by a sequence of numbers, or data values, in a
section of the computer memory and take on their correct significance
according to how they are used. Because of the symmetries described
when using real-time functions, the frequency representation only contains values corresponding to the range 0 to ν_{max}, where $\nu_{max} = 1/2T_P$. If
there are N data values, frequencies up to $\nu_{N/2}$ will be represented.

The excitation at each frequency, ν_k, is completely defined by two parameters, the amplitude $F(\nu_k)$ and phase $\phi(\nu_k)$, and is most conveniently stored as two orthogonal components, $A(\nu_k)$ and $B(\nu_k)$, where

$$A(\nu_k) = F(\nu_k) \cos \phi(\nu_k) \tag{17}$$

$$B(\nu_k) = F(\nu_k) \sin \phi(\nu_k) \tag{18}$$

When transforming to real (as opposed to complex) time functions, $B(0) = B(\nu_{N/2}) = 0$. Fourier synthesis of the complex set of values represented by $A(\nu_k)$ and $B(\nu_k)$ gives the sequence

$$f(t_j) = \frac{A(0)}{2} + \sum_{k=1}^{N/2-1} \left[A(\nu_k) \cos \frac{2\pi jk}{N} + B(\nu_k) \sin \frac{2\pi jk}{N} \right] + \frac{A(\nu_{N/2})}{2} \cos \pi j \tag{19}$$

It is clear from this expression that $f(t_{N+j}) = f(t_j)$, so that the sequence is completely defined by N values. The highest frequency is represented by two values per cycle, a relationship familiar from conventional FT experiments.

As an example, if frequencies are defined at $1/2$ Hz intervals up to 1024 Hz, 4096 data values will be required. Discrete Fourier synthesis of this set of values gives 4096 values $f(t_j)$ of the modulation function. To faithfully represent the spectrum, these values are read out of memory at the frequency $2\nu_{N/2}$ (i.e., 2048 Hz) so that it takes 2 sec to scan through the complete modulation sequence.

The set of discrete-time values is ideally suited to the modulation of pulses in a time-share modulation experiment. If the values are used sequentially to determine the amplitude of successive pulses in the train, the required spectrum and its images about zero frequency are superimposed on the set of sidebands of the pulse sequence. The influence of the higher harmonics that are generated but that are beyond the defined frequency range may often be neglected.

As an alternative to amplitude modulation of pulses of constant width, the data values may be used to determine the width of pulses with a constant amplitude. Fourier analysis[11] of a width-modulated sequence shows that, compared with an amplitude-modulated sequence, additional frequency components are generated. However, the amplitude of these additional components is usually negligible if the ratio of the maximum pulse width to the pulse interval is kept sufficiently small ($< \sim 5\%$).

The use of pulse-width modulation offers some practical advantages since the pulse widths can be accurately controlled by counting the appropriate number of cycles of a high-frequency clock signal using high-speed logic circuits. Since the pulses are all of the same amplitude, nonlinear amplifiers may be used to generate the required rf power. On the other hand, amplitude modulation requires the use of an analog modulator and a linear power amplifier to feed the rf pulses to the probe. Maintaining adequate linearity can be a problem with such a method, particularly if a complex spectrum is required.

As in a normal time-share modulation experiment, the response of the spin system may be recorded during the interval between the pulses. The receiver is gated-off for a fixed time during which the rf pulse is applied so that there is no direct detection of the excitation. Fourier transformation of the response gives a spectrum in which the amplitude and phase of each frequency component are proportional to those of the excitation as described by Eq. 1. The complete spectral response may then be calculated according to Eq. 2 or Eq. 4.

IV.   APPLICATIONS

A.   Broadband Excitation

If all the phases, ϕ_k, of Eqs. 17 and 18 are chosen to be zero, only one of the values $f(t_j)$ is nonzero, and the excitation is the familiar one of a pulsed FT experiment. The Fourier transform of the spectrometer output is the nmr spectrum directly. The only phase corrections necessary are those required to take care of phase shifts and time delays in the spectrometer receiver. The complications of the tailored excitation technique clearly offer no advantages for this experiment.

If the phases, ϕ_k, are selected randomly, the time function $f(t_j)$ is noiselike and the experiment is a form of stochastic resonance.[3,4] The steps involved in performing a stochastic resonance experiment by this method are shown in Figure 1.VIII.4. A constant amplitude excitation is defined with all the phase angles equal; for example all values $A(\nu_k)$ are set to a constant and $B(\nu_k)$ to zero. The phases are scrambled by applying a phase rotation, ϕ_k, to each component ν_k according to a value from a pseudorandom number generator. Fourier synthesis gives the stochastic excitation function which is used to modulate the rf pulses. The transform of the spectrometer output is the nmr spectrum with the phases of each frequency component rotated according to the phase of the corresponding component of the excitation. The phases are unscrambled by using the same rotations applied in the definition of the excitation, but in the opposite sense. This is equivalent to multiplying the response point by point with $E^*(\nu)$ according to Eq. 4. Once the spectrum has been properly phased, convolution operations may be performed to optimize

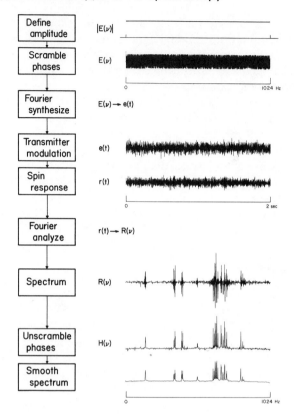

Figure 1.VIII.4
The stages of a stochastic resonance experiment by the tailored excitation method.

spectral parameters, for example, smoothing to improve the signal-to-noise ratio. For this experiment also, the method offers few advantages over other techniques of stochastic excitation. In the method of Ernst,[3] the excitation is derived from a shift register sequence and therefore requires no large memory storage for the excitation information. A shift register with n bits will generate a pseudorandom number sequence with 2^{n-1} discrete values and the spectral response will consist of an equal number of values. A fast Fourier transform algorithm is generally not suitable for use with a data set of this size but Kaiser[12] has developed a convenient method, using a Hadamard transform as an intermediate step, for determining the nmr spectrum from 2^{n-1} data values.

The power of the tailored excitation technique is that any excitation pattern may be selected. For example, as a means of avoiding problems from strong solvent peaks, an excitation function may be used in which the amplitude of the excitation is set to zero in the region of the unwanted peak and to a uniform level for the rest of the spectral region.[8] In this way the dynamic range limitations of the data system may be alleviated. However, it has not been adequately demonstrated that this method can be used while maintaining optimum sensitivity on those parts of the spectrum of interest. As the intensity of the excitation is increased, the conditions of Eqs. 14 and 15 are violated, resulting in distortion of the resonance line shapes. A simulated example is shown in Figure 1.VIII.5 in which the response of a single line to a uniform stochastic

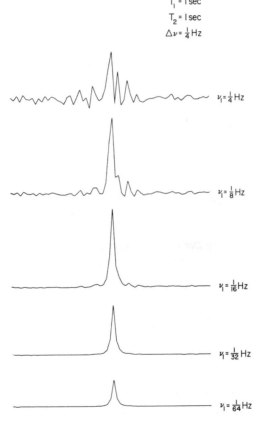

$$T_1 = I \text{ sec}$$
$$T_2 = I \text{ sec}$$
$$\Delta \nu = \tfrac{1}{4} Hz$$

$\nu_1 = \tfrac{1}{4} Hz$

$\nu_1 = \tfrac{1}{8} Hz$

$\nu_1 = \tfrac{1}{16} Hz$

$\nu_1 = \tfrac{1}{32} Hz$

$\nu_1 = \tfrac{1}{64} Hz$

Figure 1.VIII.5
Calculated lineshapes for a resonance with $T_1 = T_2 = 1$ sec excited by a stochastically modulated pulse sequence with a uniform power spectral density. The frequency components of the excitation have an intensity ν_1 and are separated by $\Delta \nu = 1/4$ Hz. The transformed and phase-corrected spectra are shown as a function of the intensity ν_1.

excitation is calculated using the Bloch equations. The frequency components of the excitation have equal intensity, but are randomly phased, and are uniformly spaced at 1/4-Hz intervals across the spectral region. Provided that Eq. 14 is satisfied, the line shape is faithfully reproduced but, as the intensity v_1 of the individual frequency components approaches their separation, the spectrum becomes badly distorted so that optimum sensitivity may not be realized.

B.   Selective Excitation

The technique of tailored excitation offers considerable flexibility as a method of selective excitation. As shown in Figure 1.VIII.3 discrete frequency components of the excitation are determined by the modulation frequency. For a sequence in which the pulse widths are given by

$$\tau = \tau_0 \cos 2\pi v_m t \tag{20}$$

the effective intensity of the irradiation at the first sideband v_m is approximately given by

$$H_1' = \frac{H_1 \tau_0}{2 T_P} \tag{21}$$

where, as before, H_1 is the intensity during the individual pulses and T_P is their separation.

Such selective frequencies may be used as a continuous irradiation, for example, in a decoupling experiment, for saturating a particular resonance by repetitive use of the pulse sequence, or as a selective pulse by using a restricted length of the sequence. Provided that any other frequency components of the excitation are far removed, the spin system will respond as if there were only one excitation component. Thus, the flip angle α induced by such a selective pulse follows the rules for a normal *monochromatic* excitation, namely

$$\alpha = \gamma H_1' T_S \tag{22}$$

where T_S is the length of the pulse sequence, that is the effective pulse length. The condition that a frequency component may be considered independently from its neighbor is the same as that stated earlier in connection with broadband excitation, namely

$$\frac{\gamma H_1'}{2\pi} \ll \Delta v \tag{23}$$

where Δv is the frequency separation from a neighboring component.

An unmodulated pulse sequence can be used in the same way by choosing the pulse repetition frequency in such a way that a sideband of the excitation falls at the required frequency. However, the effective field strength of the excitation is a function of the pulse repetition frequency (see Eq. 13). The use of a modulated sequence provides an excitation in which the intensity is independent of frequency and allows more freedom for selecting the frequency of irradiation without other harmonics of the sequence falling within the nmr spectral range. Several frequency components may be applied simultaneously by multiple modulation provided Eq. 23 is satisfied for each component. The results of applying two selective 180° pulses to a proton spectrum are shown in Figure 1.VIII.6. Use of selective pulses of this sort has been successfully applied to the measurement of selective relaxation times in proton spectra.[13,14]

A number of applications of selective excitation in ^{13}C spectra have been described by Morris and Freeman[15] using unmodulated pulse sequences. A particularly interesting example is the recording of an individual proton-coupled multiplet in a spectrum in which several multiplets overlap. This is possible provided that the proton-decoupled carbon lines are sufficiently resolved. With the decoupler on to establish an Overhauser enhancement and to collapse the spectrum, a selective pulse is applied at the frequency of the chosen resonance. The decoupler is gated

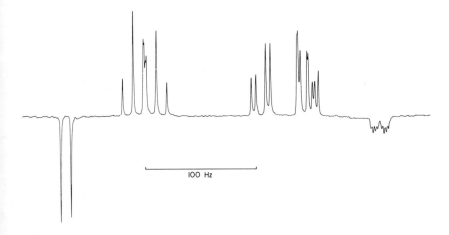

100 Hz

Figure 1.VIII.6
The proton spectrum of 3,4,6-tri-*O*-acetyl-1-*O*-benzoyl-2-chloro-2-deoxy-β-D-glucopyranose recorded by Fourier transforming the FID following a selective 180° —τ— nonselective 90° sequence. The selective 180° pulse contained modulation components chosen to invert *two* groups of proton signals.

off and the free-induction signal from the coupled line is recorded and transformed. The experiment may be repeated for each resolved line in the spectrum, or, by using multiple excitation frequencies, several nonoverlapping multiplets may be recorded simultaneously. This simple excitation method is an alternative to the two-dimensional FT methods described in Chapter 2 for chemical shift sorting.

In a modulated pulse sequence, the phase of the effective rf field is directly proportional to the phase of the modulation. Experiments involving rf phase shifts may therefore be implemented by shifting the phase of the modulation. An example of such an experiment is a selective spin-locking sequence. Figure 1.VIII.7 shows a pulse sequence for such an experiment and the effective fields in the coordinate frame rotating at the modulation frequency. The length of the first part of the sequence is selected to provide a 90° pulse. The second part of the sequence, following the 90° phase shift provides the locking field and is applied cyclically for as long as required. At the end of the locking sequence, a normal free-induction signal from the selectively excited resonance is

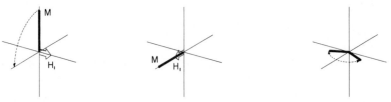

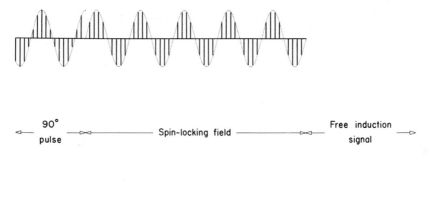

Figure 1.VIII.7
A pulse modulation sequence, and the effective fields in a frame rotating at the modulation frequency, suitable for a selective spin-locking experiment.

recorded and transformed. To ensure that the recorded spectrum has a constant phase for different spin-locking times, the times should be chosen to be integral multiples of the modulation period. The results of applying this method to the measurement of the spin-spin relaxation times of the protons in 2,4,5-trichloronitrobenzene are shown in Figure 1.VIII.8.[16]

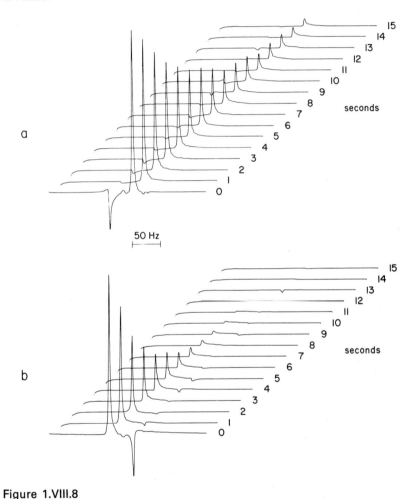

Figure 1.VIII.8
Selective spin-spin relaxation data for the protons of 2,4,5-trichloronitrobenzene. The protons were excited in turn by a selective spin-locking sequence and the subsequent free induction signals were transformed. The spectra are plotted as a function of the duration of the spin-locking sequence.

V. CONCLUSION

By modulating the pulses in a time-share modulation experiment, a spin system can be excited with any required frequency spectrum. The technique of tailored excitation provides a method of generating and storing the modulation information for an arbitrarily complex excitation spectrum. However, the usefulness of a particular excitation is determined by the response of the spin system. Because of the nonlinear behavior of a nuclear spin system, the response to an excitation is only proportional to the excitation at low levels. In recording a spectrum for optimum signal-to-noise ratio, it is necessary to drive the spins partly into saturation and therefore out of the linear region. The response of the spins is then dependent on the details of the excitation. For example, in a broadband excitation scheme, the response depends on the relative phases of the closely spaced frequency components. As a method of broadband excitation, the normal pulsed excitation in which the phase of each component is proportional to its frequency, has fewer disadvantages than more complex forms of excitation, and is very simple. For these reasons it seems unlikely that stochastic excitation will supplant the normal pulsed FT experiment for most applications.

On the other hand, in selective excitation experiments the frequency components are widely separated and can be used to drive the spins into saturation or to excite a transient response without problems. The use of the tailored excitation technique for this type of experiment allows convenient control over all the parameters of the excitation, namely frequency, amplitude, phase, and duration. Implementation of the experiments by modulation of the widths of pulses in a sequence is ideally suited to a normal Fourier transform spectrometer since the required variables are all under computer control. Selective excitation experiments therefore appear to be valuable complements to the nonselective pulsed FT method.

REFERENCES

1. I. J. Lowe and R. E. Norberg, *Phys. Rev.* **107,** 46 (1957).
2. R. R. Ernst and W. A. Anderson, *Rev. Sci. Instrum* **37,** 93 (1966).
3. R. R. Ernst, *J. Mag. Res.* **3,** 10 (1970).
4. R. Kaiser, *J. Mag. Res.* **3,** 28 (1970).
5. J. Dadok and R. F. Sprecher, *J. Mag. Res.* **13,** 243 (1974).

6. R. K. Gupta, J. A. Ferretti, and E. D. Becker, *J. Mag. Res.* **13,** 275 (1974).

7. J. Delayre and J. J. Dunand, *Abstracts of the 17th Experimental NMR Conference* (1976).

8. B. L. Tomlinson and H. D. W. Hill, *J. Chem. Phys.* **59,** 1775 (1973).

9. R. Bracewell, *The Fourier Transform and its Applications,* McGraw-Hill, New York, 1965.

10. R. W. Harris and T. J. Ledwidge, *Introduction to Noise Analysis,* Pion Ltd., London, 1974.

11. R. D. Stuart, *An Introduction to Fourier Analysis,* Methuen, London, 1961.

12. R. Kaiser, *J. Mag. Res.* **15,** 44 (1974).

13. R. Freeman, H. D. W. Hill, B. L. Tomlinson, and L. D. Hall, *J. Chem. Phys.* **61,** 4466 (1974).

14. L. D. Hall and H. D. W. Hill, *J. Am. Chem. Soc.* **98,** 1269 (1976).

15. G. A. Morris and R. Freeman, *J. Mag. Res.* **29,** 433 (1978).

16. H. D. W. Hill, XIXth Congress Ampere, Heidelberg, 1976.

2 PHYSICAL CHEMICAL APPLICATIONS OF ^{13}C SPIN RELAXATION MEASUREMENTS

DAVID A. WRIGHT
DAVID E. AXELSON
GEORGE C. LEVY

CONTENTS

I. INTRODUCTION

^{13}C spin relaxation measurements have formed a ubiquitous methodology in applications as diverse as structural and conformational analysis of natural products on the one hand and evaluation of theories for statistical mechanics on the other. Reviews of ^{13}C spin relaxation have been introductory,[1,2] or have dealt largely with organic chemical applications.[3–5] A few reviews dealing with physical chemical applications are available[6,7] but those are now dated. The excellent series of NMR Specialists Reports gives brief summaries of recent studies.[8]

In the last several years, theoretical and experimental studies have set the stage for a major assault on the concepts of molecular dynamics in liquids, previously the domain of statistical mechanicians. In addition, significant advances in the theory of magnetic relaxation have been forthcoming, largely from the laboratories of David Grant in Utah and the Volds in California.

This chapter discusses the development of ^{13}C spin relaxation and its application to physical chemical problems, and especially molecular dynamics in simple and complex liquids.

II. CROSS-CORRELATION EFFECTS

When various internuclear vectors reorient as a unit (such as in methyl or methylene groups), cross-correlation interaction terms describe the average correlation between the orientation of one internuclear vector at a given moment and the orientation of a different internuclear vector at a later time. The effects of such a correlation have been considered by numerous investigators for almost 30 years. Much of the theoretical[9–26] and experimental groundwork was done in these earlier years.

For various reasons, cross-correlation effects have been generally neglected from consideration in the interpretation of relaxation data. The effect is often assumed to be small or nonexistent. Recent theoretical advances, however, have shown that the effect may be more prevalent than anticipated, and that useful information may be derived from an explicit consideration of cross-correlation spectral densities.[27–59] An excellent detailed review of the theory of cross-correlation effects has been published recently by Werbelow and Grant and should be consulted for a comprehensive survey of the subject.[28] The problems of three spin 1/2 nuclei located at the corners of an equilateral triangle or four spaced at the corners of a regular tetrahedron with[20,23,26] and without[24,25] cross correlations included have been discussed. Within the extreme narrowing approximation, the case of a methyl group (with internal rotation) attached to a symmetric[12] or asymmetric[59] top molecule has also been evaluated. It has also been shown that the T_1 and T_2 decays are identical for any number of identical spins within extreme narrowing regardless of cross-correlation effects.[56] A general evaluation of the relaxation behavior of numerous arrangements of three or four spin 1/2 nuclei has been published by Schneider et al.[9,10,16,18,21]

The effect is more prevalent in proton relaxation studies with nonexponential relaxation behavior having been observed for the methyl protons of tetragastrin[35] (Figure 2.1), acetrizoate[38] and the methionyl methyl

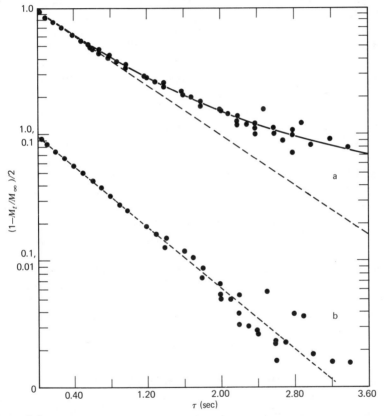

Figure 2.1
Spin-lattice relaxation in the C-terminal tetrapeptide of gastrin (0.03 M in DMSO-D$_6$, 30±1°C). (a) Methionine methyl proton relaxation (solid line denotes results of nonlinear regression analysis using the weighted sum of two exponentials, dashed line denotes initial slope calculated from the results of the double exponential fit). (b) Aromatic proton relaxation (dashed line denotes results of nonlinear regression analysis using a single exponential). (Ref. 35).

protons of methionine enkephaline[36] as well as in a number of solids. Numerous approaches have been taken in the evaluation of cross-correlation effects. The experimental technique of Grant et al.[28,31–33,37] involves selective perturbations of individual lines in coupled multiplets. These double resonance experiments on heteronuclear spin systems have shown that both auto- and cross-correlation terms may be evaluated by measuring the recovery curves following a suitably chosen selective excitation. In the Matson approach[34] to methyl relaxation the concepts of auto- and cross correlations are replaced by spectral densities associated with symmetry preserving transitions and symmetry-crossing transitions.

The detailed Redfield relaxation matrices describing the approach to thermal equilibrium of the populations of coupled nuclear spin systems of the type AX_2 and AX_3 have been tabulated[41] for relaxation by any combination of dipolar, chemical shift anisotropy, spin-rotation, and external fields (in the extreme narrowing limit), together with scalar coupling to a relaxing or exchanging spin. The presence of a significant cross-correlation interaction always serves to retard the relaxation rate while leaving unaffected the relaxation rate determined from the initial slope of the decay plot. However, multiple exponentials are required to characterize the return of the magnetization to equilibrium.[9,10,16,18,21,23–26]

The influence of cross-correlation effects, while apparent in the ^{1}H nmr results, may appear negligible in the ^{13}C results. For ^{13}C in the limit of extreme narrowing and rapid internal reorientation, the departure from exponential relaxation is within experimental error for the conditions encountered in the study of tetragastrin. This is predominantly a consequence of the difference in the angle made by the H $\cdots$ H internuclear vector and the threefold symmetry axis (90°), as compared to that made by the C—H internuclear vectors (109.47°). Noggle[13] has shown that, for ^{13}C in a methyl group, cross correlations between the ^{13}C—H', H'—H", and ^{13}C—H relaxation vectors fortuitously cancel out (to first order) even though the order of magnitude of the spectral density for cross correlation is the same as for autocorrelation. This has been confirmed by Kuhlmann et al.[57]

Most significantly, the influence of multispin dipolar cross-correlation effects has been shown to extend to the decoupled inversion-recovery experiment[32] potentially producing biexponential recovery. The influence is predicted to be more substantial if the extreme narrowing condition is violated, if other relaxation mechanisms compete with the dipolar mechanism, or if large motional anisotropies are present. This finding is of great practical importance to the experimentalist. Brondeau and Canet have shown that the longitudinal ^{13}C (or ^{15}N) relaxation curves in

$\{A\,B\,C\,D\cdots\}$ X spin systems are exponential (with the nuclear Over-hauser effect (NOE) having its usual significance) provided that (a) there is no symmetry or equivalence within the proton spin system and (b) all allowed proton transitions are saturated. The calculations apply only in the limit of extreme narrowing.[29a]

Inclusion of cross-correlation effects into the theory of multiple internal rotations results in the following predictions.[58]

1. Deviations from more conventional treatments will be most marked near the region of the NOE minimum ($\omega^2\tau^2 \sim 1$).

2. Cross correlations generally lead to an increase in the NOE over values obtained if these terms were neglected, for $D_i = 10^{10}$, $10^{11}\,\text{sec}^{-1}$.

3. They affect the NOE near its minimum in the case of rotational diffusion coefficients (D_i) equal to $10^{11}\,\text{sec}^{-1}$ by approximately a factor of 3 (i.e., raising it from the expected value).

4. The deviation produced by such terms does not continuously increase as the overall rate of molecular diffusion becomes much greater than the rates of internal reorientation ($D_0 \gg D_i$). The deviations are greatest near the NOE minimum.

5. The deviations decrease monotonically for carbons further along the chain.

These results are demonstrated in Figure 2.2.

In summary, the effect of cross correlations on NOE values is small except for the region where the "isotropic" diffusion constant for the restricted end of the chain satisfies the conditions $6D_0 \sim \omega_c$. In this case the effect depends on the internal diffusion coefficient, being insignificant for $D_i = 10^9\,\text{sec}^{-1}$ and more substantial for $D_i = 10^{10}$ and $10^{11}\,\text{sec}^{-1}$.

Cross-correlation terms provide a unique additional measure of the motional anisotropy in liquids. Whereas T_1 data on a planar molecule (with proton decoupling) is insufficient to determine the three principal diffusion coefficients, the inclusion of cross-correlation spectral density measurements is sufficient to analyze the motions.

Bovee[40] has applied this methodology to the AB_2 spin system of 2-amino-pyrimidine to determine the three rotational diffusion constants. Vold et al.[29,30] analyzed the ^{1}H AB_2 systems of the planar molecules 1,2,3 trichlorobenzene, 2,6-dichloroanisole, and 2,6-dichlorophenol-O-d in conjunction with the ^{13}C relaxation rates (to determine the rotational diffusion constants from proton relaxation curves alone requires complex spin systems with short proton-proton distances).

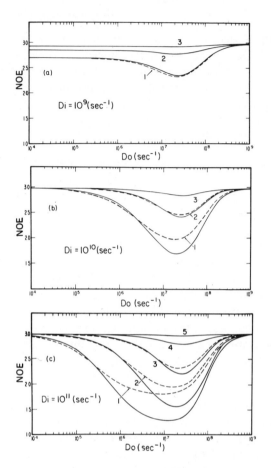

Figure 2.2

Calculations of the ^{13}C—$\{^1$H$\}$ nuclear Overhauser enhancements for a system undergoing multiple internal rotations as a function of D_0, the isotropic diffusion coefficient for the zeroth carbon atom. All angles between successive rotations and between the final rotation axis and the internuclear vector are assumed tetrahedral ($\beta = 109.47°$). Calculations assume a 23.5 kG field for which $\omega_c = 1.58 \times 10^8$ rad/sec; $\omega_H = 6.29 \times 10^8$ rad/sec. Values of D_i, the diffusion coefficient for the internal rotations assumed constant along the chain are 10^9 sec^{-1} (a), 10^{10} sec^{-1} (b), and 10^{11} sec^{-1} (c). The solid lines assume no cross correlations and are applicable to methine carbons. The dotted lines indicate the results obtained for methylene carbons including cross correlation effects.

The following conclusions were drawn by Vold et al.[29]:

1. To determine the necessary spectral density parameters the complete nonexponential recovery curves must be analyzed.
2. Relaxation of the spin system should be noticeably asymmetric.
3. Complete determination of the rotational diffusion constants for planar molecules requires the accurate determination of at least one cross-correlation term in addition to two autocorrelation terms.
4. Evaluation of the transition probability matrix in terms of the spectral densities for a given spin system need not be repeated.

Although for autocorrelation densities all weighting factors are nonnegative geometrical constants, it has also been noted that cross-correlation spectral densities need not be positive since no such restrictions are imposed on the weighting coefficients.[27]

Dynamic information from cross-correlation measurements in methyl groups can also be used to extend or complement other approaches.[27] For instance, the relationship between the internal rotation rate of the methyl group (D_{int}) and the isotropic rotational diffusion coefficient (D_0) has been the subject of consideration.[27] Autocorrelation spectral densities must decrease in amplitude with increasing internal degrees of freedom and thereby spread over a broad range of frequencies. Cross-correlation spectral densities are not limited in this sense and may decrease or increase in magnitude depending upon the relative orientation of the internuclear vectors with respect to the principal axes of diffusion.[27] Thus the quantity $\xi = (D_{int}/D_0) + 1$ is highly variable over a very large range of values for the relevent diffusion coefficients. Values of ξ greater than those that could be determined from standard theories are accessible from consideration of cross-correlation effects.

In addition, certain qualitative aspects of the relaxation behavior in a coupled methyl quartet can be useful. If the spectral density J_{HCH} is negative, which occurs when $\xi \leq 9$, the normalized intensity of the outermost lines is greater than the normalized intensity of the innermost lines. The reverse is true for J_{HCH} positive.

In an AMX system, for instance, one may qualitatively determine the presence of cross-correlation effects. Following total inversion of any of the multiplets, all four lines within any one quartet will not return to equilibrium at the same rate if cross correlation effects are significant. The recovery of the lines in the absence of cross-correlation effects would be identical. Other such relationships can be ascertained and used to practical advantage.[28]

The correlation function for an asymmetric rotor has been derived by Grant and Werbelow[31] and employed in an experimental study by Vold et al.[29] These are the resultant equations:

$$J_{abcd} = \tfrac{3}{10}\gamma_a\gamma_b\gamma_c\gamma_d\hbar^2(r_{ab}^{-3})(r_{cd}^{-3})\tau_{abcd} \tag{1}$$

$$\tau_{abcd} = \sum_{i=1}^{5}\frac{c_i}{\lambda_i} \tag{2}$$

$$\lambda_1 = 4D_x + D_y + D_z \tag{3}$$

$$\lambda_2 = D_x + 4D_y + D_z \tag{4}$$

$$\lambda_3 = D_x + D_y + 4D_z \tag{5}$$

$$\lambda_4 = 6D_s + 6(D_s^2 - L)^{1/2} \tag{6}$$

$$\lambda_5 = 6D_s - 6(D_s^2 - L)^{1/2} \tag{7}$$

$$D_s = \frac{D_x + D_y + D_z}{3} \tag{8}$$

$$L = \frac{D_xD_y + D_xD_z + D_yD_z}{3} \tag{9}$$

$$c_1 = \tfrac{3}{4}\sin 2\theta_{ab}\sin 2\theta_{cd}\sin\phi_{ab}\sin\phi_{cd} \tag{10}$$

$$c_2 = \tfrac{3}{4}\sin 2\theta_{ab}\sin 2\theta_{cd}\cos\phi_{ab}\cos\phi_{cd} \tag{11}$$

$$c_3 = \tfrac{3}{4}\sin^2\theta_{ab}\sin^2\theta_{cd}\sin 2\phi_{ab}\sin 2\phi_{cd} \tag{12}$$

$$c_4 = \tfrac{1}{4}A(3\cos^2\theta_{ab} - 1)(3\cos^2\theta_{cd} - 1) \tag{13}$$
$$+ \tfrac{3}{4}B\sin^2\theta_{ab}\sin^2\theta_{cd}\cos 2\phi_{ab}\cos 2\phi_{cd} + E$$

$$c_5 = \tfrac{1}{4}B(3\cos^2\theta_{ab} - 1)(3\cos^2\theta_{cd} - 1) \tag{14}$$
$$+ \tfrac{3}{4}A\sin^2\theta_{ab}\sin^2\theta_{cd}\cos 2\phi_{ab}\cos 2\phi_{cd} - E$$

$$A = \frac{1}{2} + \frac{(2D_z - D_x - D_y)(D_s^2 - L)^{-1/2}}{12} \tag{15}$$

$$B = \frac{1}{2} - \frac{(2D_z - D_x - D_y)(D_s^2 - L)^{1/2}}{12} \tag{16}$$

$$D = \frac{3^{1/2}(D_x - D_y)(D_s^2 - L)^{-1/2}}{6} \tag{17}$$

$$E = 3^{1/2}D\{(3\cos^2\theta_{ab} - 1)\sin^2\phi_{cd}\cos 2\phi_{cd} \tag{18}$$
$$+ (\cos^2\theta_{cd} - 1)\sin^2\phi_{ab}\cos 2\phi_{ab}\}/8$$

where γ_a is the magnetogyric ratio of spin a and r_{ab} is the internuclear distance between spin a and b. The principal axis system for the $^{13}CH_3$ group is shown in Figure 3 of Ref. 37a.

III. RELAXATION MECHANISMS

A. Dipolar Interaction with Nuclei

The majority of papers treating the subject of dipole-dipole relaxation with nuclei have recently focused on the complexities introduced by cross correlation and/or multiple internal group rotation. This work is discussed in Sections II and V.B. However, even the very simple AX system with only one correlation function may yield new information about dipole-dipole relaxation. Using the techniques of selective pulses originally due to Solomon,[60] Mayne et al.[61] obtained the four spectral densities describing relaxation of ^{13}C—H in the formation. Four experiments were required: (a) a selective ("soft") proton pulse followed by observation of the carbon magnetization, (b) a nonselective ("hard") proton pulse followed by carbon observe, (c) inversion- recovery of the ^{13}C with proton decoupling, and (d) the dynamic NOE experiment.[62] The authors describe their numerical technique for estimating the relaxation parameters by solving a set of coupled linear homogeneous differential equations and show that the resultant spectral densities are consistent with the extreme narrowing assumption. The same technique was applied to the AX_2 system in $^{13}CH_2I_2$ to determine the four dipolar and two random-field spectral densities, and from the former, four dynamical and structural parameters of methylene iodide could be calculated,[33] including the motional anisotropies $D_{xx}/D_{zz} = 0.205 \pm 0.033$ and $D_{yy}/D_{zz} = 0.079 \pm 0.057$ and the bond angle $<HCH = 104 \pm 2°$.

Determination of NOEs in simple systems continues to receive attention. Opella et al.[63], Harris and Newman,[64] and Canet[65] clarify the experimental requirements for measuring equilibrium intensities in the "decouple only" sequence, showing that waiting times of $\gtrsim 10\ T_1$ are necessary. Fagerness et al.[66] show that "selective" NOEs[67] (selective excitation of the proton resonances) can be used to obtain relative values of the relaxation rates describing coupled relaxation. A single time-dependent study can then scale these experimental values.

Dipole-dipole relaxation between two ^{13}C nuclei has been reported for doubly labeled sodium acetate[68] and diethyl malonate.[69] In both cases the ^{13}C—^{13}C dipolar mechanism produced an additional contribution to the total relaxation rate (no cross relaxation) that was negligible for protonated carbons and ~20% of the total dipolar rate for nonprotonated carbons.

B. Relaxation in Paramagnetic Solutions

When a solution contains paramagnetic nuclei the relaxation of spin-1/2 nuclei is strongly affected. Dipole-dipole interactions that occur during transient or longer-lived complexes are of the order $\gamma_n^2\gamma_e^2$ (where n denotes nuclear and e denotes electronic magnetogyric ratios) rather than $\gamma_n^2\gamma_n^2$ and these dominate the relaxation process for electron concentrations ranging from 10^{-4} to 10^{-2} M, depending upon the presence or absence of specific substrate-relaxation reagent complexes. Paramagnetic relaxation reagents (PARR) which do not induce large chemical shifts, such as $Cr(acac)_3$, $Cr(dpm)_3$,[70] $Fe(acac)_3$, or $Gd(dpm)_3$[71] can be used to probe the nature of interactions between the PARR and the organic molecule being relaxed. The analysis may be simpler than that performed with lanthanide shift reagents (LSR).[72] The advantages and disadvantages of PARRs for spectral analysis have been discussed,[73] as well as the different types of interactions possible between the PARR and an organic molecule. Levy and coworkers[74] discuss five possibilities: (a) no interaction; (b) specific steric or motional effects only; (c) polar or electrostatic effects; (d) weak bonding interactions; (e) direct metal–ligand complexation. In an investigation using $Cr(acac)_3$ and several organic molecules the authors found evidence for the first four possibilities. The steric effect, which seems to yield different relaxation times for the central versus outer carbons in both neopentane and isooctane (in contrast to an initial report[75]) may be complicated by a translation-rotation coupled enhancement of T_1^{-1} for noncentral carbons.[76] Steric and polarization effects are observed for the planar molecule benzene. In $Cr(acac)_3$ containing solutions of benzene and cyclohexane, the average PARR-substrate distance is shorter for the planar molecule. The benzene/cyclohexane ^{13}C and ^1H T_1 ratios indicate that the benzene preferentially orients edge-on to the PARR.[74]

The variation of carbon T_1 with distance has been used for structure determination both with LSRs[77,78] and PARRs.[79] The fact that the concentration of LSR or PARR is much lower than that of the organic molecule means that the relaxation rates observed for the organics depend on translational diffusion of either the substrate, the relaxation reagent, or both. The translational diffusion of CCl_4 was studied by this technique.[80] (CCl_4 is a small, "inert" spherical organic molecule. This avoids the complexities of molecular anisotropies.) The authors compared the CCl_4 ^{13}C relaxation data at two field strengths and as a function of temperature with the predictions of a small step model and a jump diffusion model. The small step model predicted T_1s that are a factor of 2 longer than observed, with too small an activation energy. The jump model explained the data, with mean jump length comparable to the CCl_4 diameter. (The relaxation is assumed to be predominantly affected by translation of the smaller CCl_4 molecule. Rotational diffusion does not affect the centralized carbon of CCl_4.)

Relaxation times may be measured for carbons in ligands bound to paramagnetic transition metals. Doddrell et al.[81] reported data for a series of acetylacetonate complexes $M(acac)_3$. They found that spin-spin relaxation of the carbon is dominated by hyperfine coupling with the unpaired electron, even in cases where proton spin-spin relaxation is dominated by a dipole-dipole interaction. Spin-lattice relaxation, on the other hand, is largely dipole-dipole[82] but the naive assumption that the T_1s depend on the sixth power of the distance to the metal nucleus is incorrect. Instead, a fair correlation between the ratios of T_1s and the ratios of spin densities at different carbons is found.[82,83] (The spin density may be obtained from the proton hyperfine coupling constant using the McConnell relationship.[84]) This effect is not unreasonable when one considers that a very small amount of spin density located in the carbon p-orbital can be as effective as a full electron several angstroms away. Accordingly, much more care must be taken in direct ^{13}C T_1 structural studies of paramagnetic molecules in solution as compared to studies of molecules weakly associating with paramagnetic substances.

C. CSA Relaxation

At the high fields presently employed, relaxation through anisotropy of the chemical shift is no longer a novelty. Measured T_1s and NOEs for several large molecules at two field strengths (different by a factor of 3) are shown in Table 2.1.[85] The observed differences in relaxation rates fit the expected quadratic H_0 dependence nicely and result in reasonable values of shielding anisotropy, in all cases but one. In this context one result[86] deserves emphasis: when cross correlations are included in calculations of relaxation rates, it is seen that the CSA mechanism may dominate even when its interaction constant is smaller than the dipole-dipole interaction, under certain conditions of a nonprotonated carbon undergoing anisotropic rotation in the nonextreme narrowing limit.

D. Spin Rotation Relaxation

In a generalization that has perhaps more application to ^{19}F relaxation, Wang[87] has extended his treatment of spin-rotational relaxation in molecules undergoing anisotropic rotational diffusion to nuclei located off

Table 2.1

^{13}C Spin Lattice Relaxation Times (T_1s) and Nuclear Overhauser Effects (NOEs) Measured at 38° at High and Low Fields[a]

Carbons	67.9 MHz[c] T_1(sec)	67.9 MHz[c] NOE(η)	22.6 MHz[d] $T_{1(sec)}$	22.6 MHz[d] NOE(η)	T_1^{CSA}[e] (sec)	$\Delta\sigma^f$ (exptl) (ppm)
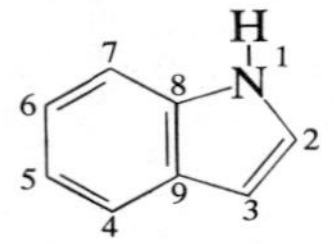 2	5.8	2.0	5.5	2.0	—	—
8	49	1.1	83	1.6	109[g]	220
9	47	1.0	85	1.5	94[g]	230
1, 2, 6[i]	1.0±0.05	1.9±0.1	1.1±0.2	1.9	—	—
11, 14, 5[i]	1.8±0.1	1.9±0.1	1.9±0.2	2.0	—	—
13	1.34	2.0	1.5	2.0	—	—
4	16.2	2.0	16.0	1.9	—	—
10	13.8	2.0	15.0	1.9	—	—
7	18.0	0.9	34.5	1.8	33[j]	220
8	25.0	0.8	48.0	1.7	42[j]	200
9	18.0	0.9	29.0	1.9	33[j]	220
12	22.0	1.1	33.0	1.8	49[j]	180

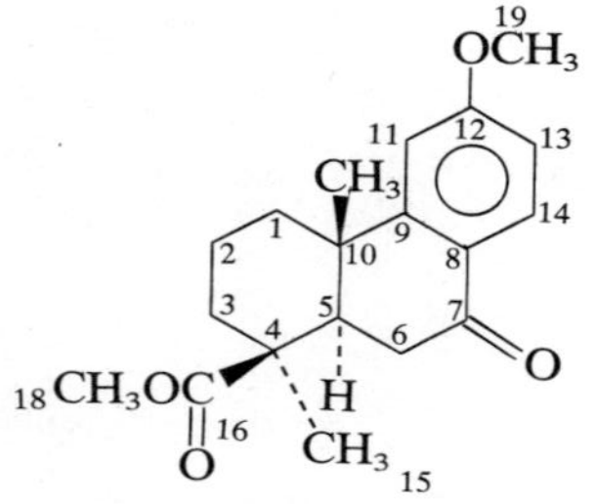

II[h]

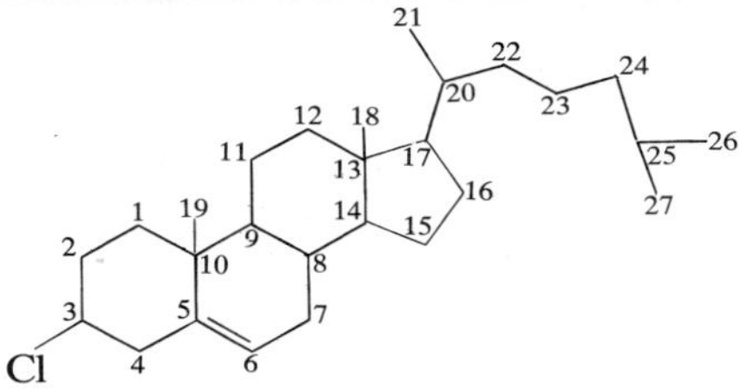

III[k]

Table (carbon positions and corresponding values):

Carbon	NT_1	NOE	NT_1	NOE			τ_{eff}
1, 2, 4, 7, 15, 16, 11, 12[i]	0.37±0.03	1.9±0.15	0.39±0.06	1.9±0.2	—	—	—
3, 6, 8, 9, 17[i]	0.70±0.10	1.9±0.15	0.70±0.16	1.9±0.2	—	—	—
10, 13[i]	4.2±0.10	1.9±0.15	4.5±0.20	1.9±0.2	—	—	—
5	3.2	0.8	1.6	5.6	—	5.3	340

[a] G. C. Levy and U. Edlund, *J. Am Chem. Soc.* **97**, 5031 (1975).

[b] Indole (4 M) in acetone-d_6; degassed by three freeze-pump-thaw cycles.

[c] T_1s and NOEs have internal estimated errors less than 10%. Several separate runs for each sample produced deviations less than 5–10%. The T_1 measurements were performed using the fast inversion-recovery sequence (FIRFT) and/or the unmodified IRFT sequence.

[d] T_1s are accurate to 5–15%, and the accuracy of the NOEs is 5% (compd I), 15% (compd II), and 10% (compd III).

[e] For the calculation of τ_{eff} the following values were used: $\gamma_C = 6720$, $h = 1.05 \times 10^{-27}$, $r_{CH} = 1.09 \times 10^{-8}$ cm, $\gamma_H = 26{,}700$.

[f] Estimated accuracy ±10%; maximum error ±20% (est) due to the relative low accuracy of the NOEs.

[g] $\tau_{eff} \approx 8.1 \times 10^{-12}$ sec/rad using $NT_1 = 5.8$.

[h] Me-OMe-Podocarpate (0.8 M) in acetone-d_6; degassed by three freeze-pump-thaw cycles.

[i] The stated values represent the range observed for all carbons in the group.

[j] $\tau_{eff} \approx 2.6 \times 10^{-11}$ sec/rad using $NT_1 = 1.8$.

[k] Cholesteryl chloride (1 M) in benzene-d_6, undegassed. $\tau_{eff} \approx 6.6 \times 10^{-11}$ sec using $NT_1 = 0.72$.

the symmetry axis. As before, the assumption is made that cross correlations of angular position and angular velocity vanish, cross correlations of angular velocity are ignored, and the angular velocity autocorrelation function is taken to be exponential. Since very few spin-rotation tensors are known at present, Wang's expression for T_{1SR} will most likely be used, along with knowledge of the τ_J's, to obtain the tensors, as in the study of CH_3Br.[88]

To this end, even empirical relations giving τ_J's are helpful, such as the proportionality between τ_J^* and the square of the absolute temperature noted by several authors.[89] Maryott et al.[90] have collected τ_J values for many small molecules (Table 2.2) to compare with the predictions of hard sphere and cell models of fluids. The authors relate τ_J^* to the time between collisions τ_{coll}^*, with an effective collision number Z as $\tau_J^* = Z\tau_{coll}^*$. Both models indicate that the collision efficiency is high (collision number between 1 and 2), with little difference in fitting the hard sphere model ($\tau_J^* \propto V_m$) or the cell model $[\tau \propto (V_m)^{1/3}]$, where V_m is the molar volume. Either model may be used to predict τ_J^* from the molecular mass, the moment of inertia, and the volume expansion of the liquid. In

Table 2.2
Molecular Properties, Effective Collision Number Z, and Reduced Angular Momentum Correlation Times τ_J^*, at the Boiling Point and the Critical Temperature[a]

	$10^{40}I$ (g cm^2)	$I_{\parallel}/I$	σ_{L-J}^{b} (Å)	C_a^{c} (kHz)	C_d^{c} (kHz)	Z_{CM}	Z_{HS}	τ_J^* at b.p.	at c.t.
$^{13}CS_2$	258	0	4.48	(-1.41)	(-2.11)	1.08	2.03	0.14	1.1
C^1H_4	5.33	1	3.76	10.4	18.5	2.10	2.85	1.0	5.6
$C^{19}F_4$	149	1	4.66	-6.85	(0)	1.05	2.36	0.26	1.6
$^{13}CCl_4$	490	1	5.95	(0,45)	0	1.11	2.35	0.19	1.4
$S^{19}F_6$	314	1	5.13	-5.21	(0)	1.24	1.98	—	1.8
					(9.2)	1.02	1.68		1.4
C^1HD_3	8.53	1.24	(3.76)	6.6	12.7	2.25	3.35	0.9	5.0
$^{31}PH_3$	6.29	1.15	3.98	-115.6	1.5	0.97	1.65	0.54	3.4
$^{31}PD_3$	12.1	1.20	(3.98)	(-59.2)	ca.0	1.51	2.34	0.56	3.8
$ClO_3^{19}F$	160	0.93	—	(-17.6)	(-28.2)	0.75	1.41	0.17	1.1
$CCl_3^{19}F$	344	1.43	5.44	(-3.66)	(-3.22)	1.20	2.13	0.21	1.5
$^{31}PCl_3$	321	1.77	ca.5.24	(4.22)	(0)	1.06	2.12	0.20	1.4

[a] For individual references see ref. 90.
[b] Lennard–Jones distance parameters from gas viscosity.
[c] $C_a = 2(C+C_{\parallel})/3$, $C_d = C - C_{\parallel}$.

small molecules such as these, care must be taken in assuming rotational diffusion since for several, $\tau_c^*\tau_J^* = 1/4$ at the melting point.

While it is common to estimate barriers to methyl rotations from dipole-dipole relaxation rates, Zens and Ellis[91] derive an equation so that spin-rotational relaxation rates may be used similarly. Using the equation

$$R_{1SR} = \frac{8\pi^2(kT+V)^{1/2}I_\|^{3/2}C_\|^2 n}{3h^2} \tag{19}$$

where all terms have their usual meaning, V is the potential barrier and n is an empherical parameter, the authors found a linear relationship between R_{1SR} and barrier height. This relationship is useful, however, only in relatively small molecules where R_{1SR} is significant.

Spin-rotational correlation times for the inertial reorientation model of Steele[92] have been calculated[93] and show that in the limit of large rotational diffusion constants the spin-rotational relaxation rate increases as $T^{1/2}$.

E. Scalar Relaxation

1. Contributions to Spin-Lattice Relaxation (T_1)

Studies of scalar contributions to ^{13}C spin-lattice relaxation continue to be largely studies of ^{13}C—Br interactions, because of the fortuitously small value of $\Delta\omega = 2\pi[\nu_L(^{13}C) - \nu_L(Br)]$ especially for ^{79}Br, at typical magnetic fields. Recent measurements[94] of the nonexponential ^{13}C relaxation in the series CH_nBr_{4-n}, $n = 0, 2$ have given the remaining unknown ^{13}C—Br coupling constants in the bromomethanes, summarized in Table 2.3. It is significant to observe that the discrepency in J_{CBr} values for

Table 2.3
Coupling Constants Between ^{13}C and ^{79}Br in Bromomethanes

Molecule	$J_{^{13}C-^{79}Br}(Hz)$	Reference
CD_3Br	30	88
CD_2Br_2	95 ± 2	94
	58.7 ± 1.4	96
$CDBr_3$	127 ± 2	94
	105	95
CBr_4	151 ± 2	94
$CFBr_3$	124.2 ± 7.0	96

CD_2Br_2 and $CDBr_3$ arise from different experimental values for the ^{13}C and ^2H relaxation times. In contrast to the Yamamoto et al.[94] work, the results of Wright et al.[95] and Huang et al.[96] were obtained from the initial slope of the ^{13}C relaxation curve with the assumption that $\tau_J^* = 0.7$ at the critical point and was adequately described by an Arrhenius-type equation. The area of greatest disagreement lies with the ^2H relaxation data or equivalently the nature of $D_\perp$. Yamamoto et al.[94] find $D_\perp = 3.49 \times 10^{12} \exp(-2500/RT)$; Wright et al. give $D_\perp = 5.37 \times 10^{11} \exp(-1610/RT)$. A Raman and Rayleigh study[97] gives $D_\perp = 6.62 \times 10^{11} \exp(-1700/RT)$, supporting the latter results. The different values for the deuterium quadrupolar coupling constant and r_{C-D} chosen in the two studies cannot, of course, account for the different temperature dependences. It may be noted that insufficient care in restricting sample diffusion[95] gives a temperature dependence very close to that of Yamamoto et al.[94] A similar discrepancy may be noted in the case of CD_2Br_2.[94,96]

2. Contributions to Spin-Spin Relaxation (T_2)

In view of the numerous possibilities for error in the spin-echo measurement of T_2,[98] little work has been reported concerning carbon spin-spin relaxation. Lilley and Howarth[99] investigated some problems unique to ^{13}C T_2s, namely that cross relaxation between carbon and proton spins[100] decreases T_2. Also, incomplete proton decoupling leaves some J modulation in the echo train. It was found that long (>0.3 sec) T_2s were seriously in error, but values less than 100 msec were accurate to 25%. Thus relative T_2s or confirmation of the T_1/T_2 ratio for macromolecules can be determined as in the study by Miziorko et al.[101] of an enzyme-bound substrate. Although T_2 broadening of resonance lines will be much greater than instrument-controlled linewidths for T_2^* in the range 10–100 msec, the overlap of many spectral lines in macromolecules can be expected to render linewidth measurements nearly useless for many of those systems.

According to Kumar and Ernst[102] heteronuclear J modulation of the echo train, which will occur whenever the proton in question is strongly coupled to other magnetically nonequivalent spins, can be used to obtain coupling constants, even when the H_0 field homogeneity is not sufficient to resolve the spectrum. The theory also states that the maximum oscillations in the ^{13}C echo train occur for $|J_{AX} - J_{A'X}|/2 = |J_{AA'}|$, which explains the unexpected result that the strongest oscillations are not observed for the olefinic carbon of maleic anhydride but rather the carboxyl carbon, where the J_{CH} coupling is weak enough to be comparable to J_{HH}.

IV. MOLECULAR MOTION IN SIMPLE LIQUIDS

In this section we will focus our attention on molecular motions in more or less rigid molecules. We will initially discuss primarily dipole-dipole relaxation and its associated correlation time τ_c (nmr), which is traditionally related to the viscosity-dependent correlation time given by

$$\tau_c(\text{Debye}) = \frac{4\pi r^3 \eta}{3kT} \tag{20}$$

where $4\pi r^3/3$ is the molecular volume, η is the macroscopic shear viscosity, k is Boltzmann's constant, and T is the absolute temperature.

A. Generalized Diffusion

In this section we will discuss diffusional motions (both rotational and translational) that have been analyzed in greater detail than provided for in the standard Debye relationship. This includes both variations and adaptations of the Gordon extended diffusion model, the recently popularized "slip" rotational diffusion model,[103] and others.

The resurgence of hydrodynamic analyses of relaxation data is due largely to the treatment by Hu and Zwanzig[103] of rotational diffusion in which the particle studied does not "drag" along the solvent "continuum" during rotation. Instead the particle is considered perfectly smooth. Consequently a spherical molecule experiences no rotational torques, and a symmetric top experiences no torque for rotation about the symmetry axis. Generalizing this approach we have[104]

$$\tau_c = \frac{4\pi\eta r^2 \kappa}{3kT} \qquad 0 < \kappa \leq 1 \tag{21}$$

where κ is a function of the axial ratio for slip boundary conditions. The parameter $\kappa = 1$ for sticking boundary conditions, and is interpreted as the degree of coupling between rotational and translational motions. At this point, the theoretical treatment is functionally equivalent with the technique of applying a microviscosity correction, but in this case the adjustable parameter has been given a molecular interpretation.[104c] Since in the limit of no coupling τ_c cannot reasonably be expected to equal zero, Bauer et al.[105,106] analyze relaxation data with the equation

$$\tau_c = c\eta + \tau_0 \tag{22}$$

where $c = (4\pi r^3/3kT)\kappa$, $\tau_0 \sim \tau_{fr} = (2\pi/9)(I/kT)^{1/2}$, and τ_{fr} is the correlation time for a free rotor. Relaxation data for 15 small molecules were compared with the predictions of the slip and stick models, and it was found that the slip model works well for organic solvents while aqueous solutions were intermediate between slip and stick, depending on the strength of H-bonding interactions. Fury and Jonas[107] reanalyzed old data (not ^{13}C relaxation) in terms of Eq. 21 and found that the parameter κ for neat liquids is independent of density and temperature. Previous work[104a,108] had shown that for a given solute-solvent pair κ is independent of temperature and pressure; thus this appears to be useful and easily obtained parameter.

In fact, Fury and Jonas[107] were able to find a relationship between κ and the molecular shape. Defining an axial ratio ρ_i (not to be confused with other axial ratios mentioned elsewhere) as the ratio of the volume of the molecule to the volume swept out by rotation about the ith axis, the authors were able to find a good correlation between κ and $\sum \rho_i^2$, shown in Figure 2.3, where the solid line is a plot of the relation $\kappa = 0.165$ $(\sum \rho_i^2)^{-2} - 0.021$. For methyl iodide and $CDCl_3$, the experimental techniques used (Raman and ^{2}H nmr, respectively) do not measure reorientation about the symmetry axis (for which $\rho = 1.0$), therefore for Figure 2.3 the summation is limited to ρ_x and ρ_y.

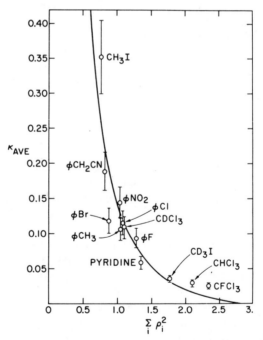

Figure 2.3
Empirical correlation between average experimental values of κ and ths sum of the squares of the axial ratios. The solid line denotes the best linear least square fit; the error bars denote the estimated experimental error limits ($\pm 15\%$).

The results of the extended diffusion model of Gordon[109] have been derived from a hard sphere binary collision model by Chandler.[110] It was assumed that the correlation functions in dense fluids are dominated by the repulsive part of the intermolecular potential. Further real systems may be approximated by a system of rough hard spheres* with the autocorrelation functions determined solely by uncorrelated binary collisions with nearest-neighbor molecules. A microscopic expression for τ_J (which was introduced in the Gordon model as the time between collisions) was derived, and predictions of the constant density and constant temperature dependence of both τ_θ and τ_J were made (which should be readily checked by experiment).

In an experiment typical of many performed on molecules with internal degrees of freedom, Ford[111] found that reorientation of the inner carbons of triethyl n-hexylammonium n-hexylbromide in aqueous solution could be described by the slip model in low viscosity solutions but not at high viscosity, where presumably internal rotations begin to compete. Other studies of methylene chain segmental motion have concentrated on the relationships between overall reorientational velocity (and mode) and observed internal motions. Chain segmental motions are not apparent from ^{13}C spectra when overall reorientation is rapid.

A combination of ^{13}C nmr and depolarized Rayleigh measurements were used to study reorientation in solutions of benzene, mesitylene, nitrobenzene, and toluene.[112,113] In agreement with the slip model, reorientations about the symmetry axis (reorientations that required no solvent displacement) were found to have little or no viscosity dependence, whereas correlation times for other motions were linearly dependent on viscosity. However, in the study of nitrobenzene by Stark et al.,[114] the zero viscosity intercept for motion about all axes was found to be well above the free rotor limit given by Eq. 22.

As was noted above, the cycloalkanes[115] show a fair relationship between R_{1DD} and η, with κ varying between 0.03 and 0.02 for stick. The ratio slip/stick for an axial ratio of 0.5 (here used in the sense of the ellipsoidol axes) appropriate to the cycloalkanes is 0.24,[83] therefore κ(slip) becomes 0.13 to 0.08. The observation that there is no concentration dependence for the cycloalkane T_1s upon dilution in C_6H_{12} shows that, as indicated by the small τ_c^* values (Table 2.4) reorientations are near the inertial limit and are dominated by the τ_0 term in Eq. 22.

Studies sufficiently detailed to separate dipole-dipole and spin-rotational relaxation over a reasonable temperature range continue to be

---

* In hard spheres, collisions occur instantaneously; in rough hard spheres, collisions cause changes in the angular (as well as translational) momentum of the particles.

Table 2.4

NMR-Derived Data Pertaining to Molecular Reorientation

Molecule	$Ea_\perp$ (kcal/mole)	$Ea_\parallel$ (or Ea_α)^a (kcal/mole)	$D_\parallel/D_\perp$ (or $D_\alpha/D_\perp$)^a	$I_\perp/I_\parallel$ (or $I_\perp/I_\alpha$)^a	$C_\parallel$ (or C_α)^a (kHz)	$C_\perp$ (kHz)
CH_3Br^b	1.7	0.9 ± 0.2	8.48^c	15.9	21.2	1.2
CH_3CN	1.67^d	0.73^d	$4.6^{c,d}$	10.1	(20.4 ± 16.6)	$(4.95\pm0.95)^e$
$CFBr_3^g$	1.42	—	—	0.59	—	—
$(CH_3)_4Sn^h$	2.25	0.80	7.9^c	61	15.4	0.25
$C_6H_5CH_3^i$			13^c		-17.3	

^{a} The subscript α refers to rotations about the methyl axis.
^{b} All data from Ref. 88.
^{c} At the boiling point.
^{d} D. E. Woessmer, B. S. Snowden, and E. T. Strom, *Mol. Phys.* **14**, 265 (1968).
^{e} Ref. 118.
^{f} Ref. 117.
^{g} Ref. 96; isotropic motion assumed.
^{h} Ref. 119.
^{i} Ref. 116.

rare. Recent results[88,95,96,116–119] obtained at least partly by ^{13}C relaxation are tabulated in Table 2.4. The general conclusions of these studies are becoming familar yet continue to be valuable. Thus in CH_3CN and CH_3Br the diffusion tensor is highly anisotropic, as is the inertia tensor, yet there is no direct relationship between the two. Similar observations can be made about nitrobenzene and the relation of methyl to overall rotation in tetramethyltin. These points have been discussed in more detail elsewhere.[120] In these molecules, and also $Sn(CH_3)_4$ the ^{13}C spin relaxation is dominated by rotation about the figure axis, as was found in CH_3I.[121] In most cases these studies require measuring relaxation rates of more than one type of nucleus, as for example Lassignes and Wells[119] investigation of $Sn(CH_3)_4$ in which they measured ^{119}Sn, ^{13}C, ^{1}H, and ^{2}H relaxation.

B. Nonassociating Liquids

Berger et al.[115] report T_1s and NOEs for cycloalkanes C_3H_6 through $C_{10}H_{20}$, enabling dipole-dipole and spin-rotation contributions be be separated (Table 2.5). The molecules form an interesting series since they should well represent noninteracting oblate ellipsoids undergoing anisotropic motion. (Rough calculations show the surprising fact that the ratio $I_\parallel/I_\perp \simeq 2$ for all but C_3H_8, where $I_\parallel/I_\perp \simeq 1.7$.) However the complexities of the motional anisotropy do not occur in the ^{13}C relaxation, since all

Table 2.5
Dynamical Properties of Liquid Cycloalkanes $C_n H_{2n}$

n	$^{13}C\ T_1$(sec)	NOE	T_{1DD}(sec)	T_{1SR}(sec)	τ_c(nmr)(sec)	τ_c^{*a}	τ_c(debye)
3	36.7	2.0	72	74	0.33	1.2	
4	35.7	2.4	51	121	0.46	1.2	
5	29.2	2.52	38	124	0.62	1.2	19
6	19.6	2.9	20	450	1.2	1.8	45
7	16.2	2.96	16	>1000	1.5	1.8	82
8	10.3	3.0	10.3		2.3		130
10	4.7	3.0	4.7		5.0		

carbons are equivalent. Several points may be made from inspection concerning the relaxation data. First, spin-rotation is negligible for $n > 5$. This is due partly to large moments of inertia and partly to measuring T_1 at lower temperatures (relative to the critical point). Secondly, T_{1SR} is roughly proportional to I. Since theory predicts $T_{1SR} \propto I^{-1}$, this is an effect of the lower relative temperatures for larger molecules. Finally, T_{1DD} is roughly proportional to η^{-1}. To make further observations, correlation times must be calculated, and this is done using Eq. 1 and 20, with a reduced correlation time τ_c^* calculated from τ_c(nmr) using the relation $\tau^* = \tau_c(kT/I)^{1/2}$. Considering τ_c^* first, we note that cyclopropane through cycloheptane are close to the inertial limit ($\tau_c^* = 0.7$). Since these relaxation times are all at temperatures well below the critical temperature, even for cyclopropane, the conclusion reached is that, as was found for other small molecules, reorientation over much of the liquid range deviates from small-step diffusion.

We next compare nmr correlation times with viscosity-derived ones and see that the τ_c (Debye) is 30 to 50 times larger than τ_c (nmr). This is the usual result for non-hydrogen-bonding fluids as was discussed previously. Several nearly isotropic molecules were studied by Grandjean and Laszlo,[122] who found the coupling parameter κ to be 0.07–0.10 for norbornadiene, 2-methylene-norbornadiene, α-pinene, and β-pinene.

The solid state of plastic crystals shows an nmr spectrum very similiar to the liquid state; therefore, relaxation studies in plastic crystalline solids will be discussed as an extension of liquid phase work. The existence of plastic solid phases is common for spherical molecules such as the above mentioned bicyclic compounds, and relaxation studies across the melting point have in the past shown rotation to be more free in the solid than the liquid, a point that arises in molecular dynamics simulations.[123] ^{13}C relaxation offers the same advantages for these solids studies as in ordinary liquid-phase relaxation measurements. Since ^{13}C line widths in the plastic solid phase are on the order of 10 Hz, at least near the melting point, relaxation measurements on multiline systems present no particular problems. Nevertheless, this area is largely untouched with very few measurements[124] of high-resolution ^{13}C relaxation times in solids using conventional FT pulse techniques.

C. Intermolecular Interactions in Liquids

The presence of weak association (usually hydrogen bonding) in liquids can be inferred from a variety of techniques (for example infared spectra or nmr chemical shift changes) but expecially dramatic effects are seen in molecular reorientation. This can be easily seen by comparing T_{1DD}s for the cycloalkanes (Table 2.5) with the corresponding alcohols[125,162] (Table 2.6). Two effects can be seen in the comparison. First, T_{1DD} in a cyclic alcohol is 30–50 times shorter than in the corresponding cycloalkane. This follows neatly from the large increase in viscosity of the alcohol. For example, in cyclohexanol, H-bonding restricts most of the mobility of C-4 which relaxes roughly 63 times faster than the CH_2 carbons of C_6H_{12}, while the ratio of viscosities is 68 (at 20°). A second observation is that the lower symmetry of the alcohols permits detection of anisotropic motions from variations in the carbon T_1s. Isotropic motion would require that in a given molecule (assuming a rigid structure) all methylene T_1s should be equal and one-half the T_1 for the methine carbon. This is clearly not true for most of the carbons in Table 2.6, and while conformational motions may complicate the picture in the larger systems and in the smallest, they are certainly not significant for cyclohexanol, where anisotropic motion is clearly indicated.

Table 2.6
^{13}C T_1s in Some Cyclic Alcohols[a]

Compound	C-1	C-2	C-3	C-4	C-5	C-6	C-7
Cyclobutanol[c]	3.2	3.1	4.3				
Cyclopentanol	2.5	4.1	3.3				
Cyclohexanol	0.72	0.59	0.60	0.32			
Cyclooctanol	0.32	0.28	0.24	0.35	0.24		
Cyclododecanol[b]	0.81	0.48	0.55	0.62	0.53	0.53	0.50

[a] Refs. 125 and 162. T_1s in seconds, taken at 35 ± 2°C; dipolar relaxation assumed.
[b] 83°C,
[c] NOEs complete (2.0η).

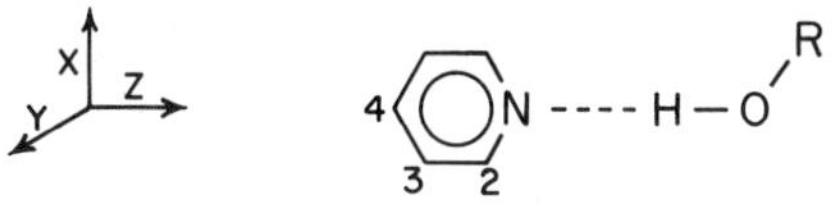

Figure 2.4
Coordinate system for the interaction of an hydroxyl group with the probe molecule, pyridine.

Quantifying anisotropic reorientation in terms of a parameter $R \equiv T_1$ (ortho + meta)/$2T_1$(para) was first introduced by Levy et al.[126] R can probe molecular aggregation as in the dimerization of benzoic acid, which yields a value of 2.0, whereas in methyl benzoate R is less than 1.3. This is due to the nearly linear geometry of the long-lived benzoic acid dimer producing a highly anisotropic complex.[127]

Observing weak interactions directly for self-associating systems can be difficult; an alternative is to use a probe molecule. Campbell et al.[128] proposed using pryridine as a "motional anisotropy probe." This choice is governed by the fact that neat pyridine does not self-associate, as evidenced by equal (within experimental error) relaxation times for all carbons. In solution with another molecule possessing an hydroxyl group, hydrogen bonding will occur at the nitrogen (Figure 2.4), hindering motion about axes X and Y. As a result, the (C-4)—H vector will not reorient as rapidly as the C-2 and C-3 C—H vectors. Using the parameter R introduced above, Campbell et al.[128] obtain for 1:1 mixtures of monohydric alcohols the data shown in Table 2.7. The axial reorientation ratio a/b is calculated from the experimental data using Woessner's[129] equations for rotational anisotropic diffusion in the extreme narrowing limit. The ratio a/b describes the size of the complex, with the major axis

Table 2.7
^{13}C Spin-lattice Relaxation Times (sec) of Pyridine Measured in Equimolar Mixtures with Some Monohydric Alchohols[128]

Alcohol	C-2	C-3	C-4	R	a/b
Methanol	14.6	14.8	11.1	1.32	2.0
Ethanol	13.6	13.5	9.6	1.41	2.3
n-Propanol	12.1	12.2	8.0	1.52	2.5
Allyl alcohol	11.9	12.3	7.6	1.59	2.7
n-Pentanol	10.0	10.0	6.0	1.67	2.9
Propargyl alcohol	10.1	10.1	5.8	1.74	3.1
Cyclohexanol	6.3	6.3	3.4	1.85	3.4
n-Decanol	7.0	7.4	3.8	1.89	3.4
n-Dodecanol	6.0	5.9	3.4	1.75	3.2

passing through the N $\cdots$ H—O bond and the (assumed degenerate) minor axes perpendicular to it. Aside from the expected decrease in all T_1s as the alcohol becomes larger there is an almost monotonic increase in the axial ratio. Glycerol, as a trihydric alcohol, can be expected to remain self-associated even as it hydrogen bonds to pyridine, and indeed a more complex behavior was observed[128] with R going through a maximum ($R \approx 2.1$ at ~ 0.50 mole fraction glycerol) and then decreasing to ~ 1.2 at high glycerol concentration. It was suggested that at high-viscosity reorientation is slowed to such an extent that H-bond breaking becomes the dominant relaxation mechanism, with the pyridine reorientating isotropically to a new H-bonded position.

Reorientation of glycerol itself is quite complex, as Burnett et al.[130] describe. Attainable temperatures suffice to reach both sides of the T_1 minimum, giving data similar to that seen for molecules relaxing through multiple internal rotations and requiring a distribution of relaxation times. The authors were able to fit the data using Woessner's[129] anisotropic rotational diffusion model (which contains five correlation times for fully anisotropic motion), yielding $D_1 = 5D_2 = 200D_3$, where the axes of the diffusion tensor are shown in Figure 2.5. The temperature dependence was taken to be $D = D^0 \exp[-E_a/R(T-T_g)]$, where T_g is a temperature near the glass transition, and $E_{a,1} = E_{a,2}$ ($E_{a,3}$ is unimportant because of negligible reorientation about the D_3 axis). Since this fitting procedure yields $E_{a,1} = 2.41 + 0.05$ kcal/mole, identical to $E_{a,\text{trans}}$[131] and to the energy required to break a single hydrogen bond, the authors propose the six-membered ring structures shown in Figure 2.5 and a simultaneous rotation-translation step in which one of the intermolecular H-bonds at the bottom of Figure 2.5 break as the molecule jumps to a new H-bonded orientation. In other work,[132] variable frequency ^1H and ^{13}C relaxation

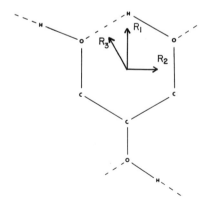

Figure 2.5
The coordinate system for the diffusion tensor and a possible hydrogen-bonded structure for glycerol.

data appear to indicate that reorientation of glycerol does require a complex description including a nonexponential autocorrelation function (see Section V.B). Thus in spite of numerous studies on glycerol, it has yet to be determined whether a distribution of correlation times (a nonexponential correlation function) is required to describe the reorientation.

In a recent ^{13}C and ^{15}N relaxation study,[133] it was shown that the $^+NH_3$ substituent, sterically identical ("iso-steric") with-CH_3, could undergo essentially unhindered internal rotation, in analogy with the situation for toluene. However, with the- $^+NH_3$ substituent, solute-solvent or counter ion interactions prevent free- $^+NH_3$ spinning, except in extremely acidic and nonnucleophilic environments. In fact, internal- $^+NH_3$ rotation was slow for substituted anilinium trifluoroacetates in $CDCl_3$ solution. The joint use of ^{13}C and ^{15}N relaxation data proved to be quite powerful in evaluation of the details of molecular dynamics of these systems.

Studies of motional anisotropy can also indicate the absence of specific interactions. Bovey et al.[134] measured relaxation times in a mixed solution of $(CF_2CH_2)_n$ and polymethylmethacrylate. From the lack of change in T_1s in the mixed solutions versus the isolated solutions they concluded that the unusual compatability of the two polymers is not due to complex formation.

The dominance by directly bonded protons of ^{13}C relaxation may often obscure interesting effects and a number of studies have focused on nonprotonated carbons. The carboxyl carbon in acetic acid was studied by Gust et al.[135] who found dipole-dipole contributions from the methyl and hydroxyl protons roughly equal, with spin rotation also important. The necessity of careful chemical preparation for these longer T_1s was demonstrated by varying the pH of a $1 M$ solution with and without added paramagnetics. The pure solution showed no change in T_1 and NOE, but the adulterated solution's values were markedly pH dependent. An intermolecular dipole-dipole contribution to the carboxyl was shown by the difference in NOE in H_2O and D_2O solutions to be given by $R_{1D,inter} \simeq 100 \; sec^{-1}$.

t-Butanol, which is known to form n-mers, was studied in $C_{16}D_{34}$ solution.[136] Since relaxation is much more efficient in a slowly reorienting n-mer than in the monomer, the observed quaternary carbon relaxation rate is sensitive to relatively small concentrations of the larger aggregates. Assuming a viscosity determined τ_c with isotropic reorientation and no internal rotation, one can argue that $\tau_c^{(n)} = n\tau_c'$, where $\tau_c^{(n)}$ is the correlation time for an n-mer. Although a model allowing for the existance of monomers, trimers, and all higher aggregates (i.e., the 1-3-∞ model) is more realistic, the author found that the simpler 1-3-6 model also fit the

Table 2.8

Analysis of Relaxation Data for $(CH_3)_3{}^{13}COH$ in $C_{16}D_{34}$ by the 1-3-6 Association Model[136]

| Concentration (M) | % Contribution | | | | | | R_1(calculated), sec^{-1} | R_1(observed), sec$^-1$ |
| | 1-mer | | 3-mer | | 6-mer | | | |
	C^a	R_1	C^a	R_1	C^a	R_1		
0.060	80	56	19	39	1	5	0.016	0.020
0.126	59	29	33	49	8	22	0.023	0.021
0.209	44	17	39	44	17	39	0.029	0.033
0.403	28	8	39	34	33	58	0.037	0.040
1.017	15	3	31	22	54	75	0.047	0.042
10.5^b	2	<1	12	7	86	93	0.071	0.071

^{a} Percentage of monomer units.
^{b} Neat liquid.

data (Table 2.8). In 0.06-M solution, 80% of the t-butanol exists as monomer, but monomer contributes only 56% of the total relaxation. This contribution drops off rapidly at higher concentrations.

D. Molecular Structure

Bond lengths and angles can in principle be calculated very accurately from nmr dipolar relaxation rates and NOEs, because of the r^{-6} dependence. Experimentally, however, this has proven difficult to implement in proton nmr, since rarely has it been possible to separate the dipolar contribution due to a single group of equivalent spins. Although ^{13}C relaxation, which is usually dominated by dipolar interaction with directly bonded protons, is experimentally more straightforward (neglecting S/N considerations) for determining molecular structure, applications have not been forthcoming Theroretical advances in the understanding of coupled relaxation of both carbon-proton[33,137] and proton-proton[30,138] systems will likely give impetus to further structural studies using both nuclei. These were discussed in Section II. One problem that the increased accuracy of T_1 measurements has made more obvious is the lack of accurate carbon-proton bond lengths. For example, in norbornadiene[122] where some r_{C-H} values are accurately known, the r_{C-H} variation alone is sufficient to account for the variation in dipolar relaxation rates. An additional complication arises in that the appropriate internuclear distance is not the equilibrium bond length r_e but rather an average r_{eff} over the displacements due to various vibrational modes, given by[139]

$$r_{eff} = r_e + \langle z \rangle + K$$

where $\langle z \rangle$ is the component of the vibrational amplitude along the bond axis and K depends on the geometry and the mean square displacements. If these are known, use of r_{eff} rather than r_e yield correlation times $\sim 10\%$ longer with corresponding variations in other molecular parameters derived from relaxation data.[140] One area in which this effect can cause systematic error is the extraction of nuclear quadrupole coupling constants (QCC) from relaxation times. A number of workers have combined

Table 2.9
Deuterium Quadrupole Coupling Constants Determined from NMR Relaxation Times

Molecule	Measured ^{2}H QCC (kHz)	Literature Value	Reference
CD_3COOH	176	168 ± 10	141
$C_6D_5NO_3$	190 ± 3	192 ± 4	114
CD_3OH	181 ± 14	155 ± 5	144
CD_3NH_2	177 ± 11	166 ± 10	144
CD_3COONa	184 ± 18	194 ± 5	144
CD_3CH_2COONa	223 ± 28	164 ± 10	144
C_6D_{12}	183	174 ± 2	141
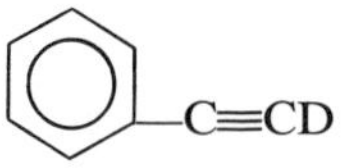 (phenyl)—C≡CD	254	215 ± 5	141
	227		142
 (phenyl)	172	186 ± 1.6	141
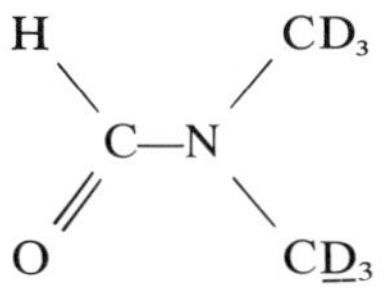 H, CD_3 / C—N / O, C$\underline{D}$_3	186		142
	160		141
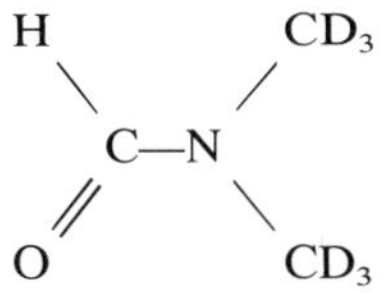 H, CD_3 / C—N / O, CD_3	168		141

^{13}C and ^{2}H relaxation data for the ^{13}C—^{1}H and ^{12}C—^{2}H moieties, respectively, to determine deuterium QCCs.[114,141–143] However, recent work[144] has shown a systematic descrepancy between QCCs measured by this relaxation method and those obtained by orienting the molecule in a liquid crystal, at least for methyl groups when the moieties measured are ^{13}CH$_3$ and ^{12}C^{2}H$_3$. The authors suggest that an isotope effect on the rotational correlation data, which predicts larger QCCs from relaxation data, could be responsible for the difference. Table 2.9 summarizes these experiments. Using the same technique for ^{13}C and ^{23}Na, Kintzinger et al.[145] have determined the QCC to range from 1.0 to 2.2 MHz for ^{23}Na complexed to various cryptate ligands. Stark et al.[114] as part of an analysis of rotational diffusion in nitrobenzene, were able to assign the principal field gradients for the nitrogen electric field gradient.

V. MODELS FOR MOLECULAR MOTION

A. Anisotropic Rotational Diffusion

The fact that the nuclear magnetic resonance relaxation parameters depend on the exact details of molecular movement through the fluctuating local magnetic fields provides a unique opportunity to study dynamic interactions in solution.[129,146–163]

An unwarranted assumption of isotropic reorientation can result in the loss of valuable information while a full analysis can be used as proof of assignment and as a check of the internal consistency of the experimental data. The presence of segmental motion or internal rotation or some other form of large amplitude molecular motion can be ascertained from a failure to fit the data to the rigid anisotropic rotor model.[150–153]

The appropriate equations required are as follows[129]

$$K_0 = \tfrac{4}{5} \tag{23}$$

$$K_1 = \tfrac{2}{15} \tag{24}$$

$$K_2 = \tfrac{8}{15} \tag{25}$$

$$D = \tfrac{1}{3}(D_1 + D_2 + D_3) \tag{26}$$

$$L^2 = \tfrac{1}{3}(D_{1,2} + D_{1,3} + D_{2,3}) \tag{27}$$

$$\delta_i = \frac{(D_i - D)}{(D^2 - L^2)^{1/2}} \tag{28}$$

$$\frac{1}{\tau\pm} = 6[\pm(D^2 - L^2)^{1/2}] \tag{29}$$

$$C_+ = d - e \tag{30}$$

$$C_- = d + e \tag{31}$$

$$C_1 = 6m'^2 n'^2 \tag{32}$$

$$C_2 = 6l'^2 m'^2 \tag{33}$$

$$C_3 = 6l'^2 m'^2 \tag{34}$$

$$\frac{1}{\tau_1} = 4D_1 + D_2 + D_3 \tag{35}$$

$$\frac{1}{\tau_2} = 4D_2 + D_1 + D_3 \tag{36}$$

$$\frac{1}{\tau_3} = 4D_3 + D_1 + D_2 \tag{37}$$

$$d = \tfrac{1}{2}[3l'^4 + m'^4 + n'^4) - 1] \tag{38}$$

$$\begin{aligned} e = \tfrac{1}{6}[&\delta_1(3l'^4 + 6m'^2 n'^2 - 1) + \delta_2(3m'^4 + 6l'^2 n'^2 - 1) \\ &+ \delta_3(3n'^4 + 6l'^2 m'^2 - 1)] \end{aligned} \tag{39}$$

$$\begin{aligned} J_h(\omega) = K_h\{&C_+[\tau_+/(1 + \omega^2\tau_+^2)] + C_-[\tau_-/(1 + \omega^2\tau_-^2)] \\ &+ C_1[\tau_1/(1 + \omega^2\tau_1^2)] + C_2[\tau_2/(1 + \omega^2\tau_2^2)] \\ &+ C_3[\tau_3/(1 + \omega^2\tau_3^2)]\}. \end{aligned} \tag{40}$$

where l', m' and n' are the direction cosines of the C—H vectors used in the calculation and the D_i's represent the rotational diffusion coefficients.

The geometric factors (direction cosines in the formalism presented here) required for the calculation are referenced to a coordinate system fixed in the molecule.[68] The choice of this coordinate system requires some care.[150–153] Alternately, the coordinate system describing the principal axes of the rotational diffusion tensor may be systematically varied in addition to the diffusion coefficients in order to obtain a proper fit.[150–154]

The orientation of the diffusion tensor principal axis system will be the result of a number of forces that act in concert on the molecular system. The diffusion tensor will reflect the overall shape of the molecule as well as the various intermolecular interactions. To a first approximation one chooses the principle axes of the moment of inertia tensor as the diffusion tensor principle axis system. This assumes that the overall molecular symmetry is the primary factor governing the diffusion.

This approximation has generally been supported by experimental results.[154,155] However, to minimize unwanted complications the molecule to be studied should not contain any heavy atoms or highly polar groups. Electric dipole moment effects have been found to be significant in certain circumstances.[156] If the overall symmetry of the molecule is not high, it may not even be possible to diagonalize the diffusion tensor.[152] A practical consequence of the theoretical formulation for anisotropic rotational diffusion is the fact that the center of the diffusion axis system cannot be located since the angles that appear in the equations arise from a rotational transformation. Therefore, translational shifts of the center of diffusion do not affect the relaxation rates.[152]

As it is assumed that the chosen coordinate system diagonalizes the diffusion tensor, at least three linearly independent, geometrically unique ^{13}C—H vectors are necessary to determine the diffusion coefficients. For this reason the full determination of the anisotropy of motion in planar molecules cannot be accomplished from consideration of ^{13}C relaxation rates alone. The results must be combined with Raman[157] or depolarized Rayleigh scattering[158] results to obtain sufficient parameters or cross-correlation spectral density measurements as noted in Section II. Alternatively, quadrupole relaxation rates alone may be used if at least one quadrupolar nucleus in the planar molecule possesses a large asymmetry parameter.[152]

One must also test for the applicability of the rotational diffusion equations. Many criteria exist that can be employed to this end; these range from comparison with hydrodynamic theory or dielectric relaxation times, to the χ test. This latter parameter, first introduced by Wallach and Huntess,[159] involves comparison of the experimental correlation time τ_{ci} with the theoretical reorientation time of the free gas molecule at the same temperature τ_{fi}. For rotational diffusion in the ith motion it was proposed that τ_{ci} be much greater than τ_{fi}, the numerical condition being

$$\chi_i = \left[\frac{5}{18D_i}\right]\left[\frac{kT}{I_i}\right] \gg 1 \qquad (41)$$

where D_i is the rotational diffusion coefficient for the ith axis, I_i is the moment of inertia about the ith axis, k is Boltzmann's constant, and T is the absolute temperature. For $\chi \gg 1$ (diffusion limit) the motion can be said to be characterized by a random walk of small angular steps. For $\chi \lesssim 5$ the molecule undergoes large angular steps between collisions and the inertial limit is attained. The choice of χ_i representing the inertial limit is a somewhat arbitrary choice. χ_i may also be related to the number of collisions required to reorient a molecule one radian and the mean angle turned per collision.

The results derived for the various molecules listed in Table 2.10 can generally be interpreted in one of the following ways. Differences in the moments of inertia are quite important in the ordering of the diffusion coefficients. If the molecule is affected by inertial reorientation then there will be a $1:1$ correspondence between the moments of inertia and the diffusion coefficients: namely the lower I_i, the greater D_i. Grant et al.[160] have defined a parameter known as the ellipticity ε. It represents the ratio of the maximum separation of atoms along the two axes perpendicular to the rotation axis. If molecular shape dominates the motion then the slower motions will correlate with the axis of greater ellipticity.

Other factors are important. Motions that sweep out the least amount of volume (solvent) should be favored and hence reorientation of a substituent is not favored. Similarly, motions that reorient an electric dipole are hindered.

The ^{13}C T_1s of a number of cyclic amino acids and peptides have been analyzed in terms of overall anisotropic reorientation of a rigid body.[154] The low-molecular-weight, rigid cyclic compounds were chosen to restrict the number of allowed conformations and minimize or eliminate interactions resulting from electrostatic effects. In addition, X-ray diffraction structural data were available from which reasonable geometric parameters could be derived and incorporated into the analysis. The principal axes of the moment of inertia and rotational diffusion tensors were assumed coincident as a first approximation, but the three Euler angles (α, β, γ) that characterized the nonalignment were allowed to vary also in the least-squares fitting. All optimum fits required little or no change in the Euler angles suggesting that the overall shape of the molecules determines the anisotropy of the motion.

Although the calculated relaxation rates were within experimental error for proline and acetylprolinamide the resultant correlation times were deemed physically unreasonable in that they ranged over two orders of magnitude. An interpretation based on overall isotropic reorientation with internal motion of particular molecular fragments was proposed. In the cyclo-triprolyl calculation the T_1s for the α and δ carbons were ~25% too large, whereas the β and γ carbons were underestimated by 7 and 8%. For cyclo (L-Ser-L-Tyr) the assumption of pseudorigid anisotropic motion resulted in very good agreement ($\tau_A = 1.26 \times 10^{-10}$, $t_B = 2.05 \times 10^{-10}$, $\tau_C = 1.57 \times 10^{-10}$, $\alpha = 10.5°$, $\beta = -0.1°$, $\gamma = 22°$ for the unsolvated case) except for the β carbon of the seryl residue. Similar results were obtained for cyclo (Gly-L-Tyr) except that the α carbon of glycine

Table 2.10
Selected Rotational Diffusion Constants for Some Small Molecules[a]

	D_x	D_y	D_z	Reference
Trichlorobenzene[b]	2.17(.06)	2.7(.2)	3.8(.1)	29b
Trichlorobenzene[c]	3.38(.06)	4.5(.4)	5.4(.1)	29b
Dichloroanisole[d]	3.61(.06)	3.1(.2)	4.4(.1)	29b
Dichlorophenol[e]	3.1(.4)	7.2(2.2)	3.7(.7)	29b
Trans retinal[f]	11.7(1.9)	1.3(0.5)	<0.5	153a
13-*cis*-retinal[f]	11.3(2.3)	2.2(0.8)	<0.7	153a
11-*cis*-Retinal[f]	4.6(1.6)	2.3(0.7)	0.9(0.5)	153a
Methylcyclohexane	48(12)[g]	3(2)	8(3)	155
	19(2)[h]	6	13(0.6)	153b
Methoxycyclohexane[g]	12(4)	2.5(1.5)	7.5(2.5)	155
Cyclohexanol	0.6(0.1)[g]	0.1	0.8(0.2)	155
	11(4)[i]	<2	10(2)	155
Methylcyclopentane[h]	95(28)	16	24(19)	153b
Methoxycyclopentane[g]	36(5)	20(4)	<1	155
Cyclopentanol[g]	19(2)	2(1)	<1	155
Methylcyclobutane[h]	25(3)	<4	61(15)	153b
Cyclopentene[j]	17(4)	29(7)	21(5)	160
Chlorocyclobutane[i]	39(5)	11(2)	9(2)	155
Bromocyclobutane[i]	6(3)	11(4)	16(7)	155
Cyclobutanol	18(2)[g]	<3	<2	155
	48(9)[i]	<8	<5	155
	20(6)[k]	6(14)	10(4)	155
2-Aminopyrimidine	0.94(.10)	0.97(.33)	1.04(0.9)	40
trans-decalin	11(3)	2.5(1)	6(2)	160
Adamantane	8(2)	8(2)	8(2)	160
Bicyclooctane	17(4)	9(2)	9(2)	160
Norbornane	17(4)	10(3)	7(2)	160
$(CH_2)_{n-1}CHCH_3$				
$n = 7$	13(5)	9(3)	8(5)	153b
$n = 8$	5(0.6)	5(1)	9(2)	153b

[a] Numbers in parentheses represent standard deviation in calculated D_j units: 10^{10} rad sec^{-1} unless otherwise noted; consult references for orientation of the principal axes.
[b] 10^{10}sec^{-1}, 0.1M, CDCl$_3$. [g] Neat liquid, 35°C.
[c] 10^{10}sec^{-1}, 0.1M, CS$_2$, 24°C. [h] 5M CDCl$_3$, 30°C.
[d] 10^{10}sec^{-1}, 0.1M, CS$_2$, 24°C. [i] 0.4M, 85% CCl$_4$-15% acetone-d$_6$.
[e] 10^{10}sec^{-1}, 0.1M, CS$_2$, 24°C. [j] Conditions not given.
[f] 10^9sec^{-1}. [k] 0.2M, 85% CCl$_4$-15% acetone-d$_6$.

appears to possess internal flexibility ($\tau_A = 1.14 \times 10^{-10}$, $\tau_B = 1.9 \times 10^{-10}$, $\tau_C = 1.09 \times 10^{-10}$.

B. Multiple Internal Rotation

Application of the theory of multiple internal rotation in its appropriate form is relevant to the study of a large number of molecular systems ranging from the n-alkanes to macromolecules including membranes, lipid bilayers, polysaccharides, and synthetic high polymers.[58,164–178] With such a diversity of systems to draw upon it is important to review the present status and recent applications of this theory.

The motion of any atom along a linear chain can be described by performing an atom-by-atom transformation of coordinates provided that one can describe the motion of the coordinate system fixed to the ith atom relative to the $(i-1)$th atom.[58,179–189] Therefore, if the matrix for the autocorrelation function for the motion of any atom in the chain is known, it is possible to calculate the autocorrelation function of any other atom given the rotational diffusion coefficient about each intervening bond (provided the motions are independent to a first approximation).

Atom 0 is usually chosen to represent the effective center of mass of the system which can be assumed to rotate isotropically (or anisotropically) with diffusion coefficient(s) D_0 (or $D_\parallel$ and $D_\perp$, respectively).[58,182,184] The resulting autocorrelation spectral densities are given by[182,184,187]

$$J(\omega) = \sum |d_{ab}(\beta_a)|^2 |d_{bc}(\beta_b)|^2 \cdots |d_{ni}(\beta_n)|^2 |d_{i0}(\beta_i)|^2 \left[\frac{\tau}{1+\omega^2\tau^2}\right] \quad (42)$$

where

$$\tau = (6D_0 + a^2 D_a + b^2 D_b + \cdots + i^2 D_i)^{-1}$$

and the $d_{ij}(\beta_i)$ are the reduced second-order Wigner rotation matrices and the angles β_a, β_b, β_c, ..., β_h are the angles between successive axes of rotation and the vector between the dipolar coupled nuclei leading to the relaxation.

London and Avitabile have shown that the computations can be reduced in complexity by taking advantage of the inherent symmetry of the problem. The resultant spectral density is given (for the case of isotropic reorientation of C_0) by[58,164]

$$J(\omega) = \sum_{abc} B_{ab} \cdots B_{hi} B_{i0} \left[\frac{\tau}{1+\omega^2\tau^2}\right] \quad (43)$$

where the B matrix is given by

B_{ij}	2	1	0
2	$\frac{1}{8}(1+6\cos^2\beta+\cos^4\beta)$	$\frac{(1-\cos^4\beta)}{2}$	$\frac{3}{4}\sin^4\beta$
1	$\frac{(1-\cos^4\beta)}{2}$	$\frac{1}{2}(1-3\cos^2\beta+4\cos^4\beta)$	$3\sin^2\beta\cos^2\beta$
0	$\frac{3}{8}\sin^4\beta$	$\frac{3}{2}\sin^2\beta\cos^2\beta$	$\left(\frac{3\cos^2\beta-1}{2}\right)^2$

In the situation in which C_0 reorients as a symmetric rotor an extra term $[B_{za}(\beta_0)]$ is added to Eq. 43 where β_0 is the angle between the Z axis of the symmetric rotor and the first internal rotation axis. The anisotropic motion of C_0 is identical to the addition of an extra internal rotation with $D_i = D_z - D_x$.

Figure 2.6 illustrates that the autocorrelation function consists of a number of exponential curves particularly for each of the the first four carbons, beyond which the function approximates very closely to a single exponential.[184]

Beyond C-5 the diffusion coefficient of the macromolecule (C-0) can be varied over a wide range without affecting the decay of the autocorrelation function provided only that $D_i > D_0$ (or $D_i > D_z$, $D_x = D_y$). Thus the relaxation times beyond C-5 are independent of the motion of the molecule (Figure 2.6). The effects of anisotropic reorientation are also illustrated in Figures 2.7, 2.8, and 2.9.[182]

One interesting prediction of the theory is that if the autocorrelation function of a C—H vector on the nth atom is nonexponential and if $(\omega/6D_i)^2 > 1$ for at least one motion in the system then $T_1 \neq T_2$ for this carbon nucleus even if the temperature dependence of T_1 is such as to suggest that the extreme narrowing condition ($\omega^2\tau^2 \ll 1$) is applicable (where normally $T_1 = T_2$).

The data obtained by Levine et al.[182] on the n-alkanes and summarized in Table 2.11 reveals the following trends:

1. The motion of the molecule about its long axis has a value (D_z) of $\sim 1-2 \times 10^{11}\ \text{sec}^{-1}$. The dependence of D_z on chain length is related to the chain length dependence of the viscosity.

2. For alkanes up to C-10 the use of equal diffusion constants for all bonds in the chain, except for the terminal methyl group explains the data.* The faster terminal bond motion was attributed to the

* Equal motion about successive bonds does not imply equal motion of successive carbons since motion of each carbon is determined by rotation of all bonds between it and the center of mass.

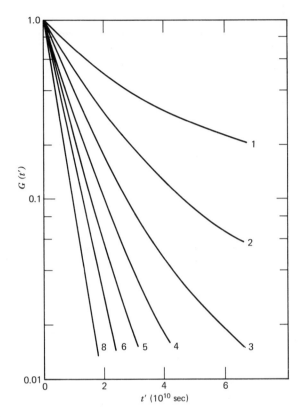

Figure 2.6
Angular autocorrelation functions $G(t')$ of the C—H vectors in methylene groups 1, 2, 3, 4, 5, 6, and 8 for a hydrocarbon chain attached to a sphere with a rotational diffusion coefficient, D_0, of 1.67×10^7 sec^{-1}. The diffusion coefficients for rotation about the carbon-carbon bonds are all the same, and equal to 1.67×10^9.

fact that this motion does not require the molecule to sweep out a volume of solution greater than the molecular volume.

3. The values of D_i decrease almost linearly with increasing chain length (Figure 2.10).[169,170,172,181,190–194]

Numerous experiments have probed the motion in lipid bilayers, particularly the fatty acid chains that constitute the hydrophobic interior (which determines such properties as premeability, flexibility of the bilayer, rate of enzymatic activity and lateral diffusion of embedded proteins).[195–200]

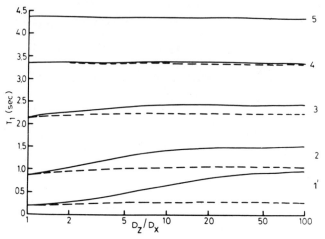

Figure 2.7
$^{13}CT_1$ values for the first five carbons as a function of β_0. For all curves $D_z = 1.67 \times 10^7 \text{ sec}$, $D_i = 1.67 \times 10^{10} \text{ sec}^{-1}$. The dashed curves are for $D_x = 0.84 \times 10^7 \text{ sec}^{-1}$, the solid curves for $D_x = 1.67 \times 10^5 \text{ sec}^{-1}$ (anisotropic ratios of 2 and 100, respectively).

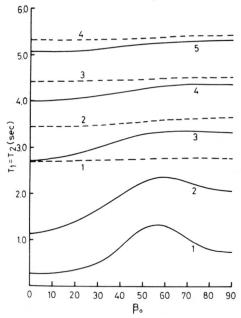

Figure 2.8
Effect of varying β_0 on ^{13}C relaxation times ($T_1 = T_2$) for the first five carbons in the chain. For all curves $D_z = 1.67 \times 10^9$, $D_i = 1.67 \times 10^{10} \text{ sec}^{-1}$. The dashed curves are for $D_x = 0.84 \times 10^9$, the solid curves for $D_x = 1.67 \times 10^7 \text{ sec}^{-1}$ (anisotropic ratios of 2 and 100, respectively).

138

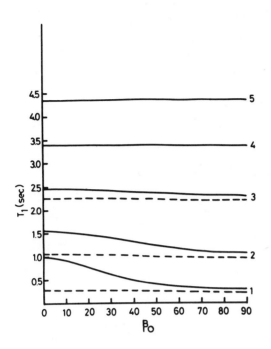

Figure 2.9
$^{13}CT_1$ values for the first five carbons as a function of the anisotropic ratio. For all curves $D_z = 1.67 \times 10^7$ sec^{-1}. The solid curves are for $\beta_0 = 0°$, the dashed curves for $\beta_0 = 90$. $D_i = 1.67 \times 10^{10}$ sec^{-1}.

Table 2.11
Rotational Diffusion Coefficients Calculated for n-Alkanes[182]

Chain Length	D_z (10^{11} sec^{-1})	$D_z\eta^a$	$D_x(=D_y)$ (10^{11} sec^{-1})	D_i (10^{11} sec^{-1})	D_ω (10^{11} sec^{-1})
6	2.6	0.71	0.35	0.2	0.9
8	1.3–1.5	0.63–0.73	0.114–0.113	0.15–0.2	0.28
10	0.93–1.0	0.73–0.80	0.059–0.064	0.110–0.119	0.3
12	(0.35–0.39)	—	(0.0.6–0.018)	0.11–0.12	0.7
14	—	—	—	0.12–0.14	0.5
16	—	—	—	0.08–0.10	0.59
18	—	—	—	0.07–0.85	0.5

^{a} η is the viscosity.

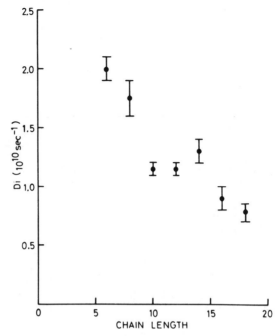

Figure 2.10
Values of D_i, the diffusion coefficient for motion about the bonds, calculated from the experimental T_1 values, as a function of chain length for n-alkanes. The values for C-6—C-10 were calculated by fitting the T_1 values for all carbons except the terminal methyl. For C-16 and C-18, D_i was obtained for the (ω-1) and (ω-2) bonds directly from the difference in T_1 values of adjacent carbons. For C-12 and C-14 both methods were used, and gave essentially identical results.

For dimyristoyllecithin[183] the data can be interpreted in terms of a rapid axial motion ($D_{\text{axial}} = 7 \times 10^9 \sec^{-1}$) together with equal motion about C—C bonds ($D_i = 1 \times 10^9 \sec^{-1}$) for carbons 2, 3, and 7. (i.e., carbons for which data was available). However, the motion has increased significantly by C-12, a result not predicted from an extrapolation of the curve for the first seven carbons. Lee et al.[183] were not able to precisely define the point in the chain at which the increase occurs. However, reasonable predictions yield $D = 1 \times 10^9$ or $3 \times 10^9 \sec^{-1}$ for bond 8 and 3×10^9 or $5 \times 10^9 \sec^{-1}$ for bond 12. Beyond C-12 the motion increased quite rapidly with values of $D_{13} = 6 \times 10^9 \sec^{-1}$ and $D_{14} = 8 \times 10^{10} \sec^{-1}$ calculated from the T_1s of C-12–C-14 by means of the simple equation

$$D_i = \frac{1}{2}\left[\frac{1}{\tau_i^{\text{eff}}} - \frac{1}{\tau_{i-1}^{\text{eff}}}\right] \tag{44}$$

where τ_i^{eff} is the effective correlation time.

Table 2.12
Comparison of Rotational Diffusion Coefficients for Dimyristoyllecithin at
52°C and n-Alkanes at 31°C[183]

	D_{axial}	D_1	$D_{\omega-1}$	D_ω
Dimyristoyllecithin	7×10^9	1×10^9	6×10^9	8×10^{10}
n-Alkanes[a]	1×10^{11}	1×10^{10}	1.2×10^{10}	6.2×10^{10}

[a] Average data for C-12, C-14, C-16 (hydrocarbons).

A comparison of the n-alkanes and hydrocarbon chains in lipid bilayers
reveals that D_{axial} and D_1 have been reduced by a factor of 15 and 10,
respectively, on going from the n-alkanes to the lipids. However, D_ω and
$D_{\omega-1}$ are very similar (see Table 2.12)

Severe computational problems arise with the incorporation of more
realistic models of chain motion in terms of the amount of computer time
necessary for the calculations. Rather than use stochastic rotational
diffusion to describe the bond rotations, a threefold jump model may be
used, with inclusion of the relaxation probabilities of the three rotamers
and the jump probabilities. Furthermore, it is likely that motions about
adjacent bonds are correlated at least to the extent that g^+g^- and g^-g^+
conformations are forbidden. This results in the rotamer and jump
probabilities becoming conditional probabilities which depend on the
instantaneous conformation about adjacent bonds. (Such calculations
have been performed using the Monte Carlo method[185,188] and subsequently extended by London et al.[164]) Some computational assistance is
provided by the fact that qualitatively, if the D_is are constant about each
bond then the plot of T_1 versus carbon number is linear from C-2
onward, the slope of the line being proportional to the value of D_i. It was
assumed in the London and Avitable[164] derivation that the C—C bonds
existed in and jumped between only three states: trans (t), gauche plus
(g^+), and gauche minus (g^-) with direct jumps between g^+ and g^- being
forbidden. The results were formulated in terms of a parameter σ defined
as the relative probability of being in a gauche versus trans state divided
by 2.

For values of σ near 0.5 plots of T_2 and T_1/T_2 versus carbon number
do not vary smoothly. Although the effect appears to become less
important for smaller values of σ the physical basis for this phenomenon
is not clear. Some representative calculations are shown in Figures 2.11
and 2.12.

The gauche-trans isomerism model also predicts that the T_1/T_2 ratio
will change very slowly for small values of σ. For D_0 (the isotropic
rotational diffusion constant) equal to 10^6 sec^{-1} and with a trans lifetime,

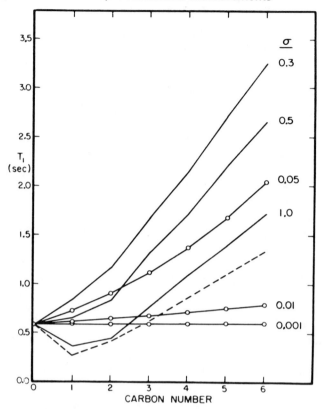

Figure 2.11
Spin-lattice relaxation times NT_1 for a six bond carbon chain undergoing isotropic rotation with a diffusion constant $D_0 = 10^6 \, \text{sec}^{-1}$ and *gauche-trans* isomerism with $\tau_t = 10^{-10}$ sec about each bond. The broken line represents the results obtained with the free internal rotation model where D_i, the internal diffusion constant, is defined to be $D_1 = \frac{1}{6}\tau_s = \frac{1}{6}\tau_t(1 + \frac{1}{2}\sigma)$ and $\sigma = 1.0$. The circles on the $\sigma = 0.001, 0.05,$ and 0.1 curves represent calculations using the approximate matrix in which $\tau_g = \tau_s$ valid for small σ.

τ_t, of 10^{-10} sec, a calculation at two different magnetic field strengths (67.9 and 25.2 MHz) reveals some interesting predictions. With $\sigma = 1$ essentially frequency independent parameters are obtained for all carbons but C_0 (with a small effect at C_1). However, an increasing frequency dependence is predicted for decreased values of σ. In the limit $\sigma \to 0$, T_1 remains constant at the value of C_0 and since the D_0 motion is so slow $T_1[C_0] \sim \nu^2$ so that the ratio of the T_1 values measured at the two frequencies is constant along the chain with a value of 7.24. It should

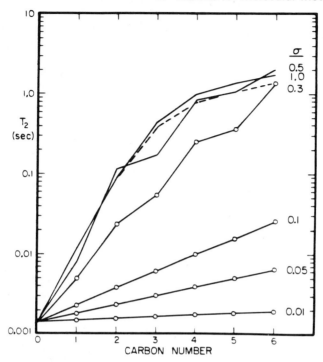

Figure 2.12
Spin-spin relaxation times for a six bond carbon chain. All parameters are the same as in Figure 2.11. The circles represent calculations using the approximate matrix described in Ref. 164. The broken line represents a free internal rotation calculation.

also be noted that the properties associated with the slowly tumbling end of the molecule (such as reduced NOE, $T_1 \gg T_2$ and a frequency dependence of T_1) propagate for into the chain.

The effect of correlated rotations is similar to the effect of restricting the motion of individual bonds. Thus if all the diffusion coefficients for internal rotation are identical, atom N in a system with correlated rotations about bonds 1 and 3 will behave like atom N-2 in an uncorrelated system and so on.

The long T_1 for the terminal methyl carbon in a straight-chain hydrocarbon has been noted previously and interpreted in terms of an abrupt increase in the diffusion rate about the last bond in the chain. The results of London and Avitabile[164] suggest that at least part of of the effect is due to a change in σ. The distinction between *gauche* and *trans* states vanishes for the final methyl rotation and $\sigma = 1$. In addition σ will increase for the penultimate bond since the unfavorable g^+g^- (and g^-g^+) states corresponding to an internal bond in the chain become essentially g^+t (and g^-t) states for the $\omega-1$ bond.

A study of the linear, intermediate-molecular-weight peptide LH-RH by Deslauriers and Somorjai[179] reveals a similarity to the previous discussion. The aliphatic sidechains appear anchored by the α-carbons of the peptide backbone. They undergo increased motion toward either end of the peptide chain. In LH-RH, arginine, proline, and leucine chains show gradations in the rates of internal motion that appear to be characteristic of the amino acid rather than of the framework of the peptide bond in which the side chain is embedded. (Because of its partial double bond character the peptide bond was simulated using an arbitrarily large correlation time). The influence of the anisotropy of motion on the leucine and arginine residues is negligible beyond the β-carbon (where the effect was less than 5%).

Gent and Prestegard[181] have noted that a model of molecular motion can be placed on a more firm basis if the model can successfully correlate the relaxation behavior of several nuclei in the bilayer. Different magnetic moments and geometric arrangements will alter the portion that each specific motion contributes to the total relaxation rate. The model they presented consisted of fast, β-coupled isomerizations of sets of carbon bonds superimposed on slow, nearly isotropic chain reorientations. (Internal bond rotations such as *trans-trans-trans* to *gauche$^+$-trans-gauche$^-$* conversions are called β-coupled *gauche* isomerizations or kink formation.)

However, the observation of frequency-dependent relaxation[189a] times for neat 1,2-decanediol cannot be explained in terms of the theories discussed above. A model of at least partially cooperative segmental motions superimposed on an overall molecular reorientation having some slow ($\geq 10^{-9}$ sec) components was proposed. The slower motions are incompletely decoupled from the methyl carbon, formally removed by 8 carbons from the sites of molecular complexation. The frequency dependence of the 1,2-decanediol T_1s is noteworthy particularly since *no part* of the molecule has *apparent motion* outside of the region of extreme spectral narrowing (T_1 for C-1 is 0.27 sec at 22.6 MHz and 54°C).[189a] Preliminary studies of 1,2-decanediol in benzene solution show additional complexities. For example, at 2 M concentration, frequency dependent T_1 values are observed at 45°C, whereas at 34°C no significant differential was observed at 22.6 and 67.9 MHz.[132a]

It is not yet known to what extent solution order (e.g., micelle formation) affects the decanediol results. In addition, intramolecular and intermolecular transmission of complex motional information to the free ends

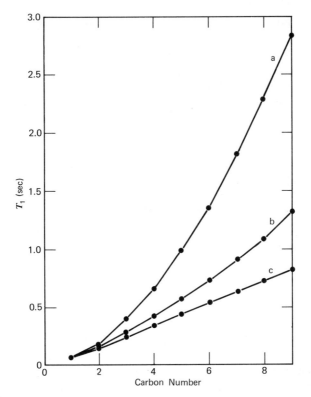

Figure 2.13
^{13}C spin-lattice relaxation times of carbons 1–9 of the chain for different models of motion. For all curves, $D_0 = 1.67 \times 10^{-7} \, sec^{-1}$. Curve c: Equal motion. All D_i values the same, and equal to $1.67 \times 10^9 \, sec^{-1}$. Curve b: Exponential increase in D_i along the chain. Values used 1.67×10^9, 1.85×10^9, 2.06×10^9, 2.29×10^9, 2.54×10^9, 2.82×10^9, 3.14×10^9, 3.49×10^9 and $3.88 \times 10^9 \, sec^{-1}$. Curve a: Linear increase in D_i along the chain. Values used: 1.67×10^9, 2.67×10^9, 3.67×10^9, 4.67×10^9, 6.67×10^9, 7.67×10^9, 8.67×10^9, $9.67 \times 10^9 \, sec^{-1}$.

of the chains are both possible; their relative contributions are not yet known.

It is readily apparent that relaxation times are sensitive to rates of rotational diffusion about the bonds (Figure 2.13). In various ^{13}C studies of the alkyl side chains of poly(n-butylmethacrylate),[189b] poly(n-hexylmethacrylate),[189b] n-hexane,[182] and amino acids,[179] it has been noted that, under certain circumstances, the relaxation time of the C-2 carbon in a linear chain is shorter than either the C-1 or C-3 carbons. In the variable-frequency/variable-temperature study of Levy et al.[189b] the effect was greatest at low temperatures and at the higher magnetic field strength. Levine et al.[182] proposed that the transformation along the chain effectively attenuates the correlation function of the molecular motion, leading to an increase in the spectral density at high frequencies and for C-2 this increase is greater than the accompanying decrease in spectral density due to the internal motions leading to a net increase in spectral density at high frequencies and thus, to a reduction in T_1. The transitory nature of the phenomenon is due to the sensitivity to the Larmor frequencies involved.

The qualitative effects of internal rotation in a large number of cycloalkanes (—CH_2—)$_n$ for n up to 44 have been reported by Fritz et al.[168] A limiting T_1 value of approximately 1.2 sec is observed with increasing ring size, this value being a factor of ~5 larger than that of a methylene group in a rigid molecule of comparable molecular weight. They concluded that the limiting value represented the T_1 of an unhindered segment of a CH_2 chain dominated by segmental motion rather than by tumbling. Other measurements on a related molecule, hexadecane, supported this argument.

The variable temperature (10–90°C) behavior of the eleven distinct carbons of 10-methylnonadecane has been reported at 20 MHz.[175] The apparent activation energies for the methine and methylene carbons ranged from 5.1–5.5 kcal mol^{-1} while the methyl carbon barriers were 1.0–1.5 kcal mol^{-1} lower. Effective correlation times for all carbons were calculated, and the results were qualitatively interpreted in terms of the various contributions of overall and internal reorientational modes.

The major constituents of chloroplast thylakoids (monogalactosyldialkylglycerol and digalactosyldiacylglycerol), which contain high concentrations of (9z, 12z, 15z)-octadeca-9,12,15-trienoic acid, have been investivated.[169,170] It was noted that the most likely secondary structure for a carboxylic acid was an inverted micelle (the carboxylic acid groups being associated with the methylene chains extending out into the solvent). Similarities with these derivatives are found in decanoic acid,[201] long-chain alcohols,[167] and long-chain quaternary amines.[202]

A model was proposed for the chain dynamics. While a $tt \rightarrow tg$ rotation in a carboxylic acid chain which causes a kink is possible, such a situation in an inverted micelle would increase intermolecular interactions with adjacent molecules. However, a double-kink coupled-rotation can occur in which simultaneous jumps occur for pairs of gauche configurations of opposite polarities at sites separated by one carbon atom. This again is the so called β-coupled rotation and results in a roughly straight configuration that minimizes unfavorable interactions.

The spin-lattice relaxation times of the thylakoid constituents[170] were measured in methanol, chloroform, and water (all deuterated) and the results were interpreted in terms of different secondary structures in different solvents, a monomeric structure in methanol, an inverted micellar structure in chloroform, and a bilayer structure in D_2O (with associated polar head groups at the water interface and a hydrophobic inner region composed of acyl chains).

Other systems studied in varying degrees of detail include tetragastrin,[177] phosphatidylcholine-cholesterol bilayers,[203] gramicidin-S,[171] various nonionic surfactants,[173] and perfluoroalkanes.[174]

In a slightly different context it is interesting to discuss some work regarding internal rotation of methyl groups. The tertiary butyl moiety has been studied numerous times by proton nmr and activation energies have been reported for rotation of the t-butyl group itself about the sp^3 bond.[204-212] Recent interest has, as noted, centered upon the dynamic behavior of the individual methyl groups in suitably sterically hindered environments.[213-222] In particular spin-lattice relaxation measurements can be particularly useful for the determination of rotational barriers of methyl groups for activation energies too low to be accessible by dynamic nmr line shape analysis. Many examples have been given.[104,155,223-229]

Previous theoretical approaches have shown that a freely rotating sp^3 methyl carbon attached to an isotropically reorienting rigid molecule satisfying the extreme narrowing condition and having only one degree of internal freedom will have a T_1 three times that of a methine carbon located on the backbone. Conversely an immobile methyl (on this fast time scale) under such conditions will have a T_1 of one-third of that of the backbone methine carbon. Relaxation rates between these extremes will be observed for intermediate rates of rotation and under these conditions energy barriers may be obtained from Eqs. 45–56.[224] Table 2.13 illustrates the agreement between barriers calculated from nmr T_1 measurements and various other methods. Despite the problems to be noted the agreement is quite good.

These calculations can be very sensitive to errors in the overall diffusion constant, methyl relaxation rate and bond angle as noted by Axelson and Holloway.[155,226] The effect of geometry was also subsequently discussed by Blunt and Stothers.[227] They noted that methyl groups subject to steric constraints are expected to increase the tetrahedral value to reduce nonbonded interactions; experimental verification is available from neutron diffraction studies.

As a consequence, for example, the increase from $\phi = 109.47°$ to $\phi = 112°$ results in a free rotation limit $T_1^{CH_3}/T_1^{CH}$ of 4 rather than 3. Thus

Table 2.13
Some Selected Barriers to Internal Rotation of Methyl Groups[223,224]

	$V_{calc.}$ (kcal/mol^{-1})	$V_{lit.}$ (kcal/mol^{-1})
1-Methylnaphthalene	2.1	2.4
1,4-Dimethylnaphthalene	2.2	
1,8-Dimethylnaphthalene	2.8	3.0
		3.2
7,12-Dimethyl benz[9]anthracene		
7 methyl	<0.4	
12 methyl	>4.4	
cis-2-Butene	0.6	0.73
trans-2-Butene	1.7	1.95
Isobutene	1.9	2.21
2-Methyl-2-butene	1.9(1–Me)	
	1.4(2–Me)	
	0.7(4–Me)	
2,3-Dimethyl-2-butene	2.1	

the barrier determined from the data must be considered in relationship to the geometry (see Figure 2.14).

$$\frac{1}{T_{1D}} = \frac{N_H \hbar^2 \gamma_c^2 \gamma_H^2}{r^6} \tau_{c,eff} \tag{45}$$

$$\tau_{c,eff} = \left[\frac{A_1}{6R_\perp} + \frac{A_2 + A_3}{6R_\perp + \rho_D} + \frac{B_1}{5R_\perp + R_\parallel} \right.$$

$$\left. + \frac{B_2 + B_3}{5R_\perp + R_\parallel + \rho_D} + \frac{C_1}{2R_\perp + 4R_\parallel} + \frac{C_2 + C_3}{2R_\perp + 4R_\parallel + \rho_D} \right] \tag{46}$$

where

$$A_1 = \tfrac{1}{16}(1 - 3m^2)^2(1 - 3n^2)^2 \tag{47}$$

$$B_1 = \tfrac{3}{4}m^2(1 - m^2)(1 - 3n^2)^2 \tag{48}$$

$$C_1 = \tfrac{3}{16}(1 - m^2)^2(1 - 3n^2)^2 \tag{49}$$

$$A_2 = \tfrac{9}{2}m^2n^2(1 - m^2)(1 - n^2) \tag{50}$$

$$B_2 = \tfrac{3}{2}(4m^4 - 3m^2 + 1)n^2(1 - n^2) \tag{51}$$

$$C_2 = \tfrac{3}{2}(1 - m^4)n^2(1 - n^2) \tag{52}$$

$$A_3 = \tfrac{9}{32}(1 - m^2)^2(1 - n^2)^2 \tag{53}$$

$$B_3 = \tfrac{3}{8}(1 - m^4)(1 - n^2)^2 \tag{54}$$

$$C_3 = \tfrac{3}{32}(m^4 + 6m^2 + 1)(1 - n^2)^2 \tag{55}$$

The directional cosine of the methyl top axis relative to the principal axis is given by m, and n is the directional cosine of the C—H bond relative to the internal methyl top axis. The energy barrier is calculated from the equation:

$$\rho_D = \rho_0 e^{-V/RT} \tag{56}$$

where ρ_0 is the mean jump rate of a freely rotating methyl group ($\rho_0 = 3/2(kT/I)^{1/2} = 1.33 \times 10^{13}$ rad sec^{-1} at 40°C, where I is the moment of inertia of the methyl group about the axis of rotation).

Another factor that may be as important in such studies is the degree of anisotropy of the overall molecular motion. As we have shown the method by which such calculations are made in Section V.A it is now

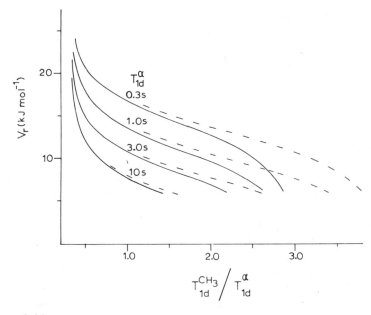

Figure 2.14
The computed values of the methyl rotational barrier (Vr) as a function of $T_{1d}^{CH_3}$ and T_{1d}^{α} for two values of θ(C—C—H). (—)$\theta = 109.47°$; (---)$\theta = 112.1°$. Each pair of curves is for the indicated value of T_{1d}^{α}. A random jump model for the methyl rotation has been assumed. $T = 308°K$. T_{1d}^{α} is the average backbone carbon relaxation time.

Table 2.14

Relationship Between the T_1 of a Locked Methyl Group Attached to an Anisotropically Reorienting Molecule and the Backbone Methine Carbon T_1s

| | | | $T_1^{b,c}$ | | | | | |
| | | | | | | | C-1 | C-2 |
$D_1{}^a$	$D_2{}^a$	$D_3{}^a$	C-1	C-2	C-3	CH$_3$	CH$_3$	CH$_3$
300	55	50	3.49	3.46	1.06	0.97	3.6	3.6
300	455	50	7.97	4.68	2.58	2.36	3.4	2.0
1230	50	50	5.67	7.00	1.24	2.10	2.7	3.3
300	50	5	2.45	2.17	0.58	0.74	3.3	2.9
300	50	105	4.11	4.70	1.50	1.13	3.6	4.2
300	50	355	5.91	8.57	3.14	1.70	3.5	5.0

^{a} Units in 10^8 rad sec^{-1}.
^{b} Ref. 155.
^{c} T_1s are in seconds, compound is 11,12-dimethylene-9-methyl-9, 10-ethanoanthracene (see Ref. 155). The C-1 carbon is the β-aromatic carbon, C-2 is the α-aromatic carbon, C-3 is one of the methylene carbons, and CH$_3$ is the methyl group attached to the bridgehead position. Thus there are four geometrically unique carbons in this rigid polycyclic compound.

informative to determine in more detail the effects on a rigid methyl group.

It is apparent from Table 2.14 that the effects of anisotropic reorientation can dramatically alter the relationship between rigidly held methine, methylene, and methyl C—H vectors. Deviations of such ratios as $T_1^{CH}/T_1^{CH_3} = 3$ in the case of isotropic overall reorientation may be considerable depending on the relative orientation of the C—H vectors with respect to the principal diffusion axes. For even relatively modest degrees of anisotropy the ratio $T_1^{C-2}/T_1^{CH_3}$ varies between 2.0 and 5.0 in the examples calculated. This is independent of the effects possible due to bond angle distortions. One need not invoke changes in geometry to account for such deviations. In fact, the effects of anisotropic rotational diffusion may be more prevalent than expected even in the case of steroids and other fairly large molecules.

VI. POLYMER RELAXATION

As indicated by the considerable coverage contained in this volume there is quite an interest in the relaxation parameters of polymers in the solid state, particularly of those compounds that have exceptionally long correlation times due to the rigid glassy nature of substances well below their glass temperatures.

As a complement to these discussions, the solution polymer area can also be described as remaining quite active. This section therefore delves into the nature of such research that is of recent origin. This discussion will also include bulk polymers studied well above their glass temperatures since spectra can be obtained in that case with scalar decoupling only.[230-277]

A number of papers have appeared reporting the relaxation parameters of polysulfone polymers.[230,233,234] In most cases the T_1 values of the polysulfone main chain methine and methylene carbons were 75 ± 10 msec and 43 ± 8 msec, respectively, both exhibiting reduced NOE factors (NOEFs) of approximately 1.0.[230] The similarities observed suggested that the dominant relaxation process of these carbons was a segmental motion that did not involve rotation about the C—C bond. The data from poly(cyclopentene) sulfone would be expected to be very different from the other polymers if such a C—C rotation were significant. This was not the case. (It was also noted that the β-CH$_2$ carbon of the cyclopentene ring was evidently affected by a ring-puckering motion since the T_1 of this carbon was much larger than that of the α-CH$_2$.) These results however, were in complete disagreement with dielectric relaxation data, which indicated that relaxation proceeded through whole molecule rotation. The differences in the techniques were emphasized by a calculation of the methine T_1s on the basis of a rigid molecular structure (from the dielectric relaxation measurements). Calculated methine T_1s for molecular weights of 2.4×10^5 and 4.5×10^4 were 30 and 2.5 sec, respectively, close to two orders of magnitude longer than observed. The variable temperature behavior of poly(but-1-ene sulfone) in CDCl$_3$ (10–50°C) indicated that $\omega^2 \tau^2 > 1$ since the T_1 decreased with increasing temperature. A calculation employing the conformational jump model of Valeur et al.[231] yielded an overall rotational correlation time (τ_0) much longer than the conformational jump correlation time (τ_D) ($\tau_D/\tau_0 \leq 1$ with $\tau_D \geq 10^{-8}$ sec). Thus segmental motion is faster than overall tumbling according to these results.

The deduction that internal segmental rearrangements within the backbone carbons was unusually slow and therefore could not complete with overall rotatory diffusion was based on comparison with other systems that are well documented.[232] In styrene polymers the transition from tumbling to local relaxation is almost complete at molecular weights of the order of 4×10^4 and a lower molecular weights for polyethylene oxide and polydimethylsiloxane. Nonetheless, the nmr data was unambiguous in that segmental motion dominated. It therefore remained for a suitable model of molecular motion to be invoked to resolve the conflict between the dielectric relaxation results and those obtained from nmr.

Some suitable, commonly adopted dynamic models for chain motions in these polymers include the 5-bond crankshaft as well as similar 3- and 4-bond rearrangements.[233] Polyethers, however, are considered to reorient according to the 4-bond process since neither of the 3- and 5-bond crankshaft motions can produce electric dipole reorientation in polyoxymethylene.

Stockmayer et al.[232] suggested that, with a sulfur atom at the apex, a 4-bond motion in polysulfones need only involve *trans* conformations at the C—C bonds both before and after rearrangements. There is no change of the electric polarization since the dipole moment about the two *trans* C—C bonds is small, and the central SO$_2$ group does not change its orientation. Cais and Bovey[234] suggested that five backbone bonds and six main chain atoms (C—S—C—C—S—C) are involved with connected segmental transitions occurring about two C—S bonds allowing interconversion of the three conformational states *ttt*, g^+tg^- and g^-tg^+. The backbone C—C bond always remains *trans* and the C—H vectors reorient while the sulfone dipoles do not. Thus the apparent conflict between the nmr and dielectric relaxation results can be resolved.

Ghesquiere et al.[235] have studied poly (4-vinylpyridine) in CD$_3$OD between 220 and 350°K by both ^{13}C (25 MHz) and ^{1}H (100 and 250 MHz) spin-lattice relaxation experiments. The anisotropic motion of the pyridyl group was treated by considering oscillations of limited amplitude (50–80°C) about the N—C4 axis and a complete derivation of the equations was given:

$$A(\alpha, \gamma) = \tfrac{2}{3}(1 - 3\cos^2\gamma) + \sin^2 2\gamma(1 + \cos\alpha) + \sin^4\gamma(1 + \cos 2\alpha)$$

$$(57)$$

$$B(\alpha, \gamma) = \sin^2 2\gamma(1 - \cos\alpha) = \sin^4\gamma(1 - \cos 2\alpha) \tag{58}$$

$$T_1^{-1}(I) = \tfrac{3}{80}\gamma_I^2\gamma_s^2\hbar^{-2}r^{-6}\{A(\alpha, \gamma)h(\tau_R) + B(\alpha, \gamma)h(\tau_\tau)\} \tag{59}$$

$$T_2^{-1}(I) = \tfrac{3}{160}\gamma_I^2\gamma_s^2\hbar^{-2}r^{-6}\{A(\alpha, \gamma)k(\tau_R) + B(\alpha, \gamma)k(\tau_\tau)\} \tag{60}$$

$$\text{NOE} = 1 - \frac{\gamma_s A(\alpha, \gamma)l(\tau_R) + B(\alpha, \gamma)l(\tau_t)}{\gamma_I A(\alpha, \gamma)h(\tau_R) + B(\alpha, \gamma)h(\tau_t)} \tag{61}$$

$$h(\tau) = \frac{\tau}{1 + (\omega_I - \omega_s)^2\tau^2} + \frac{3\tau}{1 + \omega_I^2\tau^2} + \frac{6\tau}{1 + (\omega_I + \omega_s)^2\tau^2} \tag{62}$$

$$k(\tau) = 4\tau + h(\tau) + \frac{6\tau}{1 + \omega_s^2\tau^2} \tag{63}$$

$$l(\tau) = \frac{\tau}{1 + (\omega_I - \omega_s)^2 \tau^2} - \frac{6\tau}{1 + (\omega_I + \omega_s)^2 \tau^2} \tag{64}$$

$$\tau_t = \tau_R^{-1} + \tau_G^{-1} \tag{65}$$

γ = angle between an internuclear vector and rotation axis
α = angular amplitude of the oscillation
τ_R = isotropic correlation time of polymer segment
τ_G = correlation time of oscillation

Pure oscillational motion is given in the limit $\tau_R \gg \tau_G$. The relaxation data of the main chain was interpreted in terms of an isotropic segmental motion assuming a temperature dependent distribution of correlation times.

Relaxation data for some poly (phenylthiiranes)[236] as well as poly (methylthiirane) have been reported. As for the polysulfones, there was no discernable molecular-weight effect over the range studied. Different tacticities were not reflected in the relaxation data.

The T_1s of polyisobutylene (PIB) and poly (propylene oxide) (PPO) in chloroform have been measured as a function of molecular weight, temperature, and concentration with little dependence in molecular weight (900 vs 2900) being observed.[237] T_1 increased with increasing dilution in solvent, however, reaching a limiting value at about a molar ratio [CDCl$_3$/monomer] of 5. In PIB the methyl reorientation rate increased with the backbone reorientation rate illustrating that the two motions are interlinked, whereas in PPO the methyl rate was independent of the backbone motion.

Several papers on the relaxation parameters of PMMA and related[189b,238-240] methacrylates investigating the effects of tacticity, temperature, concentration and magnetic field strength, have been published. The n-alkyl acrylate and methacrylate polymers provide interesting examples of the effects of a distribution of correlation times (and probably of conformational effects) within the aliphatic sidechain. Syndiotactic and isotactic PMMA in pyridine and CDCl$_3$ solutions at 38 and 100°C have been examined by ^{13}C relaxation experiments. The T_1s of the backbone and sidechain carbons were longer in the isotactic configuration. In addition a broader distribution of correlation times was required to fit the syndiotactic PMMA results. The methyl group attached to the backbone was found to be less constrained in the isotactic chain.

The most interesting phenomena observed[189b] are (a) frequency-dependent spin-lattice relaxation times for carbons with NT_1 values of up to approximately 20 sec and (b) less than maximum NOEFs for these same carbons. Two approaches have been developed that can be useful in treating the data; one is based in part on the fact that a distribution of correlation times is required to describe the relaxation data of the backbone carbons in these polymers and the other (described in section V.B) explicitly accounts for *trans, gauche plus,* and *gauche minus* conformations. Multiple internal rotation theory, coupled with an assumed independence of the bond rotations that themselves are described by rotational diffusion, is generally quite adequate, but it has become obvious from detailed experimental and theoretical studies that the dynamic behavior of such apparently simple systems as methylene carbon chains can require rather involved calculations.

It should be *emphasized* that the single correlation time, extreme narrowing, approximation can be inadequate for describing not only the dynamic behavior of polymer backbone carbons but also alkyl (or related) *side chain substituents.*[189b] In fact, carbons which are, strictly speaking, eight bonds removed from the polymer backbone, can exhibit frequency-dependent relaxation times and less than maximum NOEFs even though the NT_1 values are very large. Frequency-dependent relaxation times and less than maximum NOEs can be predicted and can, at least qualitatively, explain the experimental results just noted.[189b] The situation in polymeric substances, howver, results in a considerable number of ambiguities. For instance, the London and Avitabile approach[164] requires an identifiable carbon (at C_o, the center of mass) characterized by a long correlation time to generate frequency-dependent relaxation times throughout a long aliphatic sidechain. The degree to which the T_1s are frequency dependent depends in part on the ratio of the *trans* to *gauche* configurations in the chain. For a distribution of correlation times, the average overall isotropic diffusion constant may be within extreme narrowing. The London and Avitabile approach *would not predict* a frequency dependence or reduced NOEFs in this situation if measurements were made at one frequency and thus they could not account for the presence of slow-motional components in such a case. The Levy-Axelson[189] approach, while remaining predictive, does not explicitly consider configurational effects but is characterized by a tractable number of parameters. In many polymers, particularly those studied in the pure (bulk) state, intermolecular interactions may be as significant as the intrachain interactions. This further complicates the matter. Thus any approach to remain useful will most likely be deficient in some manner.

A few polymer systems are known to be compatible both in solution and in solid state including poly(alkyl methacylates) and poly(vinylidene fluoride) (PVF). A ^{13}C nmr study has been reported[72] attempting to ascertain whether some specific interaction can be discerned. No differences were observed for either PMMA in dimethylformamide or PVF as compared with a mixture of the polymers in the same solvent.

In a dilute solution of a homopolymer in a single solvent it is expected that the average composition of the polymer environment is equal to that of the solution as a whole. For block copolymers in certain solvents micelle formation (inter- or intramolecular) within which the chemical composition is very different from bulk solution can occur. The micellar structure arises when the solvent chosen is "good" for one component of the copolymer and "bad" for the other. With this in mind a styrene/butadiene/styrene triblock copolymer (SBS) in a number of selective and nonselective solvents was studied and compared with the *cis*-polybutadiene hompolymer.[241] In the nonselective solvents, $CDCl_3$ and toluene, there is little difference between the T_1s of the two polymer samples (styrene blocks have very small effect on butadiene segmental mobility). There was a distinct difference between polymers in dimethoxyethane (DME), a nonsolvent for polystyrene, and ethylacetate (a nonsolvent for polybutadiene).

While many polymers at room temperature can be considered as true solids, characterized in an nmr sense, by rotational correlation times longer than $\sim 10^{-5}$ sec, a vast number of polymers behave in a more liquid-like manner at comparable temperatures. It has been generally observed[256] for amorphous and partially crystalline polymers as well as other glass-forming substances that the temperature T_c at which the backbone carbon resonance(s) collapse, that is, become too broad to be resolved, occurs well above the glass temperature of each substance (T_g). Presently available data are summarized in Table 2.15. The difference between T_c and T_g range from approximately 30 to 87°C, with no fixed trend apparent. The values of T_c in general will depend on instrumental resolution, molecular weight, degree of crystallinity, chemical shift separations, the nmr operating frequency, and in a very important way on the motional characteristics of the polymer backbone.

This table illustrates the fact that many polymers may be conveniently studied without the need of elaborate experimental techniques such as those discussed in Chapters 4 and 5 of this book. The qualitiative differences in T_g and T_c can also serve as a guide to the feasibility of observing backbone carbon resonances in a given polymer if the glass temperature is known. Of course, the presence of substituents attached to the polymer backbone can be more readily observed by ^{13}C nmr, since these moieties will generally have a much more dynamic character, the greater degrees of freedom leading to narrower lines and shorter correlation times.

Table 2.15

Relation Between Glass Temperature and Temperature at which ^{13}C Spectra Collapse

Substance	$T_c{}^a$	$T_g{}^b$	Reference
Isotactic PMMA	110	52	256
Poly (isopropyl acrylate)	47	−11	256
Poly (n-butylacrylate)	0	−54	256
Atactic polypropylene	30	−20	256
trans-Polyisoprene	−28	−58	256
cis-Polybutadiene	−65	−102	256
cis-Polyisoprene	−30	−70	264
Polyisobutylene	−10	−70	264
Glycerol	−6	−93	139
Polyethylene oxide	−38	−70	256

^{a} Temperature at which backbone carbon resonances are too broad to observe.
^{b} Glass temperature: these values must be considered as approximate only due to the nature of the phenomenon and the varied experimental methods employed in the measurements.

VII. ^{13}C RELAXATION AND OTHER SPECTROSCOPIC METHODS

Quite a number of studies have correlated nmr data with the results of Raman, ir, Rayleigh, or dielectric measurements using well-known relationships[278] between the various relaxation times. However, only a few of these studies have relied on ^{13}C data.[112-114,279] Reorientation in nitrobenzene has been studied by ^{13}C relaxation and Rayleigh scattering by Alms et al.[113] Also ^{13}C, ^{14}N, and ^{2}H relaxation has been measured[114] with a discussion of possible explanations for the discrepencies between the two studies. A difficulty with Rayleigh scattering is the contribution due to correlations between pairs of molecules (not present in Raman scattering and, of course, not in the ^{13}C relaxation of ^{13}CH$_x$ fragments). This appears in the relationship between nmr (or Raman) correlation times τ_c and correlation times from Rayleigh light scattering τ_{ls} as

$$\tau_{ls} = \frac{1 + fN}{1 + gN} \tau_c \qquad (66)$$

for diffusional reorientation, where N is the solute number density, f is a measure of the static orientational correlation between pairs of solute molecules, and g is a measure of the time-dependent pair correlation of

angular momenta. It has been argued[280] and shown[113] for nitrobenzene
($gN = 0.1 \pm 0.1$) and chloroform ($gN = 0.0 \pm 0.01$) that dynamic correlations should be small ($gN \sim 0$). On the other hand, recent work on the
xylene[281] has yielded for o-xylene $fN = 0.03 \pm 0.13$, $gN = -0.27 \pm 0.09$;
for m-xylene $fN = -0.31 \pm 0.10$, $gN = -0.46 \pm 0.09$; and for p-xylene
$fN = 0.68 \pm 0.15$, $gN = -0.04 \pm 0.11$. Theoretical work has shown that
from symmetry arguments[282] $gN = 0$ for symmetric tops, while other
work[283] employing different starting assumptions has shown the reverse
and presented an expression for the calculation of gN. The quantity f
may be determined from the concentration dependence of the integrated
signal intensity; thus in principle τ_{ls} may be related to τ_c. Two experimental studies[97,284] of the relationship between τ_{ls} and τ_c show that caution
should be exercised in comparing the two quantities, a conclusion borne
out by the nitrobenzene work. As Stark et al.[114] discuss, the diffusion
tensor obtained from joint nmr-Rayleigh experiments[112] is smaller in all
components than the strictly nmr-derived tensor. Additionally the ratios
of D_x/D_z and D_y/D_z are smaller. Since the reported ^{13}C relaxation times
differ, Stark et al. reanalyze the light-scattering data using the longer ^{13}C
T_1s to get a third D tensor in which the ratios D_x/D_z and D_y/D_z are even
more removed from the pure nmr data.

Goulon et al.[279] have studied reorientation in several dicarbonyl compounds using infrared, dielectric relaxation, and ^{13}C relaxation techniques. Among other results, the authors determine the ratio τ_c (nmr)/τ_c
(dielectric). For several molecules the authors find this ratio to be equal to
3, the result expected for isotropic rotational diffusion, but for one
molecule they find τ_c(nmr)/τ_c (dielectric) $\cong 2$, and discuss the possible
reason for this result.

VIII. CONCLUSIONS

It is clear that ^{13}C spin-relaxation parameters have significant potential
for unraveling complex physical chemical problems in statistical dynamics
and other fields. There are two main approaches being developed today.
In one of these methodologies rigorously correct spin physics is probed
with selective pulse experiments to extend our theoretical understanding
of simple molecular systems. In the second approach, approximately
correct derivations are used as a basis for probing complex dynamics in
large molecular systems. The two approaches are not mutually exclusive,
but nevertheless they generally are applied separately. This is partly due
to the different goals of the various researchers involved, but partly this
separation results from the current impracticability of exact treatments

for complex systems. Increased understanding of the fundamental processes of molecular dynamics coupled with further development of rigorously
defined spin relaxation manifolds will jointly advance our ability to utilize
^{13}C relaxation data to probe physical chemical systems.

REFERENCES

1. G. C. Levy, and G. L. Nelson, *Carbon-13 Nuclear Magnetic Resonance for Organic Chemists*, Wiley-Interscience, New York,
   1972. Chapt. 9.
2. F. W. Wehrli and T. Wirthlin, *Interpretation of Carbon-13 NMR
   Spectra*, Heyden, New York 1976, Chapt. 4.
3. F. Wehrli, *Topics in Carbon-13 NMR Spectroscopy*, Vol. 2, Wiley-
   Interscience, New York, 1976, Chapt. 6.
4. G. C. Levy, *Acc. Chem. Res.* **6**, 161 (1973).
5. E. Breitmaier, K. H. Spohn, and S. Berger, *Angew. Chemie, Int. Ed.*
   14, 144 (1975).
6. J. R. Lyerla, Jr. and D. M. Grant, *International Reviews in Science—
   Physical Chemistry Series.* Vol. 4, C. A. McDowell, Ed., Medical
   and Technical Publishers, 1972, Chap. 5.
7. J. R. Lyerla, Jr. and G. C. Levy, *Topics in Carbon-13 NMR
   Spectroscopy* Vol. 1, Wiley-Interscience, New York, 1974, Chapt.
   3.
8. Specialist Periodical Reports, *Nuclear Magnetic Resonance*, (Chem.
   Soc., London) Vols. 4–6.
9. J. S. Blicharski and H. Schneider, *Ann. Phys.* **22**, 306 (1969).
10. H. Schneider and J. S. Blicharski, *Ann. Phys.* **23**, 139 (1969).
11. G. A. DeWit and M. Bloom, *Can. J. Phys.* **47**, 1195 (1969).
12. P. S. Hubbard, *J. Chem. Phys.* **51**, 1647 (1969).
13. J. H. Noggle, *J. Phys. Chem.* **72**, 1324 (1968).
14. M. F. Baud and P. S. Hubbard, *Phys. Rev.* **170**, 384 (1968).
15. D. Fenzke and H. Schneider, *Ann. Phys.* **19**, 321 (1967).
16. H. Schneider, *Ann. Phys.* **16**, 135 (1965).
17. D. Fenzke, *Ann. Phys.* **16**, 281 (1965).
18. H. Schneider, *Z. Naturforsch,* **19$_a$,** 510 (1964).
19. L. K. Runnels, *Phys. Rev.* **134**, A28 (1964).
20. R. L. Hilt and P. S. Hubbard, *Phys. Rev.* **134,** A392 (1964).

21. H. Schneider, *Ann. Phys.* **13,** 313 (1963).

22. G. W. Kattawar and M. Eisner, *Phys. Rev.* **126,** 1054 (1962).

23. P. S. Hubbard, *Phys. Rev.* **128,** 650 (1962).

24. P. S. Hubbard, *Phys. Rev.* **109,** 1153 (1958).

25. P. S. Hubbard, *Phys. Rev.* **111,** 1746 (1958).

26. P. S. Hubbard, *Rev. Mod. Phys.* **33,** 24a (1961).

27. L. G. Werbelow and D. M. Grant, *Can. J. Chem.* **55,** 1558 (1977).

28. L. G. Werbelow and D. M. Grant, *Adv. Mag. Res.* **9,** 189 (1977).

29. (a) J. Brondeau and D. Canet, *J. Chem. Phys.* **67,** 3650 (1977); (b) R. L. Vold, R. R. Vold, and D. Canet, *J. Chem. Phys.,* **66,** 1202 (1976).

30. D. Canet, R. L. Vold, and D. Canet, *J. Chem. Phys.* **64,** 900 (1976).

31. D. M. Grant and L. G. Werbelow, *J. Mag. Res.* **21,** 369 (1976).

32. L. G. Werbelow and D. M. Grant, *J. Chem. Phys.* **63,** 4742 (1976).

33. C. L. Mayne, D. M. Grant, and D. W. Alderman, *J. Chem. Phys.* **65,** 1684 (1976).

34. G. B. Matson, *J. Chem. Phys.* **65,** 4147 (1976); also **67,** 5152 (1977).

35. J. D. Cutnell and J. A. Glasel, *J. Am. Chem. Soc.* **98,** 7542 (1976); also **98,** 264 (1976).

36. H. E. Bleich, J. D. Cutnell, A. R. Day, R. F. Freer, J. A. Glasel, and J. F. McKelvy, *Proc. Nat. Acad. Sci., USA* **73,** 2589 (1976).

37. (a) L. G. Werbelow and D. M. Grant, *J. Chem. Phys.* **63,** 544 (1975); (b) P. E. Fagerness, D. M. Grant, K. F. Kuhlmann, C. L. Mayne, and R. B. Parry, *J. Chem. Phys.* **63,** 2524 (1975).

38. J. F. Rodriques de Miranda and C. W. Hilbers, *J. Mag. Res.* **19,** 11 (1975).

39. W. Buchner, *J. Mag. Res.* **17,** 229 (1975).

40. W. M. M. J. Bovee, *Mol. Phys.* **29,** 1673 (1975).

41. A. D. Bain and R. M. Lynden-Bell, *Mol. Phys.* **30,** 325 (1975).

42. J. W. Harrel, *J. Mag. Res.* **15,** 157 (1974).

43. S. Enid and R. A. Wind, *Chem. Phys. Lett.* **27,** 312 (1974).

44. V. A. Daragan, T. N. Khazanovich, and A. Stepanyants, *Chem. Phys. Lett.* **26,** 89 (1974).

45. J. D. Cutnell and W. Venable, *J. Chem. Phys.* **60,** 3795 (1974).

46. L. J. Burnett and B. H. Muller, *Chem. Phys. Lett.* **18,** 553 (1973).

47. L. G. Werbelow and A. G. Marshall, *J. Mag. Res.* **11,** 299 (1973).

48. M. Mehring and M. Raber, *J. Chem. Phys.* **59,** 1116 (1973).
49. I. D. Campbell, *J. Mag. Res.* **11,** 143 (1973).
50. W. Buchner, *J. Mag. Res.* **11,** 46 (1973); **12,** 82 (1973).
51. (a) P. S. Allen, A. Khazada, and C. A. McDowell, *Mol. Phys.*, **25,** 1273 (1973). (b) J. L. Carolan and T. A. Scott, *J. Mag. Res.* **2,** 243 (1970).
52. S. Albert and J. A. Ripmeester, *J. Chem. Phys.* **57,** 2641 (1972).
53. L. G. Werbelow and A. G. Marshall, *Chem. Phys. Lett.* **22,** 568 (1972).
54. K. van Putte and G. J. N. Egmond, *J. Mag. Res.* **4,** 236 (1971).
55. K. van Putte, *J. Mag. Res.* **5,** 367 (1971).
56. P. M. Richards, *Phys. Rev.* **132,** 27 (1963).
57. K. F. Kuhlmann, D. M. Grant, and R. K. Harris, *J. Chem. Phys.* **52,** 3439 (1970).
58. R. E. London and J. Avitabile, *J. Chem. Phys.* **65,** 2443 (1976).
59. P. S. Hubbard, *J. Chem. Phys.* **52,** 563 (1970).
60. I. Solomon, *Phys. Rev.* **99,** 559 (1955).
61. L. Mayne, D. W. Alderman, and D. M. Grant, *J. Chem. Phys.* **63,** 2514 (1975).
62. K. F. Kuhlmann and D. M. Grant, *J. Chem. Phys.* **55,** 2998 (1971).
63. S. J. Opella, D. J. Nelson, and O. Jardetzky, *J. Chem. Phys.* **64,** 2533 (1976).
64. R. K. Harris and R. H. Newman, *J. Mag. Res.* **24,** 449 (1976).
65. D. Canet, *J. Mag. Res.* **23,** 361 (1976).
66. P. E. Fagerness, D. M. Grant, and R. Bryce Parry, *J. Mag. Res.* **26,** 267 (1977).
67. F. Bloch, *Phys. Rev.* **102,** 104 (1956).
68. K. T. Suzuki, L. W. Cary, and K. F. Kuhlmann, *J. Mag. Res.* **18,** 390 (1975).
69. C. G. Moreland and F. I. Carrol, *J. Mag. Res.* **15,** 596 (1974).
70. G. C. Levy, U. Edlund, and J. G. Hexem, *J. Mag. Res.* **19,** 259 (1975).
71. G. N. LaMar and J. W. Faller, *J. Am. Chem. Soc.* **95,** 3817 (1973); J. W. Faller, M. A. Adams, and G. N. LaMar, *Tetrahedron Lett.* 699 (1974).
72. B. L. Mayo, *Chem. Soc. Rev.* **2,** 49 (1973).
73. G. C. Levy and R. A. Komoroski, *J. Am. Chem. Soc.* **96,** 678 (1974).

74. G. C. Levy, U. Edlund, and C. E. Holloway, *J. Mag. Res.* **24,** 375 (1976).

75. G. C. Levy, and J. D. Cargioli, *J. Mag. Res.* **10,** 231 (1973).

76. P. S. Hubbard, *Phys. Rev.* **131,** 275 (1963).

77. J. W. Faller, M. A. Adams, and G. N. LaMar, *Tetrahedron Lett.* 699 (1974).

78. G. N. LaMar and E. A. Metz, *J. Am. Chem. Soc.* **96,** 5611 (1974).

79. K. S. Bose, T. H. Witherup, and E. H. Abbott, *J. Mag. Res.* **27,** 385 (1977).

80. J. G. Hexem, U. Edlund, and G. C. Levy, *J. Chem. Phys.* **64,** 936 (1976).

81. D. M. Doddrell and A. K. Gregson, *Chem. Phys. Lett.* **29,** 512 (1974).

82. D. M. Doddrell, D. T. Pegg, M. R. Bendall, and H. P. W. Gottlieb, *Chem. Phys. Lett.* **39,** 65 (1976).

83. J. C. Ronfard-Haret and C. Chachaty, *Chem. Phys.* **18,** 345 (1976).

84. G. N. LaMar, in *NMR of Paramagnetic Molecules* G. N. LaMar, W. De W. Horrocks, Jr., and R. H. Holm, Eds., Academic Press, New York, 1973, p. 97.

85. G. C. Levy and U. Edlund, *J. Am. Chem. Soc.* **97,** 5031 (1975).

86. L. G. Werbelow and A. G. Marshall, *Mol. Phys.* **28,** 113 (1974).

87. C. H. Wang, *Mol. Phys.* **28,** 801 (1974).

88. C. R. Lassigne and E. J. Wells, *J. Mag. Res.* **27,** 215 (1977).

89. (a) K. T. Gillen, D. C. Douglass, M. S. Malmberg, and A. A. Maryott, *J. Chem. Phys.* **57,** 5170 (1972); (b) A. A. Maryott, T. C. Farrar, and M. S. Malmberg, *J. Chem. Phys.* **54,** 64 (1971).

90. A. A. Maryott, M. S. Malmberg, and K. T. Gillen, *Chem. Phys. Lett.* **25,** 169 (1974).

91. A. P. Zens and P. D. Ellis, *J. Am. Chem. Soc.* **97,** 5685 (1975).

92. W. A. Steele, *J. Chem. Phys.* **38,** 2404, 2411 (1963).

93. B. D. Nageswara Rao and P. K. Mishra, *Chem. Phys. Lett.* **27,** 592 (1974).

94. O. Yamamoto and M. Yanagisawa, *J. Chem. Phys.* **67,** 3803 (1977).

95. D. A. Wright, S. Huang, and M. T. Rogers, unpublished data.

96. (a) S. G. Huang and M. T. Rogers, private communication; (b) S. G. Huang, PhD. Thesis, Michigan State University (1977).

97. G. D. Patterson and J. E. Griffiths, *J. Chem. Phys.* **63,** 2406 (1975).

98. R. L. Vold, R. R. Vold, and R. E. Simon, *J. Mag. Res.* **11**, 283 (1973).

99. D. M. J. Lilley and O. W. Howarth, *J. Mag. Res.* **27**, 335 (1977).

100. R. Freeman and H. D. W. Hill, in *Dynamic Nuclear Magnetic Resonance Spectroscopy*, L. M. Jackman and F. A. Cotton, Eds., Academic Press, New York, 1975, p. 131.

101. H. M. Miziorko and A. S. Mildvan, *J. Biol. Chem.* **249**, 2743 (1974).

102. (a) A. Kumar and R. R. Ernst, *Chem. Phys. Lett.* **37**, 162 (1976); (b) A. Kumar and R. R. Ernst, *J. Mag. Res.* **24**, 425 (1976).

103. C. M. Hu and R. Zwanzig, *J. Chem. Phys.* **60**, 4354 (1974).

104. (a) R. E. D. McClung and D. Kivelson, *J. Chem. Phys.* **49**, 3380 (1968); (b) D. Kivelson, M. G. Kivelson, and I. Oppenheim, *J. Chem. Phys.* **52**, 1810 (1970); (c) S. Tsay and D. Kivelson, *Mol. Phys.* **29**, 1 (1975).

105. D. R. Bauer, J. I. Bauman, and R. Pecora, *J. Am. Chem. Soc.* **96**, 6840 (1974).

106. D. R. Bauer, J. I. Brauman, and R. Pecora, *J. Chem. Phys.* **63**, 53 (1975).

107. M. Fury and J. Jonas, *J. Chem. Phys.* **65**, 2206 (1976).

108. J. Hwang, D. Kivelson, and W. Plachy, *J. Chem. Phys.* **58**, 1753 (1973).

109. R. G. Gordon, *J. Chem. Phys.* **44**, 1830 (1966).

110. D. Chandler, *J. Chem. Phys.* **60**, 3500, 3508 (1974).

111. W. T. Ford, *J. Am. Chem. Soc.* **98**, 2727 (1976).

112. D. R. Bauer, G. R. Alms, J. I. Brauman, and R. Pecora, *J. Chem. Phys.* **61**, 2255 (1974).

113. G. R. Alms, D. R. Bauer, J. I. Brauman, and R. Pecora, *J. Chem. Phys.* **59**, 5310 (1973).

114. R. E. Stark, R. L. Vold, and R. R. Vold, *Chem. Phys.* **20**, 337 (1977).

115. S. Berger, F. R. Kreissl, and J. D. Roberts, *J. Am. Chem. Soc.* **96**, 4348 (1974).

116. H. W. Spiess, D. Schweitzer, and U. Haberlen, *J. Mag. Res.* **9**, 444 (1973).

117. E. von Goldammer, H. D. Ludemann, and A. Muller, *J. Chem. Phys.* **60**, 4590 (1974).

118. T. K. Leipert, J. H. Noggle, and K. T. Gillen, *J. Mag. Res.* **13**, 158 (1974).

119. C. R. Lassigne and E. J. Wells, *J. Mag. Res.* **26,** 55 (1977).

120. (a) K. T. Gillen and J. H. Noggle, *J. Chem. Phys.* **53,** 801 (1970); (b) W. G. Rothschild, *J. Chem. Phys.* **53,** 990 (1970); (c) D. Chandler, *Acc. Chem. Res.* **7,** 313 (1973).

121. K. T. Gillen, M. Schwartz, and J. H. Noggle, *Mol. Phys.* **20,** 899 (1971).

122. J. Grandjean and P. Laszlo, *Mol. Phys.* **30,** 413 (1975).

123. J. O'Dell and B. J. Berne, *J. Chem. Phys.* **63,** 2376 (1975).

124. (a) J. D. Graham and J. S. Darby, *J. Mag. Res.* **24,** 287 (1976); (b) D. A. Wright, unpublished results.

125. G. C. Levy, R. A. Komoroski, and R. E. Echols, *Org. Mag. Res.* **7,** 172 (1975).

126. G. C. Levy, J. D. Cargioli, and F. A. L. Anet, *J. Am. Chem. Soc.* **95,** 1527 (1973).

127. G. C. Levy and D. Terpstra, *Org. Mag. Res.* **8,** 658 (1976).

128. I. D. Campbell, R. Freeman, and D. L. Turner, *J. Mag. Res.* **20,** 172 (1975).

129. D. E. Woessner, *J. Chem. Phys.* **37,** 647 (1962).

130. L. J. Burnett and S. B. W. Roeder, *J. Chem. Phys.* **60,** 2420 (1974).

131. D. J. Tomlinson, *Mol. Phys.* **25,** 735 (1972).

132. (a) G. C. Levy et. al., unpublished data; (b) J. P. Kintzinger and M. D. Zeidler, *Ber. Bunsen-Gesell.* **77,** 98 (1973).

133. G. C. Levy, A. D. Godwin, and J. M. Hewitt, *J. Mag. Res.* **29,** 553 (1978).

134. F. A. Bovey, F. C. Schilling, T. K. Kwei, and H. L. Frisch, *Macromolecules* **10,** 559 (1977).

135. D. Gust, H. Pearson, I. M. Armitage, and J. D. Roberts, *J. Am. Chem. Soc.* **98,** 2723 (1976).

136. E. E. Tucker, T. R. Clem, J. I. Seeman, and E. D. Becker, *J. Phys. Chem.* **79,** 1005 (1975).

137. C. H. Wang and D. M. Grant, *J. Chem. Phys.* **64,** 1522 (1976).

138. R. R. Vold and R. L. Vold, *J. Chem. Phys.* **64,** 320 (1976).

139. (a) S. Meiboom and L. C. Snyder, *Acc. Chem. Res.* **4,** 81 (1971); (b) P. Diehl and W. Niederberger, *J. Mag. Res.* **9,** 495 (1973).

140. R. L. Vold, and R. R. Vold, private communication.

141. H. Saito, H. H. Mantsch, and I. C. P. Smith, *J. Am. Chem. Soc.* **95,** 8453 (1973).

142. L. M. Jackman, E. S. Greenberg, N. M. Szeverenyi, and G. K. Schnorr, *J. Chem. Soc. Chem. Comm.* 1974, 141.

143. A. P. Zens, P. T. Fogle, T. A. Bryson, R. B. Dunlap, R. R. Fisher, and P. D. Ellis, *J. Am. Chem. Soc.* **98,** 3706 (1976).

144. J. P. Jacobsen and K. Schaumburg, *J. Mag. Res.* **28,** 191 (1977).

145. J. P. Kintzinger and J. M. Lehn, *J. Am. Chem. Soc.* **96,** 3313 (1974).

146. T. E. Bull, *J. Chem. Phys.* **65,** 4802 (1976).

147. R. L. Vold and R. R. Vold, *J. Chem. Phys.* **61,** 2525 (1974).

148. C. Thibaudier and F. Volino, *Mol. Phys.* **30,** 1159 (1975).

149. L. D. Favro, *Phys. Rev.* **119,** 53 (1970).

150. R. S. Becker, S. Berger, D. K. Dalling, D. M. Grant, and R. J. Pugmire, *J. Am. Chem. Soc.* **96,** 7008 (1974).

151. S. Berger, F. R. Kreissl, D. M. Grant, and J. D. Roberts, *J. Am. Chem. Soc.* **97,** 1805 (1975).

152. W. T. Huntress, *Adv. Mag. Res.* **4,** 1 (1970).

153. W. T. Huntress, Jr., *J. Chem. Phys.* **48,** 3524 (1967).

154. R. L. Somorjai and R. Deslauriers, *J. Am. Chem. Soc.* **98,** 6460 (1976).

155. D. E. Axelson, Ph.D. Thesis, York University, 1976.

156. K. T. Gillen and J. H. Noggle, *J. Chem. Phys.* **53,** 801 (1970).

157. K. T. Gillen and J. E. Griffiths, *Chem. Phys. Lett.* **17,** 359 (1972); B. J. Berne and R. Pecora, *Dynamic Light Scattering,* Wiley-Interscience, New York, 1976.

158. D. R. Bauer, G. R. Alms, J. I. Brauman, and R. Pecora, *J. Chem. Phys.* **61,** 2255 (1974).

159. D. Wallach and W. T. Huntress, *J. Chem. Phys.* **50,** 1219 (1969).

160. D. M. Grant, R. J. Pugmire, E. P. Black, and K. A. Christenson, *J. Am. Chem. Soc.* **95,** 8465 (1973).

161. J. Grandjean and P. Laszlo, *Mol. Phys.,* **30,** 413 (1975).

162. R. A. Komoroski and G. C. Levy, *J. Phys. Chem.* **80,** 2410 (1976).

163. G. C. Levy, T. Holak, and A. Steigel, *J. Am. Chem. Soc.* **98,** 495 (1976).

164. R. E. London and J. A. Avitabile, *J. Am. Chem. Soc.* **99,** 7765 (1977).

165. R. Deslauriers and I. C. P. Smith, *Biopolymers* **16,** 1245 (1977).

166. R. Deslauriers, A. C. M. Paiva, K. Schaumberg, and I. C. P. Smith, *Biochemistry* **14,** 878 (1975).

167. D. Doddrell and A. Allerhand, *J. Am. Chem. Soc.* **93,** 1558 (1971).

168. H. Fritz, P. Hug, H. Sauter, and E. Logemann, *J. Mag. Res.* **21,** 373 (1976).

169. S. R. Johns, D. R. Leslie, R. I. Willing, and D. G. Bishop, *Aust. J. Chem.* **30,** 813 (1977).

170. S. R. Johns, D. R. Leslie, R. I. Willing, and D. G. Bishop, *Aust. J. Chem.* **30,** 823 (1977).

171. R. A. Komoroski, I. R. Peat, and G. C. Levy, *Biochem. Biophys. Res. Commun.* **65,** 272 (1975).

172. Y. K. Levine, N. J. M. Birdsall, A. G. Lee, J. C. Metcalf, *Biochemistry* **11,** 1416 (1972).

173. A. A. Ribeiro and E. A. Dennis, *J. Phys. Chem.* **80,** 1746 (1976).

174. J. R. Lyerla, Jr. And D. L. Vander Hart, *J. Am. Chem. Soc.* **98,** 1697 (1976).

175. J. R. Lyerla, Jr. and T. Horikawa, *J. Phys. Chem.* **80,** 1106 (1976).

176. J. D. Robinson, N. J. M. Birdwall, A. G. Lee, and J. C. Metcalfe, *Biochemistry* **11,** 2903 (1972).

177. E. Hermann, J. D. Cutnell, and J. A. Glasel, *Biochemistry* **15,** 2455 (1976).

178. E. Williams, B. Sears, A. Allerhand, and E. H. Cordes, *J. Am. Chem. Soc.* **95,** 487 (1973).

179. R. Deslauriers and R. L. Somorjai, *J. Am. Chem. Soc.* **98,** 1931 (1976).

180. D. Doddrell, V. Glushko, and A. Allerhand, *J. Chem. Phys.* **56,** 3683 (1972).

181. M. P. N. Gent and J. H. Prestegard, *J. Mag. Res.* **25,** 243 (1977).

182. Y. K. Levine, N. J. M. Birdsall, A. G. Lee, J. C. Metcalfe, P. Partington, and G. C. K. Roberts, *J. Chem. Phys.* **60,** 2890 (1974).

183. A. G. Lee, N. J. M. Birdsall, J. C. Metcalfe, G. B. Warren, and G. C. K. Roberts, *J. Chem. Phys.* **60,** 2890 (1974).

184. Y. K. Levine, P. Partington, and G. C. K. Roberts, *Mol. Phys.* **25,** 497 (1973).

185. Y. K. Levine, *J. Mag. Res.* **11,** 421 (1973).

186. Y. K. Levine P. Partington, G. C. K. Roberts, N. J. M. Birdsall, A. G. Lee, and J. C. Metcalfe, *FEBS Lett.* **23,** 203 (1972).

187. D. Wallach, *J. Chem. Phys.* **47,** 5258 (1967).

188. T. Yasukawa, D. Ghesquiere, and C. Chachaty, *Chem. Phys. Lett.* **45,** 279 (1977).

189. (a) G. C. Levy, M. P. Cordes, J. S. Lewis, and D. E. Axelson, *J. Am. Chem. Soc.* **99,** 5492 (1977); (b) G. C. Levy, D. E. Axelson, R. L. Schwartz, and J. Hochmann, *J. Am. Chem. Soc.* **100,** 410 (1978).

190. Y. K. Levine, *Progr. Biophys. Mol. Biol.* **24,** 1 (1972).

191. (a) N. O. Petersen and S. I. Chan, *Biochemistry* **16,** 2657 (1977); (b) H. A. Seiter and S. I. Chan, *J. Am. Chem. Soc.* **85,** 7541 (1973).

192. C. W. Feigenson and S. I. Chan, *J. Am. Chem. Soc.* **96,** 1312 (1974).

193. A. F. Horwitz, W. J. Horsley, and M. P. Klein, *Proc. Nat. Acad. Sci. USA* **69,** 590 (1972).

194. K. M. Keough, E. Oldfield, D. Chapman, and P. Benyon, *Chem. Phys. Lipids* **10,** 37 (1973).

195. J. DeGrier, J. G. Mandersloot, and L. L. M. VanDeenen, *Biochim. Biophys. Acta,* **150,** 666 (1968).

196. D. Papahadjiopoulous, K. Jacobsen, S. Nir, and T. Isac, *Biochim. Biophys. Acta* **311,** 330 (1973).

197. M. P. N. Gent and J. H. Prestegard, *Biochemistry* **13,** 4027 (1974).

198. C. M. Grisham and R. E. Barnett, *Biochem. Biophys. Acta* **311,** 417 (1973).

199. N. A. Machtiger and C. F. Fox, *Ann. Rev. Biochem.* **42,** 575 (1973).

200. R. James and D. Branton, *Biochim. Biophys. Acta.* **323,** 378 (1973).

201. C. Chachaty, Z. Wolkowski, F. Piriou, and G. Lukacs, *J. Chem. Soc. Chem. Commun.* **951,** (1973).

202. G. C. Levy, R. A. Komoroski, and J. A. Halstead, *J. Am. Chem. Soc.* **96,** 5456 (1974).

203. P. E. Godici and F. R. Landsberger, *Biochemistry* **14,** 3927 (1975).

204. (a) F. Suzuki, M. Oki, and H. Nakanishi, *Bull. Chem. Soc. Jap.* **47,** 3114 (1974); (b) M. Oki and M. Suda, *Bull. Chem. Soc. Jap.* **44,** 1876 (1971).

205. (a) H. I. Iwamura, *Chem. Commun.* 232 (1973); (b) G. Yamamoto and M. Oki, *Chem. Lett.,* 67 (1974).

206. M. Oki and G. Yamamoto, *Chem. Lett.* 45 (1972).

207. F. Suzuki, M. Oki, and H. Nakanishi, *Bull. Chem. Soc. Jap.* **46,** 2858 (1973).

208. M. Nakamura, M. Oki, and H. Nakanishi, *Tetrahedron* **30,** 543 (1974).

209. (a) C. H. Bushweller, G. U. Rao, W. G. Anderson, and P. E. Stevenson, *J. Am. Chem. Soc.* **94,** 4744 (1972); (b) J. E. Anderson and H. Pearson, *J. Chem. Soc. B,* 1209 (1971).

210. (a) M. Oki, *Angew. Chem. Int. Ed.* **15,** 87 (1976); (b) M. Nakamura and M. Oki, *Bull. Chem. Soc. Jap.* **48,** 2106 (1975).

211. (a) A. Ricker and H. Kersler, *Tetrahedron Lett.* 1227 (1969); (b) F. A. L. Bret, M. St. Jacques, and G. N. Chumury, *J. Am. Chem. Soc.* **90,** 5243 (1968).

212. (a) H. Kessler, V. Husoski, and M. Hanack, *Tetrahedron Lett.* 4665 (1968); (b) W. A. Gibbons and V. M. S. Gil, *Mol. Phys.* **9,** 163 (1965); (c) R. W. Frank and E. G. Leser, *J. Am. Chem. Soc.* **91,** 1577 (1969).

213. I. Morishima and T. Iizuka, *J. Am. Chem. Soc.* **96,** 7365 (1974).

214. (a) J. R. Lyerla, Jr. and D. M. Grant, *J. Phys. Chem.* **76,** 3213 (1972); (b) J. R. Lyerla, Jr. and D. L. Vanderhart, *J. Am. Chem. Soc.* **98,** 1697 (1976).

215. T. D. Alger, D. M. Grant, and R. K. Harris, *J. Phys. Chem.* **76,** 2998 (1972).

216. (a) M. Nakamura, M. Oki, H. Nakanishi, and O. Yamamoto, *Bull, Chem. Soc. Jap.* **47,** 2415 (1974); (b) V. Dave and E. W. Warnoff, *Can. J. Chem.* **50,** 2470 (1972); (c) H. Nakanishi, O. Yamamoto, M. Nakamura, and M. Oki, *Tetrahedron Lett.* **727,** (1973).

217. W. M. M. J. Bovee and J. Smidt, *Mol. Phys.* **28,** 1617 (1974).

218. J. E. Anderson and D. I. Rawson, *Chem. Com.* **830,** (1973).

219. G. C. Levy and G. L. Nelson, *J. Am. Chem. Soc.* **94,** 4897 (1972).

220. K. D. Bartle, P. M. G. Barrin, D. W. Jones, and R. L'Aime, *Tetrahedron* 911 (1970).

221. R. G. Parker and J. Jones, *J. Mag. Res.* **6,** 106 (1972).

222. (a) J. P. N. Brewer, H. Heany, and B. A. Marples, *Chem. Comm.* **27,** (1967). (b) J. P. N. Brewer, I. F. Eckhard, H. Heany, and B. A. Marples *J. Chem. Soc. C,* 664 (1968).

223. K. H. Ladner, D. K. Dalling, and D. M. Grant, *J. Phys. Chem.* **80,** 1783 (1976).

224. S. W. Collins, T. D. Alger, D. M. Grant, K. F. Kuhlmann, and J. C. Smith, *J. Phys. Chem.* **79,** 203 (1975).

225. D. E. Woessner, B. S. Snowden, Jr., and G. H. Meyer, *J. Chem. Phys.* **50,** 719 (1969).

226. D. E. Axelson and C. E. Holloway, *Can. J. Chem.* **54,** 2820 (1976).

227. J. W. Blunt and J. B. Stothers, *J. Mag. Res.* **27,** 515, 2820 (1976).

228. D. E. Woessner and B. S. Snowden, Jr., *Adv. Molec. Relaxation Processes* **3,** 181 (1972).

229. (a) T. D. Alger, D. M. Grant, and R. K. Harris, *J. Phys. Chem.* **76,** 281 (1972); (b) K. F. Kuhlmann, D. M. Grant, and R. K. Harris, *J. Chem. Phys.* **52,** 3439 (1970).

230. A. H. Fawcett, F. Heatley, K. J. Ivin, C. D. Stewart, and P. Watt. *Macromolecules* **10,** 765 (1977).

231. (a) B. Valeur, J. P. Jarry, F. Geny, and L. Monnerie, *J. Polym. Sci. Polym. Phys. Ed.* **13,** 667 (1975); (b) B. Valeur and L. Monnerie, *J. Polym. Sci., Polym. Phys. Ed.* **13,** 675 (1975); (c) B. Valeur, J. P. Jarry, F. Geny, and L. Monnerie, *J. Polym. Sci., Polym. Phys. Ed.* **13,** 2251 (1975).

232. W. H. Stockmayer, A. A. Jones, and T. L. Treadwell, *Macromolecules* **10,** 762 (1977).

233. (a) B. Valeur, J-P. Jarry, F. Geny, and L. Monnerie, *J. Polym. Sci., Polym. Phys. Ed.,* **13,** 667, 675 (1975); (b) R. H. Boyd and S. M. Breitling, *Macromolecules* **7,** 855 (1974).

234. R. E. Cais and F. A. Bovey, *Macromolecules* **10,** 757 (1977).

235. D. Ghesquiere, B. Ban and C. Chachaty, *Macromolecules* **10,** 743 (1977).

236. R. E. Cais and F. A. Bovey, *Macromolecules* **10,** 752 (1977).

237. F. Heatley, *Polymer* **16,** 493 (1975); F. Heatley and J. H. Scrivens, *Polymer* **16,** 489 (1975).

238. J. R. Lyerla, Jr. and T. T. Horikawa, *J. Polym. Sci., Polym. Lett. Ed.* **14,** 641 (1976).

239. K. Hatada, T. Kitayama, Y. Okamoto, K. Ohta, Y. Umemura, and H. Yuki, *Makromol. Chem.* **178,** 617 (1977).

240. J. R. Lyerla, Jr., T. T. Horikawa, and D. E. Johnson, *J. Am. Chem. Soc.* **99,** 2463 (1977).

241. F. Heatley and A. Begum, *Makromol. Chem.* **178,** 1205 (1977).

242. F. Heatley and I. Walton, *Polymer* **17,** 1019 (1976).

243. N. Tsuchihashi, M. Hatano, and J. Sohama, *Makromol. Chem.* **177,** 2739 (1976).

244. R. E. Cais and F. A. Bovey, *Macromolecules* **10,** 169 (1977).

245. D. E. Axelson, L. Mandelkern and G. C. Levy, *Macromolecules* **10,** 557 (1977).

246. G. C. Levy, *J. Am. Chem. Soc.* **95,** 6117 (1973).

247. F. Heatley and M. K. Cox, *Polymer* **18,** 225 (1977).

248. A. Allerhand and R. K. Hailstone, *J. Chem. Phys.* **56,** 3718 (1972).

249. J. D. Cutnell and J. A. Glasel, *J. Am. Chem. Soc.* **99,** 42 (1977).

250. F. Heatley and A. Begum, *Polymer* **17,** 399 (1976).

251. J. Schaefer, *Macromolecules* **6,** 882 (1973).

252. J. Schaefer and D. F. S. Natusch, *Macromolecules* **5,** 416 (1972).

253. Y. Inoue and T. Konno, *Polym. J.* **8,** 457 (1976).

254. Y. Inoue, A. Nishioka, and R. Chujo, *Makromol. Chem.* **168,** 163 (1973).

255. J. C. Randall, *J. Polym. Sci., Polym. Phys. Ed.* **14,** 1693 (1976).

256. D. E. Axelson and L. Mandelkern, *J. Polym. Sci., Polym. Phys. Ed.,* in press (1978).

257. R. A. Komoroski, J. Maxfield, F. Sakaguchi, and L. Mandelkern, *Macromolecules* **10,** 550 (1977).

258. R. A. Komoroski and L. Mandelkern, *J. Polym. Sci. Polym. Lett. Ed.* **14,** 253 (1976).

259. R. A. Komoroski, J. Maxfield, and L. Mandelkern, *Macromolecues* **10,** 545 (1977).

260. J. Schaefer, *Macromolecues* **5,** 427 (1972).

261. J. Schaefer, *Topics in Carbon-13 NMR Spectroscopy,* Vol. 1, G. C. Levy, Ed., Wiley-Interscience, New York, 1974, p. 149.

262. M. W. Duch and D. M. Grant, *Macromolecules,* **3,** 165 (1970).

263. J. Schaefer, S. H. Chin and S. I. Weissman, *Macromolecules* **5,** 798, (1972).

264. R. A. Komoroski and L. Mandelkern, *J. Polym. Sci. Polym. Symp.* **54,** 201 (1976).

265. (a) J. Schaefer, E. O. Stejskal, and R. Buchdahl, *Macromolecules* **10,** 384, (1977); (b) E. O. Stejskal, J. Schaefer, and R. A. McKay, *J. Mag. Res.* **25,** 569 (1977).

266. C. Chachaty, A. Forchioni, and J-C Ronfard-Haret, *Makromol. Chem.* **173,** 213 (1973).

267. S. Boileau, H. Cheradame, N. Spassky, K. Ioin, and E. Lillie, *C.R. Hebd. Seances Acad. Sci., Ser. C.* **275,** 535 (1972).

268. S. Boileau, H. Cheradame, P. Guerin, and P. Sigwalt, *J. Chim. Phys. Phys.-Chim. Biol.* **69,** 1420 (1972).

269. R. Kimmich, *Polymer* **18,** 233 (1977).

270. R. Kimmich and K. H. Schmarrder, *Polymer* **18,** 239 (1977).

271. P. G. DeGennes, *J. Chem. Phys.* **55,** 572 (1971).

272. S. Blasenbrey and W. Pechhold, *Ber. Bunsenges. Phys. Chem.* **74,** 784 (1970).

273. R. Kimmich, *Colloid Polym. Sci.* **252,** 786 (1974).

274. Y. Martin-Borret, J. P. Cohen-Addad, and J. P. Messa, *J. Chem. Phys.* **58,** 1700 (1973).

275. R. Kimmich, *Polymer* **16,** 851 (1975).

276. R. Kimmich and Doster, *J. Polym. Sci., Polym. Phys. Ed* **14,** 1671 (1976).

277. W. Gronski, G. Quack, M. Murayama, and K-F Elgert, *Makromol. Chem.* **176,** 3605 (1975).

278. R. G. Gordon, *Adv. Mag. Res.* **3,** 1 (1968).

279. (a) J. Goulon, D. Canet, M. Evans, and G. J. Davies, *Mol. Phys.* **30,** 973 (1975); (b) J. Goulon, G. Roussey, M. Hollecker, M. M. Claudon, G. W. Chantry, and E. A. Nicol, *Mol. Phys.* **33,** 377 (1977).

280. (a) T. Keyes and D. Kivelson, *J. Chem. Phys.* **56,** 1057 (1972); (b) T. Keyes, *Mol. Phys.* **23,** 737 (1972).

281. C. W. Beer and R. Pecora, *Trans. Farad. Soc.,* 78 (1977).

282. T. D. Gierke, *J. Chem. Phys.* **65,** 3873 (1976).

283. P. G. Wolynes and J. M. Deutch, *J. Chem. Phys.* **67,** 733 (1977).

284. A. M. A. Da Costa, M. A. Norman, and J. H. R. Clarke *Mol. Phys.* **29,** 191 (1975).

3 ^{13}C NMR OF NONAROMATIC HETEROCYCLIC COMPOUNDS

ERNEST L. ELIEL
K. MICHAL PIETRUSIEWICZ

CONTENTS

I. INTRODUCTION. SCOPE

The advent of powerful, commercially available instruments, in the early 1970s, for the recording of ^{13}C nmr spectra[1-3] in Fourier transform mode has led to an almost explosive outpour of publications in this field. As a result, an enormous amount of empirical information has been accumulated and used, even though the underlying theory has lagged far behind. The time has come when the spotlight needs to be turned on subsections of the field, both to maintain the maximum usefulness of the empirical data in future interpretation of spectra and to organize the data as much as possible to bring out regularities and facilitate theoretical treatment. In this chapter we focus on nonaromatic heterocyclic compounds with a view, in particular, to correlate ^{13}C parameters (chemical shifts, and, to a lesser extent, coupling constants, relaxation times, dynamic nmr parameters) with structure (ring size, configuration, conformation) wherever possible. In the process we have attempted to make a survey of spectra of nonaromatic heterocycles in general.

Systematic literature searching in this area is not easy. Much of the literature is of the last 6 years (1972–1977), which puts a heavy penalty on the lag in the secondary (abstracting) literature, and a good bit of it has been developed peripherally to other chemical work and is thus not always mentioned in title and key words. We have searched Chemical Abstracts Condensates for the years 1970–January 1977 by computer through the Lockheed search system and have endeavored to keep current through an alerting system based on CA Current Awareness Search provided by the University of North Carolina CHEM System. For the retrospective search we used combinations of ^{13}C and nmr (in the various forms of the terms) and had the computer scan Sections 22 (Physical Organic Chemistry), 27 and 28 (Heterocyclic Compounds with one and several heteroatoms), 31 (Alkaloids), 32/7 (Steroidal Alkaloids), and 33 (Carbohydrates). Unfortunately the computer reported (within a very few minutes) that there were nearly 1500 references. By excluding certain terms, such as organometallic compounds, carbocations, etc., we were able to reduce this number to 1036, which were printed off-line and scanned manually; fully aromatic heterocycles and a number of false drops were screened out.

It became obvious that not all the remaining material could be covered within the scope of this chapter. We, therefore, have omitted the discussion and referencing of natural products except in so far as their spectra serve to illustrate general principles. We have taken this step reluctantly, realizing that the major interest of saturated heterocycles is in such areas as the chemistry of alkaloids, carbohydrates, nucleic acids, etc. Fortunately, there are a number of earlier reviews of ^{13}C spectra of carbohydrates,[4-7] alkaloids,[8,9] naturally occurring coumarins,[9] nucleosides, nucleotides, and nucleic acids,[4,7,10] peptides[4,5,7,10] (including the pertinent cyclic peptides and proline derivatives) and macrocyclic compounds.[4]* There is also a review on ^{13}C nmr of santonin derivatives[18] that are representative of naturally occurring lactones. ^{13}C spectra of cyclic halonium ions are discussed in a monograph.[19]

This review, then, focuses on relatively simple, mostly synthetic heterocyclic systems of the nonaromatic type (i.e., derivatives of pyridine, furan, thiophene, and pyrrole, as well as pyrilium and thiopyrilium salts are omitted, unless they contain a second, nonaromatic ring with a hetero-atom). Monocyclic systems are taken up by size to facilitate the comparison of carbocyclic rings and corresponding rings with various hetero-atoms. This kind of arrangement has, on occasions, provided new insights. Polycyclic systems are discussed rather more briefly with emphasis on those compounds from whose spectra information of some general interest can be derived.

While we have made every effort to cover all the pertinent literature, some references have undoubtedly escaped our notice. Moreover, even in those references we have cited, there may be ^{13}C spectra of compounds not explicitly mentioned in this review. Thus the absence of a nonaromatic heterocycle from this chapter should not be taken as a definitive indication that the compound has not been studied through the end of 1977.

---

* The review on macrocyclic compounds does not include macrolides although some excellent primary publications in this area are available.[11-17]

Finally a word on chemical shifts is in order. Shifts reported prior to 1973 tend to suffer from referencing problems; a variety of standards were used and there may be a question as to the relative position of TMS and the reference chosen under the conditions of the original experiment. We have generally used shifts cited in Stothers' book[2] (and converted to TMS as a standard) for compounds from the older literature. Where we have made our own conversion, we have used the recommended[2] shifts. Even today the agreement between shifts reported by different investigators is frequently no better than 0.3 ppm and conclusions drawn from small shift differences in compounds investigated in different laboratories must be taken with a large grain of salt. Solvent changes affect shifts appreciably and even concentration changes may not be innocuous.

II. MONOCYCLIC SYSTEMS

A. Three-Membered Rings

Several systematic studies on epoxides[20-23] and aziridines[24-26] have been reported. There are also isolated reports on oxaziridines, **1**[27,28a] and their fluoroborates,[28b] diaziridines, **2**,[29] episulfide, **3**,[30] azirines, **4**,[31] thiirenium ions **5**[32] and the phenylphosphirane **6**.[33]

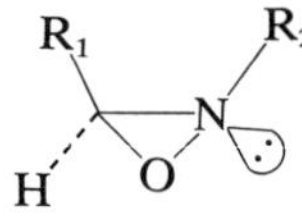

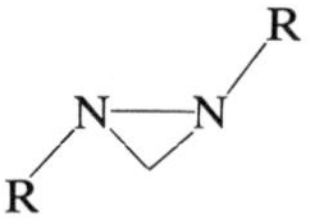

R = Me, Et, Pr   R = H, Me, Ph   5

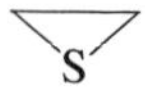

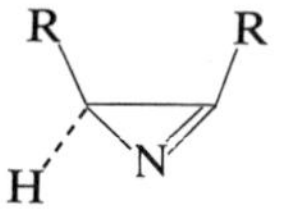

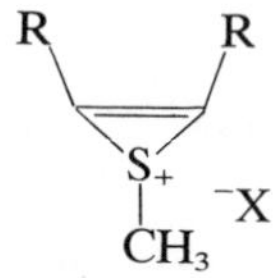

R = Me, Et

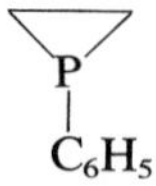

6 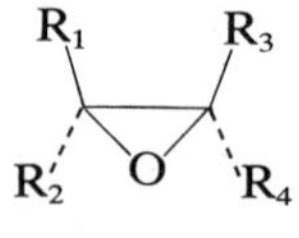   7 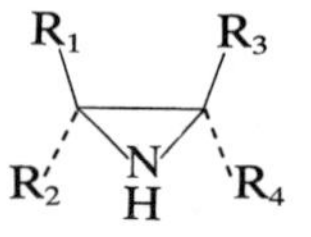   8 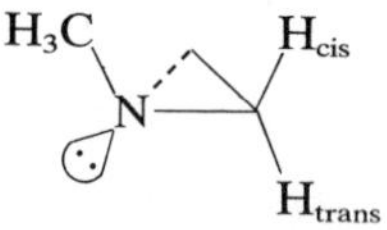   9

In the epoxide (**7**)[21] and ethylene imine (aziridine, **8**)[25] series, substituent parameters[34] have been calculated, in the case of the epoxides for alkyl substituents, in the case of the ethylene imines for methyl and phenyl substituents. These parameters are summarized in Table 3.1 and compared with corresponding parameters for cyclopropanes, calculated

Table 3.1
Alpha, Beta, Cis, and Gem Effects in Three-Membered Rings

System	Base Value[a]	Shift Parameters[b]			
		α	β	cis	gem
Cyclopropane	$(-2.9)^c$	8.0	8.7	-4.0	-1.6
Oxirane (**7**)[d]	40.8	8.3	7.1	-2.7	-4.3
Aziridine (**8**)[e]	18.2	6.9	7.6	-3.7	-1.8

[a] Shift value (from TMS) for parent system.
[b] To be added to given value for parent system.
[c] Ref. 36.
[d] Ref. 21.
[e] Ref. 25.

by hand from the limited amount of data[35-37] available. The data are similar in the three systems except for a slight reversal of the magnitude of the α and β effects and the *cis* and gem effects in the epoxide system.*

The actual shifts for cyclopropanes, epoxides, and aziridines with methyl substituents are summarized in Scheme 3.1; parenthesized shifts are those calculated using the shift parameters in Table 3.1. The agreement is only modestly good for the reasons indicated in the footnotes *b* and *c* to Scheme 3.1 on page 176.

The methyl resonances in the heterocycles (Scheme 3.1) are at the same or slightly higher field than in the methylated cyclopropanes. This is unusual,[38] the normal pattern of replacement of CH_2 by O^{40} or $N^{41,42}$ being a slight deshielding. Part of the reason for this upfield shift may be the presence of lone pairs of electrons on the heteroatoms, since in aziridines of type **8** ($R_1 = R_2$), under conditions of slow nitrogen inversion and NH exchange, there may be a difference as large as 2 ppm between the substituent (e.g., methyl) *cis* and *trans* to the pair.[25] In oxaziridines, **1**, also, it has been shown that the carbon attached to the ring in the R_1

---

* An α-effect of a substituent X in a given parent compound is the chemical shift increment due to X (relative to the parent) at the site of substitution. A β-effect is the corresponding increment at the carbon or carbons next to the substitution site and a γ effect the increment at the carbon(s) next to those. A gem effect is the *additional* increment (additional to those of the substituents themselves) when two substitutents are geminal to each other; in principle, it may be α, β, or γ, depending on the position of the geminal substituents relative to the carbon observed. A *cis* effect is the corresponding additional increment for two *cis*-located substituents and analogously for a *trans* effect. (cis and trans effects are defined only in ring systems or in chain segments of fixed conformation.)

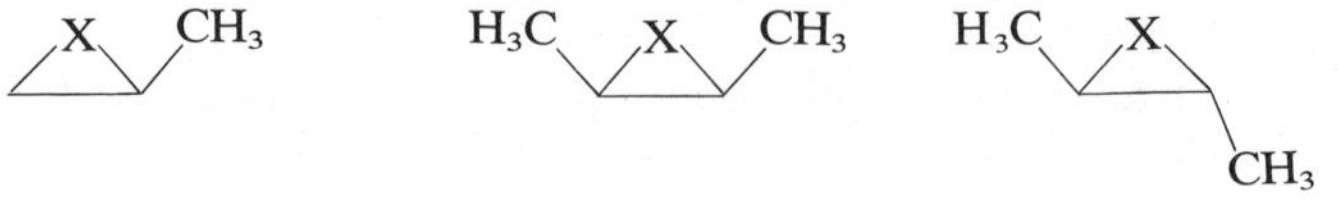

	α	β	Me	α	β	Me	α	β	Me
X = CH$_2^a$	5.3	6.0	19.6	9.8	13.6	13.0	14.2	14.6	19.0
	(5.1)	(5.8)		(9.8)	(14.5)		(13.8)	(14.5)	
X = O^b	48.0	47.8	18.1	52.4	—	12.9	55.2	—	17.6
	(49.1)	(47.9)		(53.5)			(56.2)		
X = NH^c	25.1	25.8	19.6	29.2	—	13.6	33.5	—	19.3
	(25.1)	(25.8)		(29.0)			(32.7)		

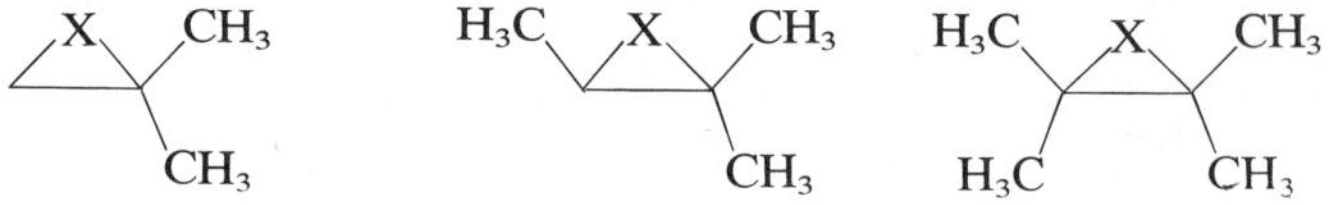

	α	β	Me	α	β	Me	α	β	Me
X = CH$_2^a$	11.5	14.1	25.7						
	(11.5)	(14.5)							
X = O^b	54.5	53.9	23.1	59.9	58.1	18.5	61.7	—	21.2
						14.2			
	(53.1)	(55.0)		(57.5)	(60.6)	$\begin{cases}24.8\\19.5\end{cases}$	(61.9)		
X = NH^c	30.2	32.5	24.1	35.1	37.8		40.3	—	22.4
	(30.2)	(33.4)		(34.1)	(36.6)	$\begin{cases}14.8\\27.3\end{cases}$	(38.0)		

Scheme 3.1

^{a} Ref. 35.
^{b} Ref. 20. The parameters (from Ref. 21) are not optimized for this data set.
^{c} Ref. 25. We have omitted the small *trans* and β-gem parameters given in this reference.

group resonates at higher field when R_1 is *trans* to the lone pair on nitrogen (Scheme 3.2a) than when it is *cis* (b).*

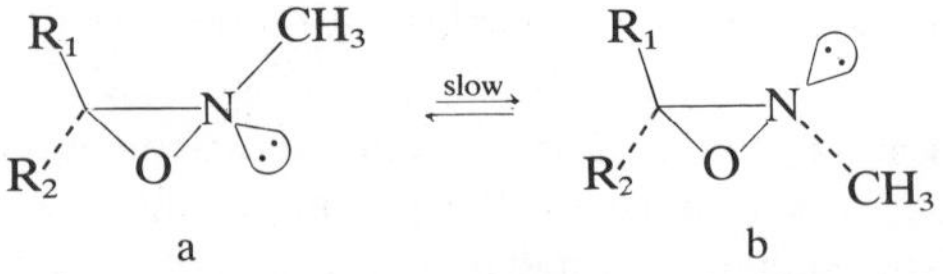

Scheme 3.2

---

* Nitrogen inversion of oxaziridines is slow on the nmr time scale, both because the three-membered ring structure impedes nitrogen inversion and because of the presence of the electron-withdrawing oxygen. The spectra of a and b thus coexist in solution; the configurations of the associated molecular species were ascertained by NOE experiments.

One could alternatively (or additionally) take the point of view that the slight upfield shift of the exocyclic methyl groups is due to their being gamma as well as beta to the ring hetero-atom. This would be in line with other observations on gamma-upfield shifting effects of heteroatoms.[43]

Derivatives of propylene oxide (7, $R_1 = XCH_2$, $R_2 = R_3 = R_4 = H$) have been investigated[22,23] and the rotational preference of the CH_2X group has been inferred from the spectra.[22] In the diaziridines 2, as in the oxaziridines mentioned above, nitrogen inversion is slow on the nmr time scale[44]; as a result the compound with $R = CH(CH_3)_2$ shows diastereotopic (anisochronous) methyl groups.[29] Nevertheless separate cis and trans isomers of 2 are not seen; it appears that the trans isomer greatly predominates at equilibrium to the point where the spectrum of the cis isomer cannot be found. The downfield shifting effect on the ring carbon of the two methyl groups in 2 ($R = CH_3$) is twice that of the single CH_3 group in N-methylethylene imine.[29]

The slow nitrogen inversion in the aziridines 10 and 11 causes the methylene groups of the aziridine rings to be diastereotopic on the nmr time scale; thus they give rise to distinct signals.

A limited amount of information on $^1J_{^{13}CH}$ coupling constants in aziridines,[26] azirines,[31] and oxaziridines[27] is available; of particular interest is the stereochemical dependence of these constants. In aziridines 9, $^1J(^{13}C/H_{cis}) = 161$ Hz whereas $^1J(^{13}C/H_{trans}) = 171$ Hz; i.e. the proton cis to the lone pair has the larger coupling constant. Similarly, in oxaziridines 1, $^1J(^{13}C/H)$ is 182–186 Hz in the cis isomer (shown, H trans to R_2) and 173–180 Hz in the corresponding trans isomer; again the proton cis to the lone pair has the larger ^1J.

Coupling constants (as well as chemical shifts) have also been reported in the tin derivatives 12, including the aziridino-, methylaziridino-, and 2,2-dimethylaziridino-substituted ones.[46] A correlation has been made[47] between vertical ionization potentials (IP), measured by photoelectron spectroscopy and $^1J_{CH}$ in the oxacycles 13, including oxirane: the smaller the ring is, the larger the IP and the larger J since both increase with increasing s-character of the C—H bond.

B. Four-Membered Rings

Information on four-membered ring systems is scarce. Even the carbocyclic parent system has not been investigated extensively; the resonance of cyclobutane is at 23.3 ppm,[2,30,37] but, although relaxation times in methylcyclobutane have been reported,[48] the signal positions have not. Data for methylated oxetanes[49] (**14**, Table 3.2) allow one to calculate the parameters shown in Table 3.3 which are unexceptional.

In the dimethyl compounds, as in the corresponding three-membered ring analogs (Scheme 3.1), the CH_3 and C_α resonances of the *cis* isomers are at higher field than those in the *trans* analogs.* This trend will be seen to recur in larger rings as well.

It has been pointed out that the α-carbon of oxetane resonates at lower field and the β-carbon at higher field than carbon atoms in corresponding acyclic compounds, such as 1,3-propanediol.[51] A closer comparison would be that between oxetane and cyclobutane: there is little difference in the shift of the β carbons (23.3 vs 23.1[49] or 23.7 ppm[51]), that is, the β-upfield shift seen in oxiranes has disappeared. But the α-carbon is

Table 3.2
^{13}C Chemical Shifts in Oxetanes[a]

Carbon	Oxetane	2-Methyloxetane	2,3-Dimethyloxetane	
			cis	*trans*
α	72.8	78.3	79.6	85.0
β	23.1	29.2	31.8	37.4
γ	(72.8)	66.6	74.5	73.5
$CH_3(\alpha)$	—	24.1	16.9	22.8
$CH_3(\beta)$	—	—	13.0	17.5

[a] In ppm from TMS, from Ref. 49. For structures see **14**.

---

* Configurational assignments[49] are based on proton shifts and are contrary to those reported in Ref. 50. Methyl shifts are assigned on the basis of the reasonable assumption that the hetero-atom exerts a gamma-upfield shift[43] and on the closeness of CH_3 (α) to that in the monomethyl compound. There seems to be a small upfield-shifting effect even in the *trans* isomer perhaps because the puckering of the four-membered ring brings the *trans*-Me groups closer together.

Table 3.3
Shift Parameters in Methyloxetanes

Parameter[a]	α-2	α-3	β-2[b]	β-3	γ-2[c]	cis-2	cis-3
Value	5.5	8.2	6.8	6.1	−6.2	−5.4	−5.6

[a] The numeral refers to the site of the ring carbon atom observed (not necessarily the site of substitution).
[b] Average of β-2 and β-4.
[c] Or γ-4.

shifted 49.5 ppm downfield—an unusually large downfield shift for a replacement of CH_2 by O,[40] larger even than in oxiranes (43.7 ppm). [The oxirane and oxetane $CH_2 \rightarrow O$ C_α displacements are closer if one takes the point of view that C_α in oxirane, but not in oxetane, also has an upfield β shift (cf. Scheme 3.1 and earlier discussion). However, one might then expect a γ-upfield shift in the oxetane where C_α is also C_γ and there is no indication for that.]

No data seem to be available for azetidines and only isolated reports for derivatives thereof, such as four-membered lactams[52] (**15**), azetines[53] (**16**), trimethylene sulfides[30,54] (**17**) (C_3H_6S: C_α, 26.0; C_β, 28.0. C_4H_8S: C_α, 37.65; C_β, 35.80; C_γ, 21.29; Me, 25.25.) unsaturated sulfones[55] (**18**) (both chemical shifts and coupling constants) and β-propiolactone.[56]

14   R_1
$R_1 = R_2 = H$
$R_1 = CH_3, R_2 = H$
$R_1 = R_2 = CH_3$
(cis, trans)

15   CO_2CH_3
H_5C_6—C_6H_5
—N
O C_6H_5

16   C_6H_5 Cl
Cl—
H_3C—N
H

17   R
R = H[30,54]
R = CH_3[54]

18   $(CH_3)_n$

19   X
—P—Y
$\frac{4}{3/2}$
$(CH_3)_n$

X = O, Y = OH, OCH_3, CH_3, $(CH_3)_3C$, C_6H_5, p-C_6H_4F, Cl, $C_6H_5CH_2$; X = S, Y = C_6H_5; X = CH_3, Y = CH_3, C_6H_5, p-C_6H_4F, $C_6H_5CH_2$; X = Y = $C_6H_5CH_2$; X = C_6H_5, Y = lone pair; C_6H_5, $C_6H_5CH_2$.

20   CF_3
F_3C—O
—P$\cdots$O—$CH(CF_3)_2$
$(CF_3)_2HC$—O CH_3

Methyl-substituted phosphetanes and their oxides, sulfides, and quaternary salts (19) have, in contrast, been studied in considerable detail.[33,57,58] The chemical shift of C(3) and $^2J(^{31}P/^{13}C\text{-}3)$ in the 3-methyl compounds are strongly configuration dependent, and the coupling constant has been used to assign configuration in 1-phenyl-2,2-dimethyl-3-vinylphosphetane oxide.[59a] A pentacoordinate phosphorous compound (20) has also been investigated.[59b]

C. Five-Membered Rings

The conformational situation in five-membered rings is more complex than that in four-membered ones[60b]; thus, although a substantial amount of ^{13}C nmr data is available for five-membered heteroatom-containing rings, the interpretation of the data is far from straightforward. The carbocyclic parent, cyclopentane (21), has been studied by Roberts and coworkers[61] and the oxygen analogs tetrahydrofuran (22) and 1,3-dioxolane (23) in our own laboratories.[62] Parameters for methyl substitution, including the shifts of the methyl group(s) are shown in Table 3.4. In the case of 22 and 23 the parameters depend on the position of the carbon atom under consideration (indicated in parentheses). In some instances of β and γ shifts, the position of the substituent is also of significance (see footnotes to Table 3.4).

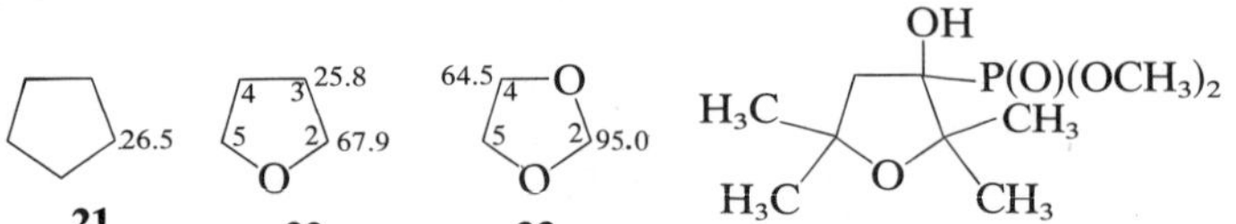

Table 3.4
Methyl Substitution Parameters for Cyclopentane (21) Tetrahydrofuran (22) and 1,3-Dioxolane (23)

Parameter	Cyclopentane	Tetrahydrofuran (22)	1,3-Dioxolane (23)
α-CH$_3$	9.3	7.4(2), 8.1(3)	6.2(2), 7.6(4)
β-CH$_3$	9.3	7.1(2), 7.6(3),[a] 8.8(4)[b]	6.3(4)
γ-CH$_3$	−0.1	0.4(4), −0.2(5),[a] 0.0(5)[b]	0.0(2), 0.4(4)
CH$_3$ shift[c]	20.5	21.1(2), 17.9(3)	19.8(2), 18.1(4)

[a] Methyl group at C(2).
[b] Methyl group at C(3).
[c] In monomethyl compound.

To begin with, it is perhaps surprising that the five-membered ring systems can be parametrized (with respect to chemical shift increments) at all, since the five-ring is notoriously conformationally mobile.[60a] Of course, the significance of a successful parametrization increases with an increasing amount of data and vice versa; thus it is particularly noteworthy that the 1,3-dioxolane set comprises 33 compounds (99 ring carbon shifts, not taking into account degeneracies) and that the standard deviations of the parameters (Table 3.4) range from 0.1 to 0.3 ppm,* despite the fact that the system is known[64] to be quite conformationally variable from compound to compound. It would appear, therefore, that the parameters are not strongly conformation dependent. This insensitivity is clearly seen in 1,3-disubstituted cyclopentanes (in which the *cis* and *trans* isomers are known to have different conformations[60a]) and the corresponding tetrahydrofurans and 1,3-dioxolanes; the data are summarized in Table 3.5. Whereas *cis-trans* differences range from 1.5 to 5.7 ppm in the six-membered ring, the range is 0.0–1.9 ppm in the five-membered ones and for corresponding positions or groups the difference is always smaller in the five-membered cycles. This, of course, is just what one would expect in view of the less marked conformational differences between *cis* and *trans* isomers in five-membered rings as compared to six-membered ones.[60a]

In all five-membered rings C(1), C(2), and C(3) resonate at lower field and C(4) and C(5) at higher field in the *cis*-1,3-disubstituted species than in the corresponding *trans* isomers.† This regularity is seen throughout Table 3.5 (except where the pertinent C is replaced by O) and may be found useful for configurational assignment; additional examples will be encountered later. (The situation is different from that in 1,3-dimethylcyclohexanes and its heterocyclic analogs where the shifts of *all* carbon atoms in the *cis* isomer are downfield from corresponding shifts in the *trans*.)

Other interesting features are the virtual identity of the α- and β-parameters throughout Table 3.4 (in six-membered rings the β-parameters tend to be approximately 4 ppm larger than the α[34]) and the near-zero value for the γ parameter. The latter, in six-membered rings, is near zero for equatorial but around -5 to -6 ppm for axial methyl

---

* Excluding parameters for the highly congested *cis*-4,5-diisopropyl- and -di-*t*-butyl-1,3-dioxolanes.

† The numbering system is that in cyclopentane. C(1) and C(3) in 1,3-dimethylcyclopentane correspond to C(2) and C(5) in 2.5-dimethyltetrahydrofuran and C(2) and C(4) in 2,4-dimethyloxolane. C(4) and C(5) in **21** correspond to C(3) and C(4) in **22** and C(5) in **23**.

Table 3.5

Calculated and Observed Ring Shifts for 1,3-Dimethylcyclopentanes[61] 2,5-Dimethyltetrahydrofurans,[62] and 2,4-Dimethyl-1,3-dioxolanes[62]

System	Carbon Atom	Chemical Shift		
		cis Isomer	trans Isomer	Calculated[a]
(1,3-dimethylcyclopentane)	1	35.5	33.6	34.8
	2	45.1	43.2	44.0
	4	34.4	35.3	34.8
	Me	21.2	21.5	—
(2,5-dimethyltetrahydrofuran)	2	76.7	76.0	75.1
	3	34.8	35.9	33.8
	Me	23.0	23.1	—
(2,4-dimethyl-1,3-dioxolane)	2	101.6	100.6	101.6
	4	72.9	71.9	72.6
	5	71.0	71.7	71.3
	Me (2)	20.2	20.2	—
	Me (4)	18.8	18.5	—
(six-membered ring)	1[b]	33.0	27.3	33.3; 27.6
	2[b]	44.7	41.3	45.2; 41.6
	4[b]	35.2	33.8	36.2; 34.2
	5[b]	26.6	20.9	27.2; 20.8
	Me[b]	23.0	20.7	—

[a] The data for the five-membered rings are calculated from the (configuration-independent) parameters given in Table 3.4. The data for the six-membered rings are calculated from the (conformation-dependent and therefore configuration-dependent) parameters in Ref. 34b.
[b] Data from Ref. 66.

groups. The absence of a γ_a effect in five-membered rings is perhaps what one would expect (there being no truly axial positions); however, it might be noted that a substantial upfield-shifting γ effect exists in 2-methyloxetane (Table 3.3) and also in methylcyclobutane.[65] This may imply a more marked conformational preference in four-membered as compared to five-membered rings.*

The methyl shift in methylcyclopentane (20.5 ppm) is approximately midway between that of equatorial (22.7) and axial (17.5) methyl in cyclohexane.[66] Surprisingly, in the 1,3-dimethyl compounds (Table 3.5) the methyl resonances are shifted *downfield* suggesting not only the

* However, 2-methyloxetane, unlike cyclobutane, is very nearly planar.[60b,c]

absence of a reciprocal γ_a but the existence of a downfield-shifting δ effect. The effect is even more marked in the 2,5-dimethyltetrahydro-furans (compare Tables 3.4 and 3.5) but is nearly absent in 2,4-dimethyl-1,3-dioxolanes. Methyl groups at C(3) in tetrahydrofuran and C(4) in 1,3-dioxolane resonate substantially upfield from those at C(2) indicating an upfield-shifting effect of the γ-oxygen; we shall return to this point later.

Comparison of chemical shifts in **21**, **22**, and **23** discloses a downfield shift of the α-carbon in **22** of 41.4 ppm, that is, somewhat less than in oxetane (p. 179). The upfield shift of the β-carbon (-0.7 ppm) is small. The downfield shift in **23** at C(4), 38.0 ppm, is somewhat less than the sum of an α- and a β-shift (40.7) and the shift at C(2), 68.5 ppm, is much less than the sum of two α-shifts (82.8) presumably because of a satura-tion of polarizability.[30]

The spectrum of 2-isopropyl-3,3,5,5-tetramethyltetrahydrofuran has been tabulated[67a] and data for various ethyl-, methyl-, and trimethyltetra-hydrofurans have been obtained.[62] The spectra ot γ-butyrolactone[47,56,68a,d] maleic anhydride,[67] succinic anhydride,[67] and a substituted succinimide[68b] are on record, as is the spectrum of compound **24**.[68c]

In γ-valerolactone (**25b**)[68a] there is a very large α effect of the methyl group (19.6 ppm) relative to γ-butyrolactone (**25a**). Compounds homologous to **25b** (with R up to n-heptyl) have also been studied and T_1 for the n-heptyl homolog has been obtained.[68a]

	Shift, ppm	C_5	C_4	C_3	C_2
25a,	R = H	57.6	22.3	27.8	177.9
25b,	R = CH$_3$	77.2	29.7	29.1	177.2

In contrast to the tetrahydrofuran system, the pyrrolidine skeleton has received little systematic study. The spectra of pyrrolidine (**26a**) itself and N-methylpyrrolidine (**26b**),[2,8] 2-pyrrolidones,[69] succinimide, nicotine, pro-line, and 3-hydroxyproline are on record,[67] as are the spectra of substi-tuted prolines and pyroglutamic acids.[70a] Relaxation times in proline have been measured.[70b,c] The spectra of exocyclic enamines derived from pyrrolidine have been studied.[71]

Lanthanide induced shifts [Eu(dpm)$_3$, Pr(dpm)$_3$] have been recorded for pyrrolidine[71a] and N-alkylpyrolidones[72b]; in the latter case (amides) contact shifts contribute. The spectrum of pyrrolidine itself (**26a**)[67,72a] shows a substantial upfield shift (relative to **21**) at C-α (21 ppm) but virtually no shift at C-β.

N-Methylpyrrolidine (**26b**) shows a downfield shift of approximately 9 ppm at C-α (β effect) and an upfield shift of 1–2 ppm at C-β (γ effect) relative to pyrrolidine (**26a**). The β effect of the methyl is in line with that shown in Table 3.4 for C-methyl groups; the γ-effect is, if anything, somewhat larger. Quaternization[73] of **27** produces two diastereomers of **28** (*cis* and *trans*). The β effect of quaternization (i.e., influence on C(2)] is of the same order of magnitude as that of replacing H by CH$_3$ (compare **27** → **28** with **26a** → **26b**) but the effect at C(3) is substantially larger, resulting in a sizeable upfield shift (4–10 ppm). The latter is clearly an effect of the charge; the resulting upfield shift is observed in other systems as well.[42]

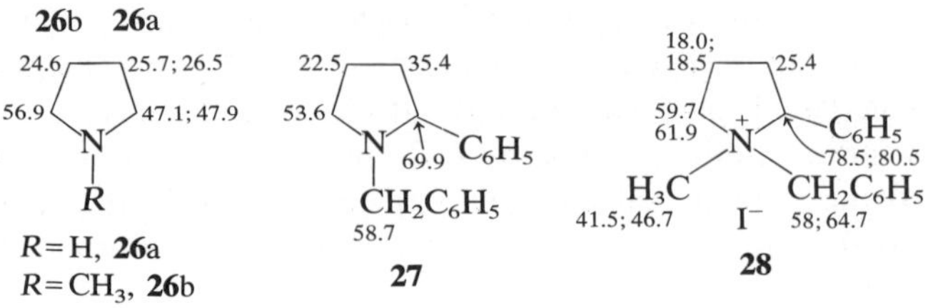

Tetrahydrothiophene (**29**)[54,67,74] and some of its methyl derivatives,[54,75] salts[54,74] and the corresponding sulfoxide and sulfone[74] have been investigated.* The shifts for **29** and its homologs are summarized in Scheme 3.3.

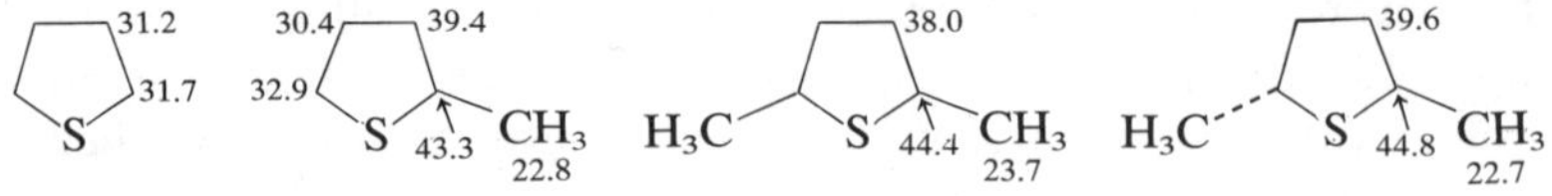

Scheme 3.3

The following findings are of note:

1. The Me(α) effect, +11.6 ppm, is larger than in cyclopentane and other heterocyclic analogs, whereas Me(β), +8.2 ppm and Me(γ), −0.8 (2 → 4) or +1.2 (2 → 5) ppm are typical of the series. (Large α effects are seen in other sulfur compounds, *vide infra*, and may be indicative of enhanced polarizability of the α carbon. The four-membered ring sulfide **17** also shows this effect.

2. The above parameters, derived from the shifts in **30,** can be adequately used to compute the shifts in both **31** and **32,** at C(α). Additivity at C(β) is slightly less good.

3. Although tetrahydrothiophene is believed to be more puckered and less easily pseudorotating than cyclopentane (**21**) or tetrahydrofuran (**22**),[60a] the spectral differences between the stereoisomers

---

* The assignments were made by specific proton decoupling: Ref. 54.

31 and **32** are small, especially at the α-positions, where they are reduced to insignificance.

4. As in other five-membered rings (cf. p. 181), the β-carbons in the 2,5-*cis*-dimethyl compounds are upfield from those in the *trans*.

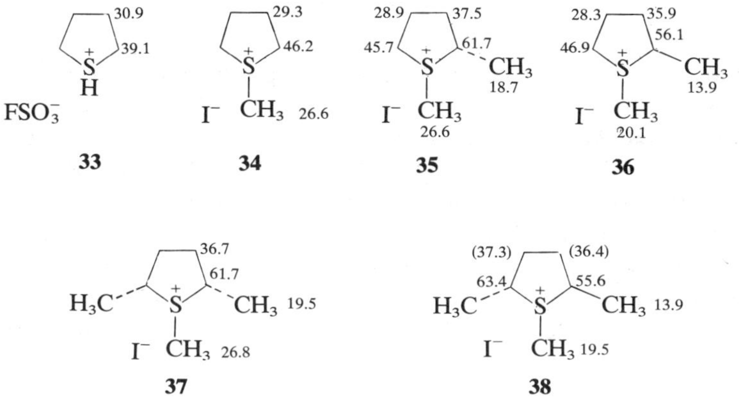

FSO$_3^-$

33   **34**   **35**   **36**

37   **38**

Scheme 3.4

The effect of salt formation and ternerization is shown in Scheme 3.4. Salt formation (compare **33** with **29**, Scheme 3) leads to a sizeable downfield shift (7.4 ppm) at C(2) and a marginally upfield one (0.3 ppm) at C(3), a situation which is similar to that in ammonium salts.[42] The effects are enhanced in a ternary salt (compare **34** with **29**, **35**, and **36** with **30**, **37** with **31**, and **38** with **32**). *Prima facie* comparison of **34** with **33** suggests a sizeable β (downfield shifting) and palpable γ (upfield shifting) effect of the *S*-methyl substituent, though there may be complications due to solvent effects (CDCl$_3$ vs. HSO$_3$F) and referencing differences (external vs. internal TMS). While this matter is in doubt, there can be no question that, unlike the parent compounds **31** and **32**, the diasteremeric salts **37** and **38** differ appreciably in chemical shifts as do **35** and **36**. Thus the *cis*-S,2-dimethyl combination leads to notable upfield shifts at C(2), C(3) and the two methyl groups compared to the trans arrangement, analogously to what has been seen earlier in smaller rings (p. 178). It is instructive to compare the *cis/trans* differences in this series with those in 1,2-dimethylcyclopentanes[61]; this is done in Scheme 3.5. The negative figures are the upfield shifts encountered in going from the *trans* to the (more congested) *cis* isomers. It is interesting that the shift at C(5) in the sulfonium salt is positive, in contrast to the shift at C(3) in the dimethylcyclopentane; the reason for this remarkable difference is not evident.

Scheme 3.5

Relative to thiolane, **29**, the corresponding sulfoxide, **39** and sulfone, **40** display[74] downfield shifts at C(2) and upfield shifts at C(3), similar to the ternary salt **34**. The order of the downfield C(2) shifts is **39 > 40 > 34** whereas the order of the upfield shifts at C(3) is **40 > 39 > 34**. Neither order corresponds to the amount of positive charge at sulfur, which is presumably least for **39** (dipole moments: **40**, 4.75D, **39**, 4.17D). The shifts at C(3) may comprise a steric (through-space) component that would presumably be largest for **40**. We shall return to this point in connection with the conformationally better-defined thiane sulfoxide and sulfone.

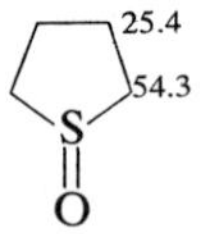

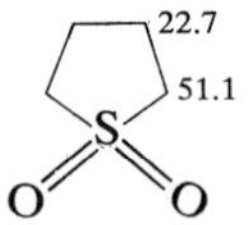

2,5-Dihydrothiophene sulfones (Scheme 3.6) have been investigated.[76] Compared to acyclic olefins[77] there are some anomalies: the β effect at C(2) in **43** (3.4 ppm) is unusually small and the γ_{π} effect at C(5) is positive rather than negative. Nonadditivity of the α effects in **44** (calculated 11.7 − 5.1 = 6.6 ppm, observed 2.5 ppm) points to a *cis*-effect, which is larger than that[77] in acyclic olefins. (Unfortunately, the more closely related methylcyclopentenes do not seem to be available for comparison.) In compound **45** the introduction of the methyl group at C(2) causes the

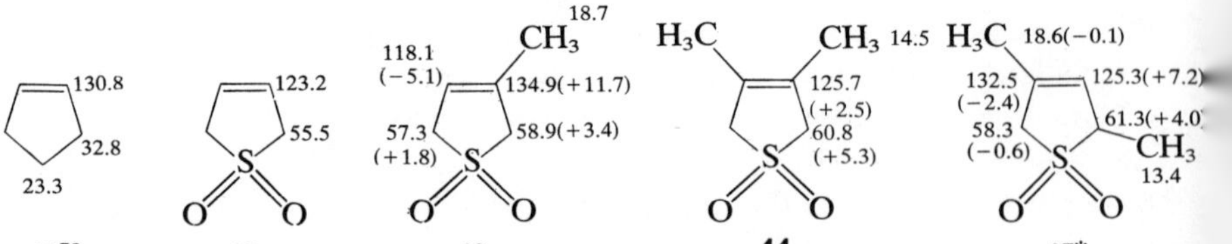

Scheme 3.6

* Comparison (in parentheses) with **43**. Other comparisons are with **42**.

expected α- and β-downfield shifts at C(2) and C(3) but slight upfield shifts at the γ positions, C(4) and C(5). The $^1J_{CH}$ coupling constants have also been recorded.[76]

An extensive investigation of eighteen ethylene sulfites (46) with substituents CH_3, CH_2Cl, or $CH_2OC_6H_5$ has been made.[79] The data were interpreted in terms of a rapidly pseudorotating five-membered ring with a fixed configuration at sulfur (i.e., without inversion of the

$$\begin{array}{c} O \\ \parallel \\ -O-S-O \end{array}$$

pyramid); the preferred conformations are those in which CH_3/O gauche interactions are minimized. The earlier-mentioned relation between shifts in 1,3-*cis* and 1,3-*trans*-disubstituted isomers holds in ethylene sulfites as well, provided the S=O oxygen is considered one of the two substituents. As applied to the sulfites, the prediction is that the α carbon should be

downfield in the *cis* isomer and the β carbon should be upfield. The pertinent data are shown in Table 3.6; clearly the shift differences for the α-carbons are larger and more dependable than those for the β carbons.

An extensive study of 1,3,2-dioxaphospholanes, 47, with X = CH_3O, $(CH_3)_3CO$, $(CH_3)_3C$, $C_6H_5CH_2$, C_6H_5 and $(CH_3)_2N$ has been carried out.[80] A number of different arguments was brought to bear on the

Table 3.6

Ring Shifts in Substituted Ethylene Sulfites 46

-Substituent A, *cis* or B, *trans*)	CH3		CH2Cl		CH2OPh		C6H5		4,5-di-CH3[b]		4,5-di-CH3[c]		4,5-di-CH3[b]	
Configuration[a]	cis	trans	cis	trans	cis	trans	cis	trans	cis	trans	cis	trans	cis	trans
ppm, C(4)	80.2	76.5	81.4	79.3	79.9	77.8	85.6	80.9	81.2	78.5	85.4	80.5	86.4	84.8
ppm, C(5)	70.9	72.9	70.4	69.5	70.0	68.9	71.4	73.5	—	—	—	—	—	—

Relative to sulfite oxygen.
Methyl or phenyl groups at 4 and 5 *cis* to each other, comparison between syn- (substituents A,) and anti- (substituent B, D) diastereomers.
Methyl groups *trans* to each other (A, D in 46), comparison between diastereotopic methyl groups.

question of the configuration and conformation of these compounds, including ^{13}C nmr spectroscopy. It was noted here (as had been pointed out above in connection with other five-membered ring systems) that in the *cis* isomer (CH$_3$ and X in **47** *cis*) C(4) is downfield and C(5) upfield relative to the *trans*. The Me-4 substituent shift also correlates quite well with configuration, being upfield in the *cis* isomer in all cases save one (the *t*-butoxy compound). The pertinent data are summarized in Table 3.7. Also in the table are the ^{31}P/^{13}C coupling constants; in the case of the methyl carbons it is seen that $^3J_{trans} > {}^3J_{cis}$, presumably because of the existence of a Karplus relationship and the fact that the *trans* methyl group is more nearly anti-periplanar to the phosphorus atom than the *cis*.

For the case where X = OCH$_3$, higher methylated homologs of **47** (diastereomeric 4,5-dimethyl as well as 4,4,5,5-tetramethyl compounds) have also been studied.[81] In the case of the dimethyl compounds, once again the ring carbons bearing substituents *cis* to the methoxyl on phosphorus resonate downfield from those bearing trans substituents. Also the methyl groups in *trans* position again have the higher ^{31}P/^{13}C coupling constants; however, the relative shifts of the methyl groups (*cis* upfield of *trans*) found for the monomethyl analogs is not maintained in this series. It appears, as already implied in Table 3.5, that the shifts of the methyl substituents are not good criteria of configuration.

In the cyclic phosphine series, phenyltetramethylenephosphine (**48**), its phenyl and benzyl quaternary salts (**49** and **50**) and the corresponding phosphine oxide (**51**) have been investigated.[33,82] The chemical shifts of

Table 3.7
Ring Shifts and ^{31}P/^{13}C Coupling Constants in 1,3,2-Dioxaphospholanes **47**

P-Substituent (X)	Configuration	$\delta_{C(4)}$ (ppm)	$J_{PC(4)}$ (Hz)	$\delta_{C(5)}$ (ppm)	$J_{PC(5)}$ (Hz)	δ_{CH_3} (ppm)	$^3J_{PCH_3}$
OCH$_3$	cis	74.2	9.8	70.0	7.6	19.0	<3
	trans	72.8	9.4	70.8	8.3	20.0	4.2
OC(CH$_3$)$_3$	cis	73.8	9.0	69.0	7.6	20.6	<3
	trans	71.2	7.8	69.6	7.6	19.6	4.5
C(CH$_3$)$_3$	cis	74.1	6.4	71.0	6.6	17.0	3.4
	trans	73.6	8.8	72.0	7.6	20.1	4.5
CH$_2$C$_6$H$_5$	cis	74.4	9.7	69.9	8.4	18.8	<3
	trans	72.5	9.8	70.6	8.0	19.8	3.7
C$_6$H$_5$	cis	75.1	9.5	70.2	8.6	18.5	<3
	trans	72.5	8.9	70.5	7.9	19.4	3.7
N(CH$_3$)$_2$	cis	72.9	7.5	69.8	7.5	17.4	3.0
	trans	71.8	8.1	70.7	9.5	20.1	4.1

the α and β carbons as well as the $^{31}P/^{13}C$ coupling constants (in parentheses) are indicated in the formulas;

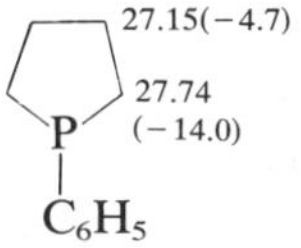

the absolute magnitude of the coupling constants leads to an immediate assignment of the two saturated carbon resonances. ^1J is negative for the phosphines but positive for the compounds with tetra-coordinate phosphorus; this is general for phosphorus compounds and can be accounted for theoretically.[33] The negative ^2J in the phosphine is believed to be due to an anti-periplanar orientation of C_β and the lone pair on P.[33] In comparison with cyclopentane the carbon atoms in **48** are shifted downfield similarly as in thiolane (**29**) but in contrast to tetrahydrofuran (**22**) and pyrrolidine (**25**) (where only C_α is shifted downfield and C_β slightly upfield, if at all) and also in contrast to the six-membered homolog.[33] Formation of quaternary phosphonium salts (**49, 50**) leads to upfield shifts at both C_α and C_β; in contrast to sulfonium (Scheme 3.4) and quaternary pyrrolidinium salts where C_α is shifted downfield.* The C_α downfield shift recurs in the phosphine oxide **51**. The spectra of several phospholenes[83] and their oxides are shown in Scheme 3.7. Here, as in other series, one finds an upfield shift of the methyl substituents in the *cis* as compared to the *trans* compounds (compare **52** to **53, 54** to **55**, and **56** to **57**). In the phospholenes (**52, 53; 54, 55**), the $^{31}P/^{13}C$ coupling

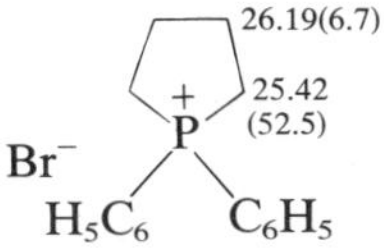

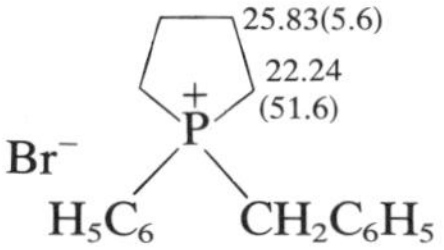

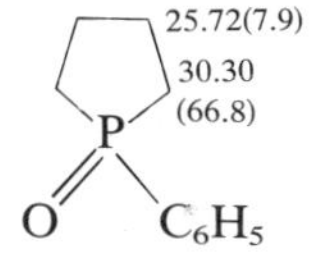

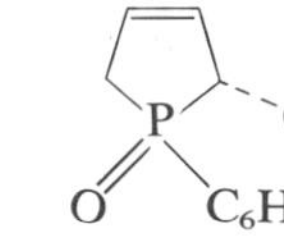

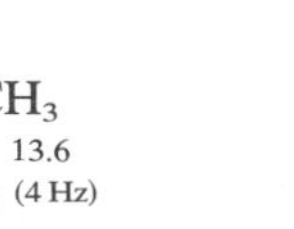

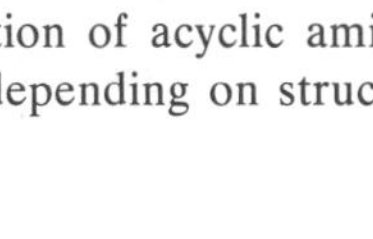

Scheme 3.7

---

* It may be noted that protonation of acyclic amines may give rise to either upfield or downfield shifts at C_α, depending on structure: Ref. 42.

constants (^2J, in parentheses) are enormously larger for the *trans* than for the *cis* isomers. Again this is likely due to the orientation of the lone pair relative to the substituent; the difference disappears in the oxides (**56, 57**).

Spectra of additional substituted phospholenes (e.g., **58**)[84] and the corresponding phospholene oxides (**59**) and sulfides (**60, 61**)[85,86] as well as phospholanes **62** and **63**[85] (Scheme 3.8; coupling constant in hertz in parentheses) have been recorded. The spectra of the phenylphospholene oxides **64** are as yet unpublished.[87]

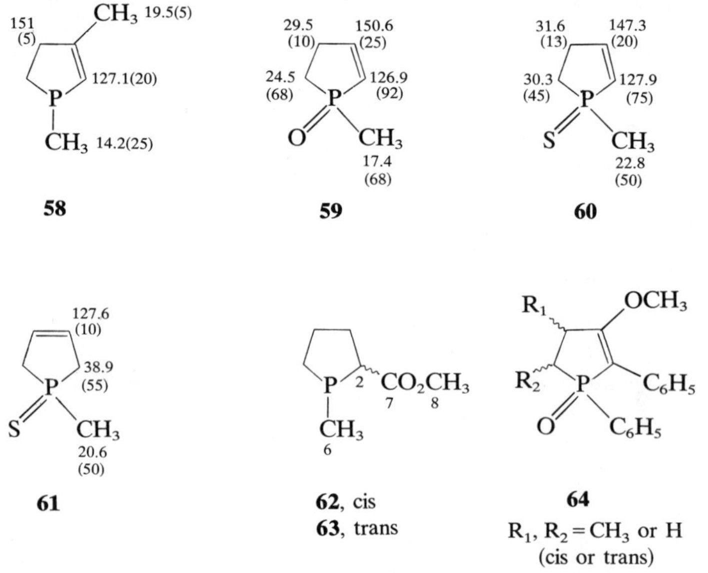

	C(2)	C(6)	C(7)	C(8)
62	43.8(23)	9.0(27)	173.7(<15)	51.9(<1.5)
63	48.3(20)	13.3(20)	177.2(17)	51.9(<1.5)

Scheme 3.8

The pentacoordinate spirophosphoranes **65**[88] and **66**[89] have been investigated in conjunction with pseudorotation phenomena.

65 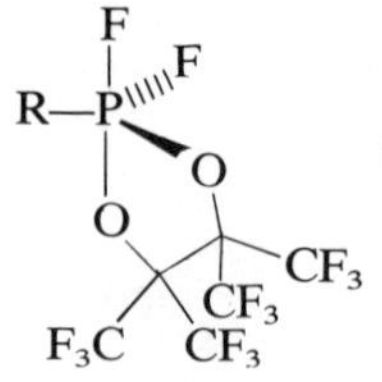

65, R = F, Me, Bu, Ph, NEt_2, $N(SiMe_3)_2$

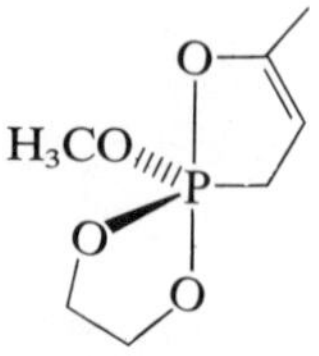 **66**

67 **68** R = H, Me, CCl₃ **69** **70**
Et

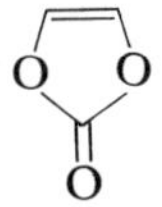

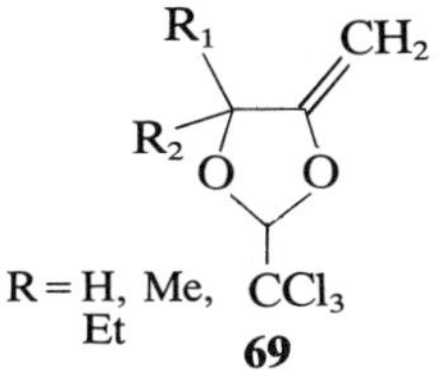

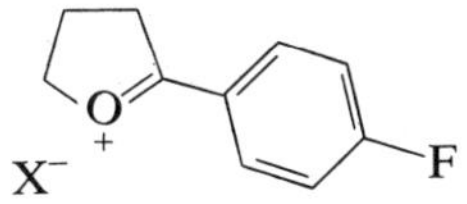

We conclude this section with a brief mention of other five-membered ring systems. Proton/^{13}C coupling constants in dioxolenes (**67**), their dimers, and ethylene carbonate (**68**)* and some of its mono-, di- and trisulfur analogs have been determined from ^{13}C satellite peaks in proton spectra.[90] Data on the methylene-dioxolane **69**[91] and the oxonium salt **70**[92] are on record. Imino-lactones **71** and -anhydrides **72** have been investigated[93,94] and the configuration of substituted 1,2-oxazolines **73** has been assigned by ^{13}C nmr.[95a] The spectrum of oxazoline **74** is on record,[95b] as is that of pyrazoline **75**.[95c]

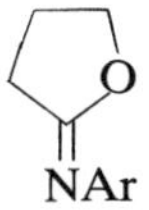

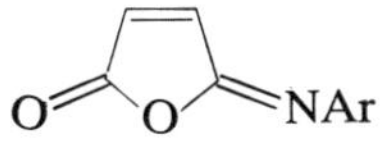

72 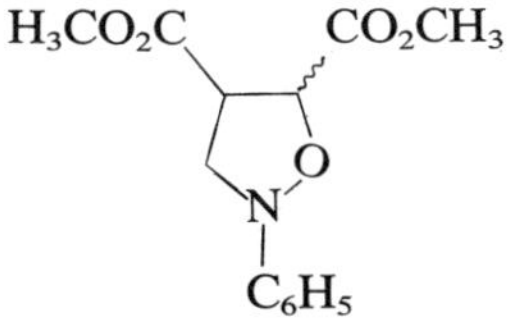 **73** 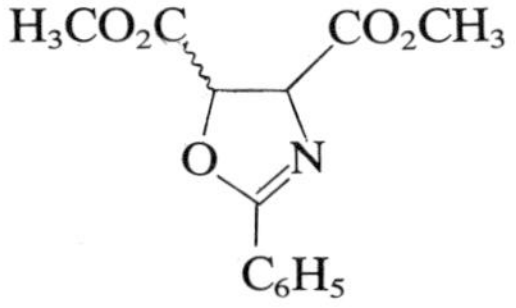 **74**

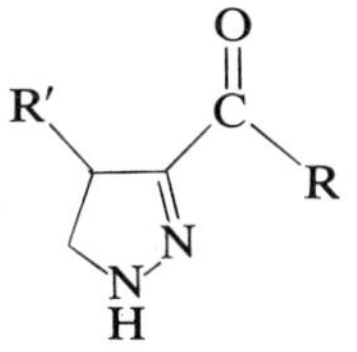

75

The structures of a number of ethylene ureas, thioureas, guanidines and their derivatives whose spectra have been recorded are summarized in Scheme 3.9. In compounds **77a** and **77b** (the $\Delta^{3,4}$ unsaturated analogs were also investigated) a curious reversal of the usual substituent effect occurs: whereas the marked carbon moves upfield as R is changed from methyl to ethyl to isopropyl, further change to t-butyl causes a downfield shift suggesting serious ring distortion.[98] Formula **78** represents the alkaloid ergothioneine.

---

* Data on the corresponding saturated carbonates are also on record: K. Pihlaja and K. Rossi, *Acta Chem. Scand.*, **B31**, 899 (1977).

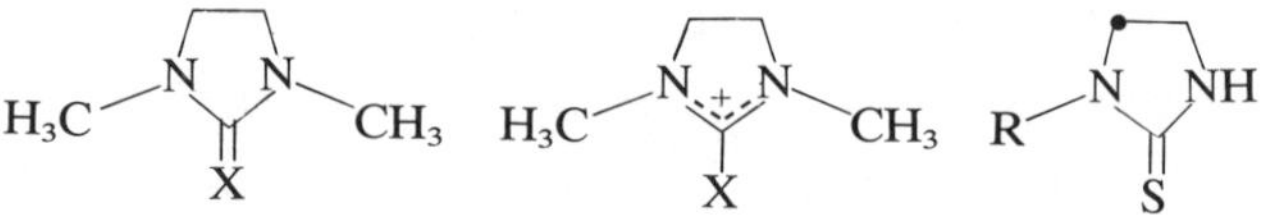

X = O, S, NR   X = Cl, O$^-$, OAr, NR$_2$

76a[96]   **76b**[96a,97]   **77a**[98]

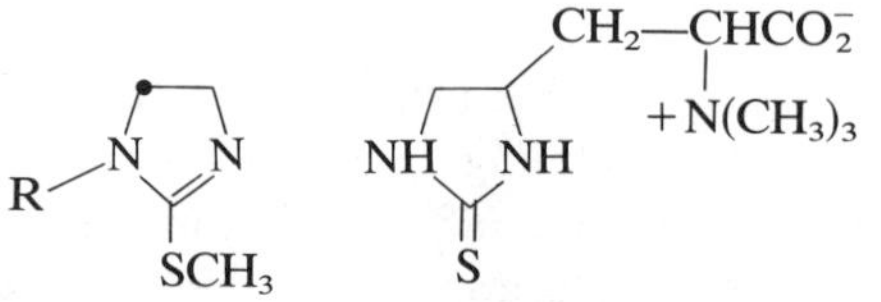

77b[98]   **78**[99]

Scheme 3.9

In the imidazolidines **79**, ^{1}H/^{13}C coupling constants have been determined by observation of the ^{13}C satellite proton spectrum.[100] ^{13}C nmr has contributed evidence[101] that the system **80**, which is pharmacologically active in lowering blood pressure, exists exclusively (or nearly so) in the imino-diazolidine form **80b**.

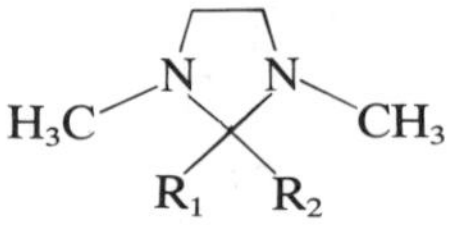

$R_1, R_2 = H, H;$
Me, Me;
Me, H;
Ph, H

79   **80a**   **80b**

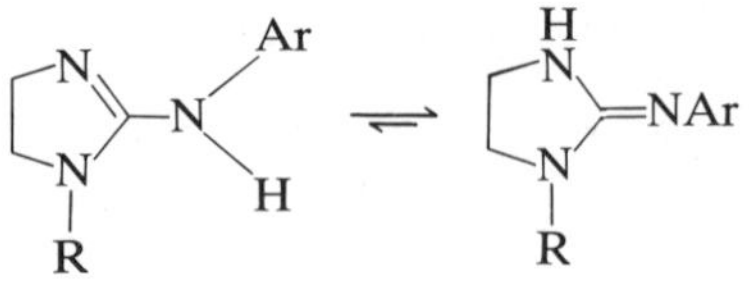

The thiazolidine system **81** has been investigated[102] as have its 2-, 4-, and 5-methyl homologs as well as the 2,2,4- and 2,5,5-trimethyl and 2,2,4,4-tetramethyl derivatives. The α, β, and γ parameters for methyl substitution shown in Table 3.8 were thus derived.[102] It is evident that the

Table 3.8
Methyl Substituent Effects in 1,3-Thiazolidines (**81**)

Effect at	Substitution at		
	C(2)	C(4)	C(5)
C(2)	11.25	−1.05	0.65
C(4)	−0.3	8.0	7.4
C(5)	2.55	6.65	11.75

$$53.1 \overbrace{}^{} 34.05$$
$$\substack{4 \quad 5}$$
$$HN \underset{2}{} S$$
$$55.65$$

81

$$HO_2C \qquad R_1$$
$$HC{-}N \quad S \cdots R_2 \qquad R_3 = H \text{ or } CH_3$$
$$\| \qquad \qquad$$
$$O \qquad R_3 \quad R_3$$

82

α effect next to sulfur is greater than that next to nitrogen (cf. Scheme 3.3 for a large α effect in tetrahydrothiophene), whereas the two β effects are of similar magnitude. The γ effects are small and may be either shielding or deshielding. In **82**, prepared from diastereomeric 2-amino-3-mercaptobutyric acids, the configuration of the methyl group at C(4) could be deduced from ^{13}C nmr in that the group *cis* to the carboxyl resonates at higher field; this, in turn, revealed the configuration (erythro or threo) of the parent acid $CH_3CH(SH)CH(NH_2)CO_2H$.[103]

Lipoic acid (**83**), a 1,2-dithiolane, and its amide have been investigated as to chemical shifts and relaxation times.[104a] Spectra of 1,3-dithiolane **84**, the corresponding $\Delta^{3,4}$-thiolene and carbenium salts thereof have

$$(CH_2)_4CO_2H$$

83

$$S \quad H$$
$$S \quad SCH_3$$

84

$$S{-}S{-}X$$

$$\rightleftharpoons$$

85

$$S{-}S \quad X$$

85

$$CH_3$$
$$|$$
$$N$$
$$\diagdown$$
$$P{-}NR_2$$
$$\diagup$$
$$N$$
$$|$$
$$CH_3$$

86

been reported.[104b] Unpublished results on 1,3-dithiolanes and 1,3-oxathiolanes are cited in Ref. 63. The aromatic structures of **85**[105] (X = O or S) have been investigated by C-13 nmr; it appears that the compound with X = S[105a] but not the one with X = O[105b] exists in the bicyclic (fully aromatic) form. The phosphorus triamides **86** have been studied by dynamic nmr,[106] and data on the thianines **87** are on record.[107a]

Structure **88** has been established on the basis of ^{13}C nmr studies;[94] the spectrum of the oxa analog (O in place of NR_2) is also known.[94] Spectra of various silicon and germanium containing rings of type **89** have been reported.[107b]

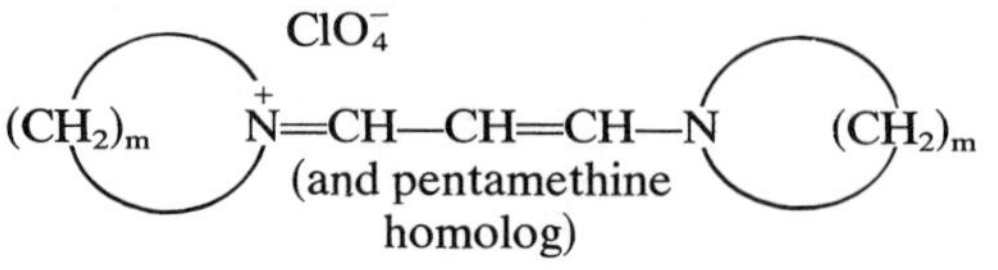

$$\text{(CH}_2)_m \quad \overset{+}{N}=CH\!-\!CH=CH\!-\!N \quad (\text{CH}_2)_m \qquad \text{ClO}_4^-$$

(and pentamethine
homolog)

87

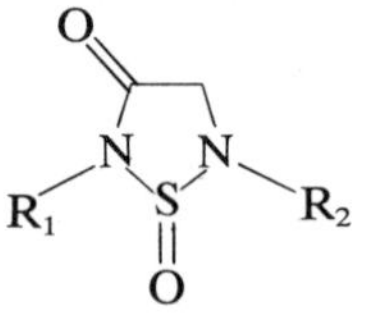

88

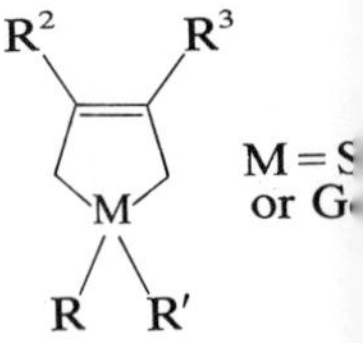

89

$M = S$
or Ge

D. Six-Membered Rings

1. General

Six-membered saturated heterocycles have been studied more extensively than those of other ring sizes, presumably because the deep potential well of the chair conformation provides a stable framework within which the effects of substituents may be considered as second-order perturbations. This situation, which pertains in monohetero- and diheterocyclohexanes* as well as in cyclohexane itself, facilitates the understanding of spectral effects (just as it had earlier facilitated the understanding of chemical effects) because the geometry of the ring is fixed and the position of the substituents is either equatorial or axial. (In contrast in a substituted five-membered ring, neither the shape of the ring nor the position of the substituents in space are obvious *a priori*.)

Most of the fundamental ^{13}C studies of chemical shifts in six-membered rings have dealt with one or other of two problems: (a) the effect of the hetero-atom or atoms on the chemical shifts of the ring carbon atoms or substituent carbon atoms; (b) the effect of substituents on the shifts of the ring carbons as a function of the nature of the heteroatom.

A comprehensive paper on the effect† of heteroatoms in monoheterocyclohexanes on the shifts of ring carbons has been published

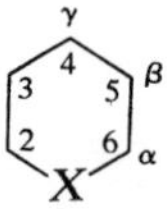

90

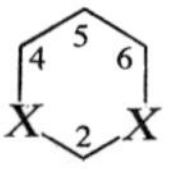

91

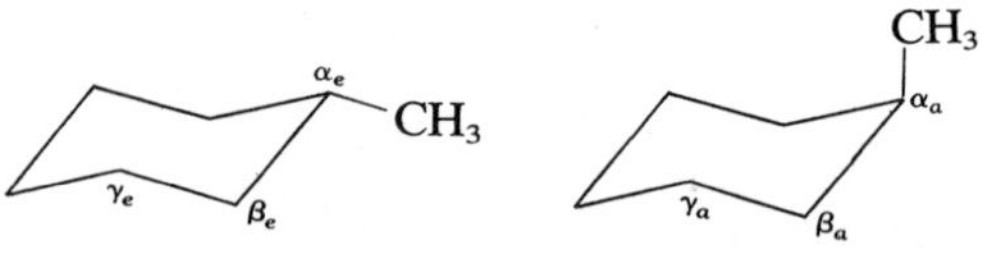

Scheme 3.10

---

* The proposition becomes increasingly uncertain as the number of heteroatoms increases. For example, 1,2,4,5-tetrathiane has chair and twist forms of comparable stability.[108] However, the barrier to chair-twist (or chair-chair) interconversion is even higher (approximately 16 kcal/mol) than the already high barrier (10–11 kcal/mol) in cyclohexane and simpler six-membered heterocycles.

† The effects refer to shifts relative to cyclohexane.

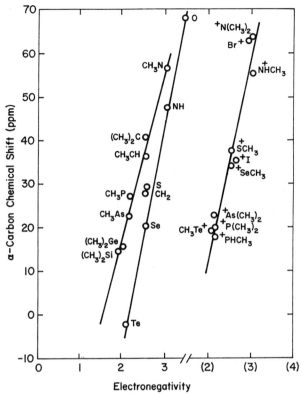

Figure 3.1.
The chemical shift of the α carbon as a function of electronegativity for the pentamethylene heterocycles of groups 4 and 5 (left), for those of group 6 plus piperidine and cyclohexane (center), and for the positively charged heterocycles of groups 5, 6, and 7 (right). From Ref. 109, by permission of the authors and publisher.

by Lambert and coworkers.[109] The general conclusions are as follows: The α shift (see **90**) is a steep function of the electronegativity of the heteroatom X with an increase in one unit in electronegativity producing a downfield shift of about 50 ppm (Figure 3.1). Data are available for C, Si, Ge, Sn, N, P, As, O, S, Se, Te, Br^+, and I^+ in various states of substitution (Table 3.9) and the only major deviation appears to occur for S, which behaves as if it were more electronegative than it actually is. Methyl substitution at the heteroatom causes a consistent downfield shift and lowers the slope to approximately 40 ppm/electronegativity unit (el.u.) Formation of an onium salt (where possible) changes the slope to

Table 3.9

^{13}C Chemical Shifts of the Pentamethylene Heterocycles ^{a}

X	Solvent	$\delta(\alpha)^b$	$\delta(\beta)^b$	$\delta(\gamma)^b$	$\delta(S)^{b,c}$
		Group 4			
CH_2	None	27.7	27.7	27.7	
$CHCH_3$	None	36.4	27.1	27.0	
$C(CH_3)_2$	None	40.4	23.2	27.3	29.4
$Si(CH_3)_2$	None	14.3	24.4	30.1	−3.3
$Ge(CH_3)_2$	None	15.4	25.9	30.6	−3.9
$Sn(CH_3)_2{}^d$	None	8.5	23.1	26.5	−9.4
		Group 5			
:NH	None	47.5	27.2	25.5	
	None	47.7	27.5	26.1	
:NCH$_3$	None	56.7	26.3	24.3	47.0
	None	56.7	26.2	24.3	46.9
	None	57.0	26.6	24.6	47.2
:N-t-C$_4$H$_9$	None	46.9	26.0	27.1	53.2, 25.2
:NCl	CH_2Cl_2	64.0	27.8	23.2	
$^+NH_2$ I$^-$	H_2O	45.8	23.2	22.4	
	H_2O	45.5	23.1	22.6	
$^+NHCH_3$ I$^-$	H_2O	55.9	24.1	21.7	44.3
	H_2O	55.2	23.7	21.8	45.1
$^+N(CH_3)_2$ I$^-$	H_2O	63.7	20.5	21.0	52.7
	H_2O	63.3	20.6	21.0	52.7
$N(O)CH_3$	C_6H_6	66.1	21.1^e	21.7^e	
:PCH$_3$	None	27.0	23.5	28.6	11.3
	None	26.7	23.4	28.3	11.3
:PC$_6$H$_5$	None	24.6	23.4	27.9	
$P(O)CH_3$	$CDCl_3$	29.9	23.6	27.1	14.4
$P(S)CH_3$	$CHCl_3$	32.7	22.4	26.2	18.5
$^+PHCH_3$ I$^-$	$CHCl_3$	17.6	22.8	24.9	5.3
$^+P(CH_3)_2$ I$^-$	H_2O	19.9	21.2	24.8	6.9
:AsCH$_3$	None	22.4	23.9	29.3	5.1
:AsC$_6$H$_5$	CCl_4	22.8	24.7	29.4	
$As(O)CH_3$	$CDCl_3$	33.0	25.5	27.8	14.6
$As(S)CH_3$	CS_2	34.0	24.9	28.2	17.5
$As(Se)CH_3$	C_6H_6	32.1	23.1	26.6	18.6
$AsCH_3Cl_2$	CCl_4	51.1	23.9	52.0	22.4
$AsCH_3Br_2$	$CDCl_3$	47.2^e	24.2	47.2^e	22.2
$^+As(CH_3)_2$ I	$H_2O/DMSO$-d_6	22.8^e	22.8^e	25.8	6.4

Table 3.9 (*Continued*)

X	Solvent	$\delta(\alpha)^b$	$\delta(\beta)^b$	$\delta(\gamma)^b$	$\delta(S)^{b,c}$
		Group 6			
O	None	68.0	26.6	23.6	
	None	69.7	27.9	25.1	
S	None	29.3	28.2	26.9	
:SO (average)	CDCl₃	49.0	19.3	25.3	
:SO-ax[f]	CD₂Cl₂	45.1	15.5	24.7	
:SO-eq[f]	CD₂Cl₂	52.1	23.3	24.7	
SO₂	CDCl₃·	52.6	25.1	24.3	
SBr₂	CH₂Cl₂	34.7	22.9	27.6	
SI₂	CH₂Cl₂	33.3	25.9		
:S⁺H FSO₃⁻	FSO₃H	31.2	24.1	21.8	
:S⁺CH₃ I⁻	H₂O	37.8	20.5	22.7	22.4
	SO₂[g]	37.8	20.2		22.1
Se	None	20.2	29.1	28.4	
:SeO (average)	CH₂Cl₂	42.1	18.6	26.3	
:SeO-ax[h]	CH₂Cl₂	39.4	16.8	25.1	
SeO₂	CH₂Cl₂	57.5	25.1	24.9	
SeBr₂	CH₂Cl₂	51.2	20.9	22.9	
SeI₂	CH₂Cl₂	29.7	25.7[e]	26.0	
:Se⁺H FSO₃⁻	FSO₃H	41.8	23.8	22.5	
:Se⁺CH₃ I⁻	H₂O	34.1	20.5	23.9	15.7
	SO₂[g]	34.0	20.1		15.2
Te	None	−2.1	29.9	30.9	
TeBr₂	CH₂Cl₂	36.9	20.3	25.9	
TeI₂	CH₂Cl₂	33.2	21.4	25.5	
:Te⁺H FSO₃⁻	FSO₃H	24.0[e]	25.0[e]	25.8[e]	
:Te⁺CH₃ I⁻	H₂O		20.7	27.7[e]	
	SO₂[g]	17.7	19.9		−0.6
⁺Br	SO₂	62.9	25.4	22.6	
⁺I	SO₂	35.4	24.8[e]	25.4[e]	

[a] From Ref. 109, by permission of the authors and publisher.
[b] In ppm, downfield from internal TMS; taken at 30°C unless otherwise noted.
[c] Substituent on X.
[d] See p. 233.
[e] Assignments are uncertain.
[f] Taken at −93°C.
[g] The material was γ deuterated.
[h] Taken at −98°C.

45 ppm/el.u. but the actual effect of salt formation on the shift is ambiguous, being generally downfield-shifting but in a few instances (e.g., As, P) either with little effect or actually upfield shifting. A methyl carbon on the hetero atom is affected similarly as the α carbon. It follows that the α carbon in heterocyclohexanes may be either upfield or downfield of that in cyclohexane (27.7 ppm; see first footnote on p. 199) depending on whether the hetero atom is more electronegative (O, N, and possibly S) or less electronegative (Si, Ge, Sn, P, As, Se, Te) than carbon.

The shift of the β carbon is a much softer function of the electronegativity of X (−2.5 ppm/el.u.) and, moreover, is reversed in sign so that the β carbon is most upfield (i.e., <27.7 ppm) for the most electronegative X (e.g., oxygen) and downfield (i.e., >27.7 ppm) for heteroatoms less electronegative than carbon. This finding suggests a charge alternation effect between C_α and C_β.[109] However, the effect on C_γ is qualitatively the same as on C_β with a slope of about −5 ppm/el.u. Thus atoms of low electronegativity (e.g., Si, Te) shift C_γ downfield and those of high electronegativity (e.g., O, N) shift it upfield relative to C. The γ effect is almost invariably larger than the β effect, showing neither alternation nor attenuation. The situation with respect to the β effect, and, to a lesser extent, the γ effect is complicated when there is a substituent at X (such as methyl); if the substituent is equatorial, there is little effect at C_β but when it is axial, C_β is shifted substantially upfield. Introduction of charge at X makes it more electronegative and thus leads to an upfield shift at both C_β and C_γ.

A systematic study of relaxation times (T_1) in the ^{13}C spectra of pentamethylene heterocycles has been carried out by Lambert and Netzel.[110] In virtually all cases it turns out that dipole–dipole relaxation[111] is dominating; thus the relaxation rates for C_α are proportional to molecular weight. (An exception is piperidine where there is a complication due to intermolecular association.) The relaxation rates for C_β and C_γ reveal some evidence for anisotropic tumbling[111] around the X—C(4) axis in those cases where X bears a substituent or substituents (e.g., Si, N, P, As) but not in those cases where it does not (O, S, Se, Te).

Much of the work dealing with heteroanalogs of cyclohexane has been addressed to the following questions: (a) effect of the heteroatom on the shifts of the ring carbons; (b) effect of substituents on the shifts of the ring carbons; (c) effect of the heteroatom on shifts of substituents. The first question is the one dealt with in the most general terms by Lambert (vide supra, compare cyclohexane with heterocyclohexanes); more specific instances will be presented in conjunction with individual ring systems. The second question has been addressed in the case of methylcyclohexanes in the pioneering work of Grant and coworkers.[34] It was found that the effect of a methyl substituent depends on its position and conformation (Scheme 3.10); relative to the shift of unsubstituted cyclohexane* (27.15 ppm^{66}) the shift parameters are α_e 5.96 ± 0.12, β_e 9.03 ± 0.08, γ_e 0.05 ± 0.07, α_a 1.40 ± 0.23, β_a 5.41 ± 0.19, γ_a -6.37 ± 0.15.[34b] It may be seen that all shifts in axially substituted cyclohexanes are upfield of those in equatorially substituted stereoisomers, but particularly striking is the upfield γ shift ("steric shift") produced by an axial substituent about whose origin there has been much controversy.[112]

The introduction of a heteroatom in general changes the shift parameters only in a minor way (Table 3.10).† Perhaps the most striking effect is the large enhancement of the α_e and α_a effects in thianes and, *a fortiori*, in thianium salts. This enhancement may result from the high polarizability of sulfur in these species. No corresponding effect is seen in oxygen or nitrogen analogs, nor is it seen at C(3) or C(4) even in the sulfur compounds. Another recurring feature is the diminution of the β_e effect at C(2), that is, in the position adjacent to the heteroatom. The β_e effect at C(4) is normal and that at C(3) seems to be somewhat reduced in some cases but not in others, depending also on the position of the substituent (2 or 4). No anomaly is seen in the β_a effects. The γ_e and δ_e and δ_a effects are small throughout except for some relatively large δ_a-3 effects, which are probably real.‡ Finally there is some indication for enhanced γ_a effects at C(2), especially in the sulfur compounds, but only if the effect is transmitted through carbon (γ_a^4-2) rather than through the

---

* Even though shifts can be measured with a precision of better than ±0.1 ppm, systematic errors (e.g., in calibration or in referencing) mar the accuracy to the point where measurements in different laboratories at different times may vary by as much as 0.2–0.5 ppm. The cyclohexane shift was originally given[34a] as 101.44 ppm from benzene, corresponding to 27.26 ppm from TMS (cf. Ref. 2), is recorded in reference books as 27.5,[1] or 27.7[2] ppm and is taken in the parametrization scheme[34b] as 27.0 ppm (101.70 ppm from benzene). For this reason the shifts recorded for cyclohexane in Table 3.9 differ slightly from the above.

† A word on notation: The Greek letter indicates whether the shift effect is caused by a substituent on the same carbon (α), on an adjacent carbon (β), etc. The subscript indicates if the substituent is equatorial or axial and a superscript, if any, indicates the position of the substituent if it is not obvious; thus a β_e effect at C(3) can be either caused by a substituent at C(4): β_e^4-3 or at C(2): β_e^2-3. The last number is the locant of the carbon under observation: 2, 3, or 4 in **90**, 2, 4, or 5 in **91**.

‡ The standard deviations of the parameters usually vary between 0.1 and 0.6 but are occasionally larger; thus the significance of small parameters is questionable.

Table 3.10

Shift Parameters in Methyl-substituted Cyclohexanes and Heterocyclohexanes

Effect	XCH_2[b]	$O^{65,c}$	$NH^{d,113}$	$S^{114,e}$	$SCH_3^{115,f}$	Effect	1,3-Dioxane[116,g]	1,3-Dithiane[117]	1,3-Oxathiane[h]
α_e-2	6.0	5.1	4.8	8.5	13.9	α_e-2	5.3 (5.1)	10.3	7.6
α_e-3	6.0	4.6	5.1	6.0	7.2	α_e-4,6	5.7 (5.9)	8.3	5.6 (9.2)
α_e-4	6.0	6.7	7.4	6.2	7.2	α_e-5	3.1 (3.6)	5.6	4.5
β_e-2	9.0	6.2	6.6	6.7	4.9 (8.3)[i]	β_e-4,6	5.8 (6.3)	7.0	6.1 (7.1)
β_e^2-3	9.0	7.2	5.8	9.2	7.1	β_e-5	7.3 (7.3)	9.0	6.5 (9.0)
β_e^4-3	9.0	8.6	9.8	8.4	8.0				
β_e-4	9.0	9.0	8.8	8.8	8.7				
γ_e^4-2	-0.3	-0.5	-0.3	0.1	0.1	γ_e^2-4,6	0.1 (-0.1)	1.2	0.3 (0.5)
γ_e^6-2	-0.3	-0.1	0.2	0.7	1.0	γ_e^2-2	0.8 (-0.4)	0.8	-0.3 (0.4)
γ_e-3	-0.3	0	0	0.7	0.4 (-1.8)[i]				
γ_e-4	-0.3	0.1	-0.1	0.5	0.9	γ_e^6-4	-0.1 (-0.3)	0.6	0.2[j] (0.2)
δ_e-2	-0.6	-0.5	-0.4	-0.5	-0.6	δ_3-2	-0.2 (-0.4)	-0.4	-0.3
δ_e-3	0.3	-0.7	-0.1	-0.9	0.4	δ_e-5	-0.8 (-0.9)	-1.3	-1.5
α_a-2	1.4	-1.1	0	4.5	10.3	α_a-2	-1.7^m (-1.0^n)	8.3	—
						α_a-4,6	0.6 (1.3)	3.2	0.4 (4.8)

α_α-3	1.4	2.2	0.9	−0.6	4.3
α_α-4	1.4	1.3	1.5	−0.6	1.2
β_α-2	5.4	4.4	5.8	6.2	6.6(2.6)[i]
β_α^2-3	5.4	3.6	6.8	5.9	7.5
β_α^4-3	5.4	5.6	5.0	5.3	6.6
β_α-4	5.4	5.7	4.3	5.5	5.9
γ_α^4-2	−6.3	−5.6	−6.4	−7.4	−5.1
γ_α^6-2	−6.3	−8.4	−4.9	−4.9	−3.0
γ_α-3	−6.3	−5.3	−4.5	−6.5	−5.4(−7.8)[l]
γ_α-4	−6.3	−5.7	−6.2	−6.0	−5.2
δ_α-2	0.2	0.3	0	0	0.5
δ_α-3	0.2	−0.8	−1.6	−0.2	−0.3
G_α-2[l]	−3.8	−1.4	—	—	—
G_α-3	−3.8	−2.9	—	−2.9	−2.2
G_α-4	−3.8	−3.6	—	−2.2	−1.9
G_β-2	−1.3	−0.8	—	−1.1	−1.1
G_β^2-3	−1.3	−0.8	—	—	—
G_β^4-3	−1.3	—	−1.4	—	—
G_β-4	−1.3	−1.5	—	−2.5	−1.2
G_γ^4-2	2.0	—	—	−1.3	1.2
G_γ^6-2	2.0	1.3	—	2.6	—
G_γ-3	2.0	1.3	—	1.7	1.6
G_γ-4	2.0	2.0	—	—	—

α_α-5	3.1(3.3)	−1.8	1.0
β_α-4,6	4.5(4.8)	6.4	4.3(6.2)
β_α-5	3.7(3.8)	5.7	3.5(5.7)
γ_α^2-4,6	−7.3[m](−7.8[n])	−5.5	—
γ_α-2	−9.0(−7.1)	−9.4	−8.6(−5.2)
γ_α^6-4	−5.3(−5.4)	−6.5	−6.4[k](−6.6)
δ_α-2	0.4(0)	0.2	0
δ_α-5	0.1[m](−1.0[n])	−1.1	—
G_α-2	(−0.5[n])	−1.6	—
G_α-4,6	−2.0(−2.3)	−1.4	−3.5(−2.9)
G_α-5	−2.3(−2.5)	3.1	−3.6
G_β-4,6	(−0.6)	−1.0	−0.7(−1.0)
G_β-5	−0.8(−0.6)	−0.9	−0.8(−1.4)
G_γ^2-4,6	−(1.0[n])	1.7	o[p]
G_γ-2	2.3(1.3)	2.1	1.2(1.2)
G_γ^6-4	1.5(1.5)	1.5	1.6[q](2.1)

Table 3.10 (Continued)

Effect	XCH₂[b]	O[65,c]	NH[d 113]	S[114,e]	SCH₃[115,f]	Effect	1,3-Dioxane[116,g]	1,3-Dithiane[117]	1,3-Oxathiane[h]
G_δ-2	0.7[r]	-0.1	—	—	—	G_δ-2	-(0.2)	-0.5	0.3
G_δ-3	0.7	1.0	—	—	—	G_δ-5	-(1.0[n])	0.4	—

[a] In ppm; to be added to experimental values for parent compounds (**90**, **91**) except where it is indicated that basic shifts were calculated rather than experimental values.

[b] α and β values from Ref. 34; γ and δ values from Ref. 66. Since some of the shifts in Ref. 66 supersede those in Ref. 34, these "combined" parameters may be slightly inaccurate.

[c] Calculated base values for C_2, C_3, and C_4 are 68.7, 26.7, and 23.6, respectively.

[d] Tentative values from hand calculation including shift measurements at low temperature; base values for C_2, C_3, and C_4 are 47.6, 27.4, and 25.3, respectively.

[e] Calculated base values for C_2, C_3, and C_4 are 29.1, 27.7, and 26.4, respectively.

[f] Calculated base values for C_2, C_3, and C_4 are 32.0, 25.2, and 22.4, respectively.

[g] Parameters in parentheses are derived from data of K. Pihlaja and T. Nurmi. Since these data were obtained with modern instrumentation, they are probably more accurate than those in Ref. 116. We thank Prof. K. Pihlaja, University of Turku, Turku, Finland, for these data. Base values to be used with these parameters are C_2, 95.3; C_3, 66.9; and C_4, 26.6.

[h] K. Pihlaja, P. Pasanen, and T. Nurmi, personal communication. The unparenthesized data, where there are two sets, are effects on or from the oxygen side of the molecule (or through oxygen); the parenthesized, on the sulfur side; base values: C_2, 71.2; C_4, 27.5; C_5, 27.0; and C_6, 69.7.

[i] Effect of methyl on S.

[j] γ_e^4-6.

[k] γ_a^4-6.

[l] Geminal effect; to be added to calculated shifts after giving effect to all pertinent α, β, γ, and δ effects. For vicinal effects the original articles should be consulted.

[m] Combined α and G effects.

[n] K. Pihlaja. T. Nurmi, G. Furst and E. L. Eliel, personal communication.

[o] Combined γ_s^2-6, γ_a^2-6, and $G_\gamma^{2,6}$:-7.8.

[p] Combined γ_e^2-4, γ_a^2-4 and G_γ^2-4:3.2.

[q] G_γ^4-6.

[r] Hand calculated.

heteroatom (γ_a^6-2). The effects in the diheterocompounds **91** (X = O, 1,3-dioxane; X = S, 1,3-dithiane) generally parallel those in the monoheteroanalogs **90** if one keeps in mind that C(4) in **91** corresponds to C(2) in **90**, C(5) in **91** to C(3) in **90**, and C(2) in **91** to C(2) in **90**, but with a double effect of the heteroatom.

We turn to a discussion of the effect of heteroatoms on shifts of substituents, for example, the effect of the nitrogen on the methyl shift in 3-methylpiperidine (19.8 ppm) compared to that in cyclohexane (23.0 ppm,[66]). To some extent these effects parallel the effects on ring carbons that have been addressed in Table 3.9. However, when one compares exocyclic carbon shifts with ring carbon shifts, a conformational component comes into play in addition to (or as part of) the electronic or electrostatic one. Thus formula **92** indicates that a methyl group at C(2) in piperidine may be either anti-periplanar (axial Me) or cyn-clinal (equatorial Me) to the lone pair, whereas the ring carbon C(3) is necessarily syn-clinal. Similarly in **93** the C(4) ring carbon is synclinal to the heteroatom but of the methyl substituents at C(3) the axial is syn-clinal but the equatorial is anti-periplanar.

Before discussing effects of heteroatoms, it is well to consider the situation in cyclohexane. An axial methyl group (δ 17.5–18.9 ppm[66]) is always upfield of an equatorial one (23.0 ppm). The effect is the reciprocal of the steric shift at C_γ caused by an axial methyl (*vide supra*). However, increased compression, as in the 1,3-synaxial situation (**94**)

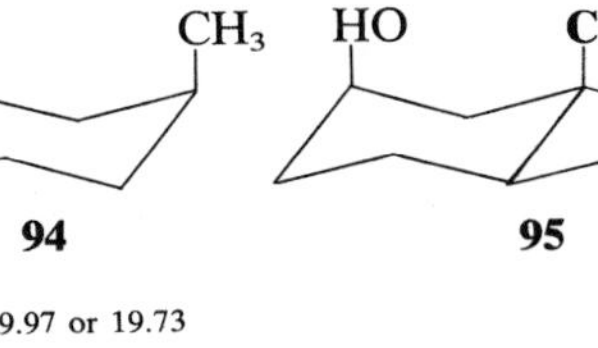

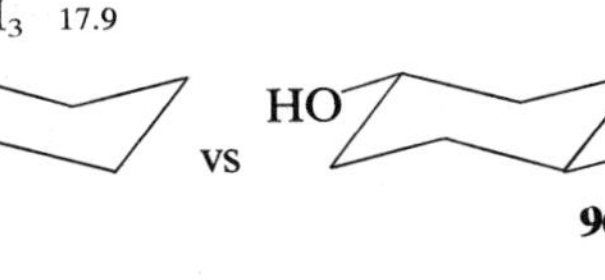

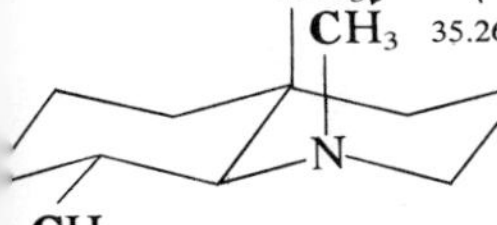

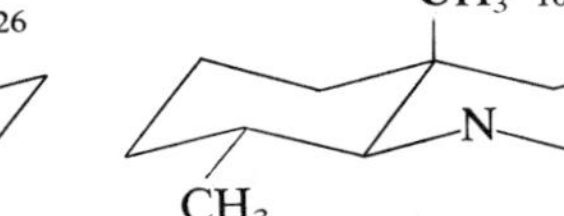

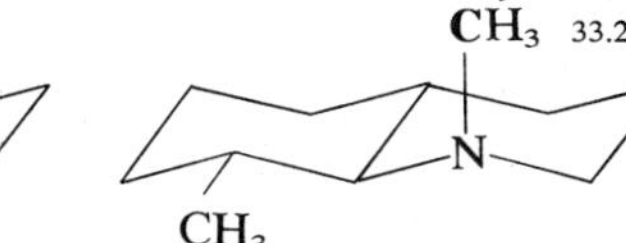

Scheme 3.11

leads to a downfield shift; examples[118–120] are shown in Scheme 3.11 (compare **95** to **96** or **97** to **98** and **99**).

The effects of heteroatoms may be discussed under four headings: (a) the γ-gauche effect,* (b) the γ-anti-effect,* (c) the effect of lone pairs, and (d) other effects. Since the γ-gauche-effect operates on the ring carbons (cf. **93**), it may be extracted directly from Table 3.9 by comparison of $\delta(\gamma)$ for the heterocyclohexane with $\delta(\gamma)$ for cyclohexane. As already explained the γ-gauche-effect of the heteroatom is then upfield shifting for atoms more electronegative than carbon (O, N) and downfield shifting for the less electronegative atoms (P, Se, Te, As). (In an earlier publication[43] it had been alleged that the effect is always upfield shifting; this misconception was due to the fact that the least electronegative atom investigated was sulfur, and sulfur, as already noted earlier,[109] feigns an electronegativity greater than that recorded for it in the literature.)

The γ-anti-effect (Scheme 3.12) has attracted considerable attention ever since it was first proposed in 1975.[43] The two situations where it is most generally observed are shown in formulations **93** (effect on equatorial CH$_3$) and **100** (effect of X on C$_\gamma$ of the ring).[121] It was originally

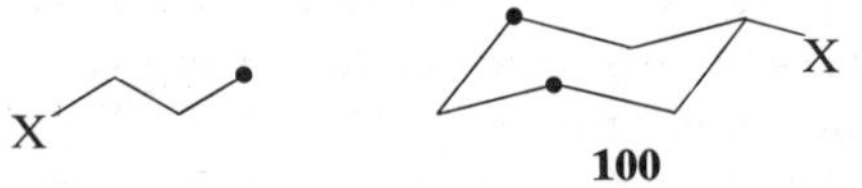

Scheme 3.12

believed[43] that the effect was upfield shifting for second-row heteroatoms (O, N, F), but that there was no effect for third-row atoms (P, S, Cl). As a result of this observation plus the finding that the effect vanished or was reversed (to a downfield shifting one) when the heteroatom was attached to a bridgehead atom in a bicyclic structure, it was postulated[43] that the effect was a hyperconjugative one. This hypothesis is no longer tenable, however. First of all it has been shown[122] that the γ-anti-effect, like the γ-gauche-effect (*vide supra*) is reversed (i.e., becomes deshielding) for atoms less electronegative than carbon (Scheme 3.13). This is probably

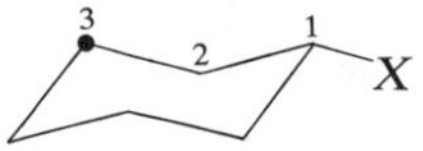

	X	H	C(CH$_3$)$_3$	Si(CH$_3$)$_3$	Ge(CH$_3$)$_3$	Sn(CH$_3$)$_3$	Pb(CH$_3$)
	$\delta_{C(3)}$	27.0	27.8	28.3	28.3	29.0	30.1

Scheme 3.13

---

*We mean here the shift effect of the heteroatom X—C—C—C$^\gamma$ on the γ carbon over and above the effect exercised by a CH$_2$ group in lieu of the heteroatom (H$_2$C—C—C—C$^\gamma$). In the case of the γ-anti-effect, since a γ-anti-carbon causes practically no incremental shift, the comparison is often made with H—C—C—C$^\gamma$.

the reason why the effect for S and P, whose electronegativity is similar to that of C and H, is small; the situation for Cl is less easily understood. Second it was found[123,124] that the upfield-shifting effect is turned into a downfield-shifting one when the intervening atoms are heavily substituted (cf. Scheme 3.14); this observation no doubt explains why bridgehead-substituted compounds (e.g., **101**)[125] show downfield rather than upfield γ-anti-shifts: not because of the bridgehead position of the electronegative substituent but because of the high degree of substitution around the bridgehead! Even with all this said the origin of the effect (upfield or downfield shifting) is still not clear; indeed there is no certainty that a single factor is involved. Recently a fresh point of view has been brought to bear on the subject of gamma effects[126]; it is postulated that the γ-gauche-effect is due in the main not to the substituent but to the removal of the hydrogen atom it replaces.

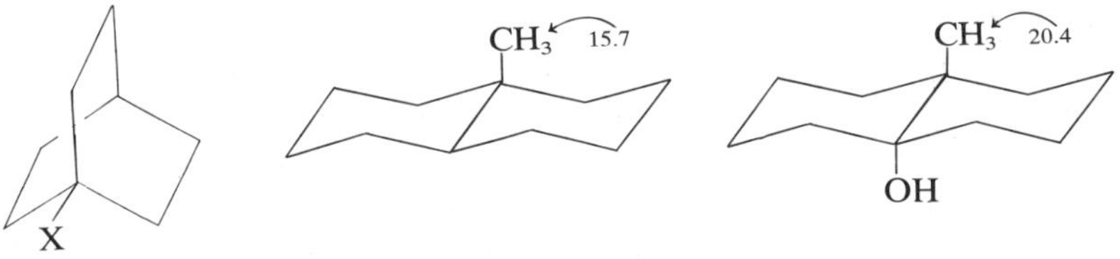

101

Scheme 3.14

Another fairly general shift effect is related to the orientation, with respect to the carbon nucleus observed, of lone pairs on heteroatoms. Such an effect is not readily discerned for atoms such as oxygen or sulfur where the orientation of the pairs cannot readily be influenced and any separate effect of the pair thus can not be distinguished from a general anisotropy or electronegativity effect of the heteroatom as such. The situation is more favorable with nitrogen, for if the nitrogen is substituted with an alkyl group, the position of the alkyl group can be changed and the orientation of the lone pair changes accordingly.* There is still the difficulty that the change in orientation of the pair is accompanied by a change in orientation of the alkyl group and sometimes it is not evident which of the two changes affects the shift of the nucleus used as a probe and to what extent.

A salient example is provided by $N,2\alpha$- and $N,2\beta$-dimethyl-*trans*-decahydroquinolines (Scheme 3.15).[119,127] The position of the N-methyl group can be influenced by placing an equatorial or axial methyl group at

---

* As yet there is no clear-cut case of a change of orientation of the hydrogen in a secondary amine, —NH— as distinct from that of an alkyl group in the tertiary amine, —NR—; *vide infra*.

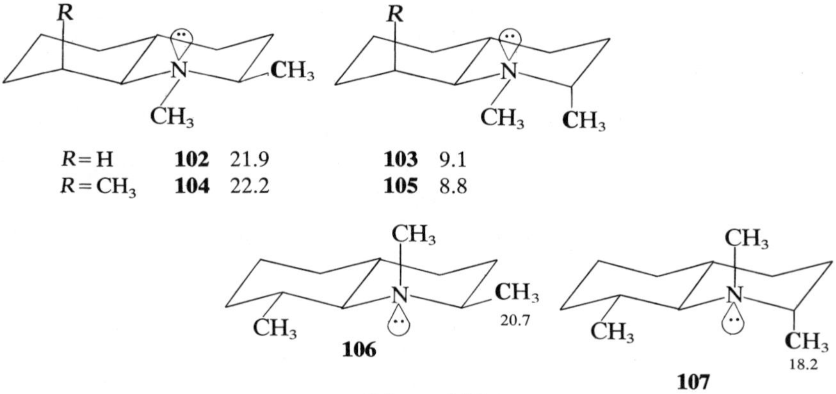

$R=H$ **102** 21.9 **103** 9.1
$R=CH_3$ **104** 22.2 **105** 8.8

Scheme 3.15

C(8). It is evident from Scheme 3.15 that the two axial methyl groups anti-periplanar to the lone pair resonate at unusually high field; another example is provided by the low-temperature spectrum[128] of *N,N,2-*trimethyl-1,3-diazane (conformer shown as **108**). Of course one must ask whether the high-field shift is not just due to a combination of the methyl group at C(2) being axial and also compressed by the (*gauche*) *N*-methyl. This possibility was assessed by converting **103** and **105** to their respective hydrochlorides. The corresponding shifts for Me-2 now were: **103**.HCl,[119] 11.0 ppm, **105**.HCl,[127b] 10.9 ppm. It is seen that a substantial part of the unusual upfield shift of Me(2a) persists in the hydrochloride and may be an indication of an unusually strong gauche interaction (due to close proximity) with the N—CH$_3$ group.* However, Me(2a) in the amines is still approximately 2 ppm upfield of its position in the corresponding hydrochlorides; considering that the β-effect of hydrochloride formation is almost uniformly upfield shifting in other cases[42,119] the conclusion that the lone pair in **103** and **105** is responsible for a substantial fraction of the remarkable upfield shift of Me(2a) remains on a

108

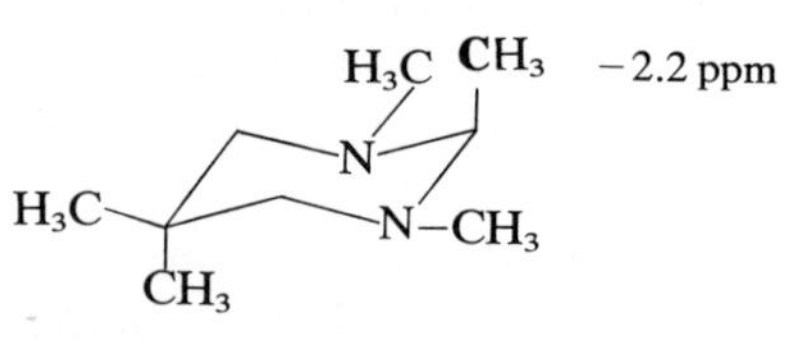

---

*Puzzlingly, the *N*-methyl group is not at unusually high field; 38.6 ppm in **103**.HCl, 38.0 ppm in **105**.HCl. This compares with 35.7 ppm in **102**.HCl and 34.9 in **104**.HCl; the axial methyl at C(2) is actually *less* shielding than an equatorial one!

firm base.* The effect is also seen in N-methyl-cis-decahydroquinolines[129] and in N-methylpiperidines.[113a] It is of interest that a similar upfield shift effect of an anti-periplanar lone pair has long been known[130] in proton nmr spectroscopy.

Among the miscellaneous effects we have already noted that β_e effects at carbons next to heteroatoms are generally smaller than in cyclohexanes, whereas β_a effects tend to be somewhat larger. The reciprocal of this effect is seen in the 2-methyl-1,3-dithianes **109** and **110**; apparently the heteroatom deshields the axial methyl group and shields the equatorial one to the point where the normally upfield position of Me_a (17.5–18.9 ppm[66]) and downfield position of Me_e (23.0 ppm) are reversed (even though one pair of p electrons on sulfur is anti-periplanar to the axial CH_3, in contrast to the situation in piperidines, $vide$ $supra$.) Apparently both sulfurs are required to effect this reversal; a single sulfur atom, as in 2-methylthiane,[114] shifts the methyl groups close to each other (Me_e, 22 ppm; Me_a, 20.5 ppm) but there is no crossover. Again there is a parallel between these ^{13}C shifts and corresponding shifts of equatorial and axial protons at C(2) in 1,3-dithianes ($\delta_e < \delta_a$).[127a]

109   **110**

2. Oxygen Heterocycles

The six-membered oxygen containing ring is the building block of the pyranose sugars, which, however, will not be discussed here as such (cf. Sec. I). Tetrahydropyran (oxane), some of its 2-alkoxy derivatives, and 2-acetoxy-, 2-hydroxy-, 2-isopropylthio-, 2-dimethylamino-, and 2- and 4-methyl derivatives have been investigated by de Hoog.[131] It was found that the substitution parameters were generally similar to the Grant parameters[34] in cyclohexane, a conclusion that was confirmed by later, more extensive studies[65] summarized in Table 3.11. The parameters are indicated in Table 3.10, column 3; one salient feature, already recognized by de Hoog[131] is the unusually large γ_a^6-2 effect. It was suggested[131] that this is due to the very close proximity of the axial substituent at C(6) to the axial hydrogen attached to C(2) (and, indeed, to the C-2 carbon). In

---

* The possibility remains that the change in the Me(2a) shift in going from the amine to the salt is a function not of the disappearance of the lone pair but of some (unrecognized) change in geometry.

Table 3.11
^{13}C nmr Chemical Shifts of Methylated Tetrahydropyrans (in CDCl$_3$; ppm)

Position of Methyl Group(s)	C-2	C-3	C-4	C-5	C-6	2-CH$_3$	3-CH$_3$	4-CH$_3$	5-CH$_3$	6-CH$_3$
None	68.78	26.91	23.80	26.91	68.78	—	—	—	—	—
2	73.91	33.92	23.84	26.13	68.44	22.25	—	—	—	—
3	74.80	31.18	32.17	26.11	68.18	—	17.47	—	—	—
4	68.07	35.20	30.19	35.20	68.07	—	—	22.32	—	—
23c	75.69	32.63	30.65	21.34	68.05	18.41	12.02	—	—	—
23t	79.89	37.68	32.77	26.90	68.34	19.76	18.22	—	—	—
24c	73.53	42.67	30.50	34.66	68.01	(22.12)	—	(22.37)	—	—
24t	68.04	39.44	24.99	32.27	62.34	21.05	—	18.73	—	—
25c	73.70	(28.86)	(29.16)	28.34	72.52	21.58	—	—	16.73	—
25t	73.52	33.82	32.72	30.81	74.84	21.92	—	—	17.23	—
26c	73.74	33.25	23.91	33.25	73.74	22.33	—	—	—	22.33
26t	66.49	31.45	18.07	31.45	66.49	19.38	—	—	—	19.38
34c	73.15	34.18	32.25	30.61	67.18	—	11.75	17.62	—	—
34t	74.25	38.17	36.99	35.20	68.46	—	14.53	19.51	—	—
35c	74.33	31.32	41.60	31.32	74.33	—	17.31	—	17.31	—
35t	74.04	27.64	38.65	27.64	74.04	—	17.53	—	17.53	—
22	71.23	36.84	20.06	26.28	61.89	26.68	—	—	—	—
33	78.48	30.39	36.80	22.89	68.41	—	25.53	—	—	—
44	64.17	39.36	28.05	39.36	64.17	—	—	28.32	—	—
226	71.41	36.12	20.26	33.58	66.28	32.07e (21.99a)	—	—	—	(22.78)
335	77.99	30.95	46.06	27.26	74.78	—	27.46e 24.15a	—	17.39	—
235eae	75.51	32.84	40.38	25.03	75.32	18.86e	12.04a	—	17.29e	—
235eee^a	79.39	37.55	41.88	31.65	74.62	19.49e	18.11e	—	17.19e	—
235eea^a	80.21	32.18	38.59	29.31	72.77	19.62e	18.28e	—	17.33a	—
345eae	67.94	35.33	36.15	35.33	67.94	—	14.68e	5.07a	14.68e	—
345eee^b	74.65	37.50	44.04	37.50	74.65	—	14.64e	15.90e	14.64e	—
345aee^b	74.33	34.32	39.65	32.42	74.33	—	11.84a	14.97e	16.64e	—

^{a} Mixture of 24% *eee* and 76% *eea*.
^{b} Mixture of 45% *eee* and 55% *aee*.

accordance with this explanation the same large effect is seen in 1,3-dioxanes at C(4) and, *a fortiori*, at C(2) (Table 3.10, column 8). It was also noted[131] that the effect of alkoxy groups at C(2) is much less than the corresponding effect of an alkoxy substituent in cyclohexane; the explanation invoked is one of saturation of polarizability (i.e., an atom that has already been substantially polarized by one substituent cannot be as effectively polarized by a second one).[30] The shifts of the parent tetrahydropyran[65] are shown in **111**. (Agreement among the various reports[30,65,131] for this compound is only moderately good.)

Shifts of various phosphonate derivatives of type **112**, especially in the sugar series, have been reported[132a] with the finding that $^1J(^{31}P/^{13}C)$ is considerably larger (172–173 Hz) when phosphorus is equatorial (**112e**) than when it is axial (**112a**, 158–160 Hz). This situation parallels that in the corresponding cyclohexyl phosphonate,[132c] but is in contrast to that where phosphorus is part of the six-membered ring (p. 229). Conformational equilibria in the α-halo derivatives of tetrahydropyran-4-ones have been studied by ^{13}C nmr.[132b]

111 **112e** **112a**

113 $X = CO_2CH_3$
114 $X = CO_2C_4H_9\text{-}n$
115 $X = CH_2OH$
116 $X = CH_2OAc$

117

The shifts of a number of 2-methoxy-6-substituted $\Delta^{3,4}$ dihydropyrans (**113–116**) have been recorded[133]; in general the ring carbon atoms in the *trans* isomers resonate slightly upfield of those of the *cis* isomers. The difference is less than in *cis*- and *trans*-2,6-disubstituted tetrahydropyrans (Table 3.11) as might be expected, since the equilibrium in the dihydropyrans involves pseudoaxial versus pesudoequatorial (rather than axial versus equatorial) substituents. Occasional reversals do occur, especially at C(5), and the only reliable carbons for configurational assignments are C(2) and C(6). A substantial amount of diaxial conformer appears to exist in the case of **113**.

An extensive study of 1,3-dioxane has been carried out by Kellie and Riddell[116ab] and some additional data are available.[116c,134,135] The shifts of the parent molecule are shown in **117** and the shift parameters for methyl substitution in Table 3.10.* C-2 in 1,3-dioxane, like that in 1,3-dioxolane

---

*The shifts for **117** are taken from Ref. 2 (there is, again, considerable discrepancy between Ref. 30, 116b and 116c). The increments in Table 3.10 will presumably be nearly independent of solvent, as long as the unsubstituted and substituted compounds are compared in the same solvent.

(p. 183) shows saturation of polarizability (compare **117** with **111**). As mentioned earlier, the shift of an equatorial methyl at C(5) occurs at an unusually high field (12.4 ppm in *trans*-2-*t*-butyl-5-methyl-1,3-dioxane)[116c] because of the presence of two anti-periplanar oxygen atoms (γ-anti-shift); thus this resonance is upfield of that of the only slightly shifted axial methyl in the *cis* isomer (at 15.9 ppm).[127a]

Information on ^{13}C nmr in the 1,4-dioxane system is surprisingly limited. The parent compound[30] resonates at 67.8 ppm.[2] The ^{13}C satellite proton spectrum has been used[136] to measure the inversion barrier in 1,4-dioxane since, as a result of ^{13}C/^{1}H coupling, it does not display the H_a/H_e degeneracy found in the ^{12}C-proton spectrum.

118   **119** (*cis*)   **120** (*trans*)

121   **122**

The spectrum of δ-valerolactone (**118**) has been recorded[56,68a] as have the spectra of its 5-alkyl,[68a] 2-methyl-2-hydroxy-[56] and the diastereomeric *cis*- (**119**) and *trans*- (**120**) 3,5-dimethyl analogs.[137] In the monoalkyl analogs[68a] the expected downfield-shifting α and β and upfield-shifting γ effects are seen. Configuration of the dialkyl derivatives (which exist in half-chair forms[137]) was readily assigned on the basis that the trans isomer, with one axial or pseudoaxial methyl group, has all of its resonances at higher field than the equatorial, a situation that, as already indicated, is typical for six-membered rings. Also, since Me(5) is nearly constant in shift in the two isomers whereas Me(3) resonates at considerably higher field in **120**, Me(5) would appear to be equatorial in **120** and Me(3) axial.[137] Finally we note that the Me(3)s in both **119** and **120** resonate at unusually low field; this is not typical of either 4-methyltetrahydropyrans (Me$_e$, 22.3 ppm; Me$_a$, ~18.7 ppm; see Table 3.11) or 3-methylcyclohexanones (Me$_e$, 22.6 ppm; Me$_a$, calculated 19.6 ppm[137]). The spectrum of 4-oxanone has been recorded.[138a]

Glutaric anhydride (**121**) and several of its 2-, 3-, and 2,4-substituted homologs have been studied.[138b] The compounds exist in "sofa" conformations (anhydride carbons and oxygen in a plane, saturated carbons out of that plane) and the substituted members generally are equilibrium mixtures of conformational isomers. In the case of the 2,4-dimethyl analogs, carbons in the *trans* isomer generally resonate at higher field than those in the *cis*, similar to what is seen in cyclohexanes. In general the α and β parameters are similar to those in cyclohexanes, even though the conformation is different.

Chemical shifts, $^{13}C/^{1}H$ coupling constants and T_1 relaxation times have been measured in the substituted methylene malonates (1,3-dioxane-4,6-diones) **122**.[139] These compounds exist in boat forms and the substitution parameters are somewhat different from those in 1,3-dioxanes. It is of interest that substitution of a methyl at C(5) has a slightly *up*field shifting α effect; this is true also for 5-axially methylated 1,3-dioxanes (Table 3.10) but not the 5-equatorially substituted ones. In the measurement of the coupling constants it was noted that the aniso-chrony of the protons at C(5) in the 2-*t*-butyl-2-methyl homolog leads to a non-first-order coupled ^{13}C signal for C(5) in chloroform-*d*. The T_1 values of the methyl substituents at C(2) are markedly conformation dependent with the bowsprit (equatorial) methyl having the shorter (approximately 1 sec) and the flagpole (axial) methyl the longer (2–3 sec) relaxation time.[139]

Senda and coworkers[140] have studied a series of chroman-4-ols (**123**) including flavanols (aromatic substituent) and also the $\Delta^{2,3}$-dihydropyran **124**. They noted that in the latter there is no shift effect of oxygen on C_β (contrary to the situation in tetrahydropyran, *vide supra*), which resonates at 23.05 ppm (compared to 23.11 ppm in cyclohexene).

The single carbon resonance in 1,3,5-trioxane is at 93.65 ppm,[65] very close to that of 1,3-dioxane (**117**); in the *cis*-trimethyl homolog the ring shift is at 98.33 ppm and the methyl shift at 20.57 ppm.[65]

123   124

3. Nitrogen Heterocycles

Because of its ubiquity in alkaloids (see Sec. I) and a number of substances of pharmacological interest, the six-membered nitrogen-containing ring has received more attention than any other. Formulas **125** and **126** display the chemical shifts[2,8] of piperidine and N-methylpiperidine, the parent compounds in this series. In discussions of

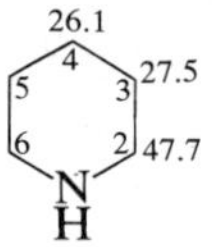

125 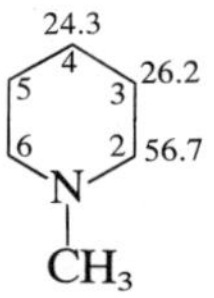   **126** 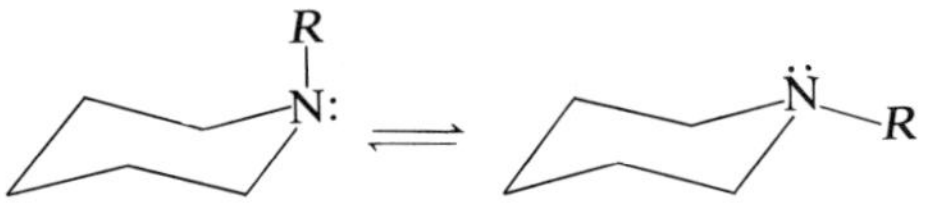   **Scheme 3.16**

spectral properties of derivatives there appears an immediate complication: both piperidine and its N-alkyl homologs exist, in principle, as two isomers,* shown in Scheme 3.16. Indeed, in piperidine itself (R = H) the two isomers contribute nearly equally and it was a matter of long controversy[144] as to which one predominates at equilibrium. There is now a fair amount of evidence (supported by very recent studies[141b]) that, at least in nonhydrogen-bonding solvents, the isomer with equatorial H predominates by a factor of less than 2 at 25°C.[141] Fortunately, with respect to ^{13}C nmr chemical shifts, the present evidence is that the position of the hydrogen (and therefore that of the lone pair) is of little consequence. Salient data are shown in Scheme 3.17[127a]; the situation here is in sharp contrast to that with the N-methyl compounds displayed in Scheme 3.15. The interpretation of the insensitivity of the shifts of Me(2a) to the conformation of Me(8) means *either* that Me(2a) is indeed unaffected by the position of the equatorial-axial lone pair equilibrium *or* that that equilibrium is virtually completely insensitive to the conformation of Me(8)† or both. In the case of proton NMR[141b] the former hypothesis has

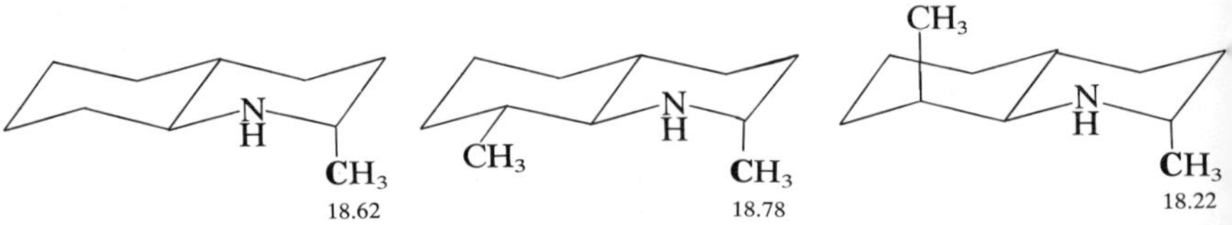

Scheme 3.17

---

* There seems to be no agreement as to whether these two isomers should be said to differ in "configuration" or "conformation" since they can be interconverted either by ring reversal (a conformational change) or by nitrogen inversion (a configurational change). Nmr spectroscopy suggests that, in general, the nitrogen inversion process is faster. ($\Delta G = 6.1$ kcal/mol for N-inversion[141] as determined by low temperature ^{13}C nmr vs 10.4 kcal/mol for ring reversal.[142]) Regarding the definition of "isomer," see Ref. 143.

† Efforts are presently underway to resolve this question. The assumption that the equilibrium of Scheme 3.16, R = H is unaffected by a syn-axial methyl group is contrary to Ref. 130b.

been shown to be the correct one, but the difference between the NH and NMe cases remains a puzzle.

In the case of N-methylpiperidine, the situation is more clear-cut, since it has recently been shown beyond a shadow of a doubt that the N-methyl group is nearly entirely (>99%) equatorial.*[145,146] One of the techniques[145b,c] used to evaluate the equilibrium (Scheme 3.16, R = CH$_3$) was irreversible quenching of a ^{13}CH$_3$-enriched sample of a conformationally anchored N-methylpiperidine (e.g., N,cis-3,5-trimethylpiperidine, **127**) followed by analysis of the mixture of diastereomeric salts **128** by ^{13}C nmr spectroscopy.

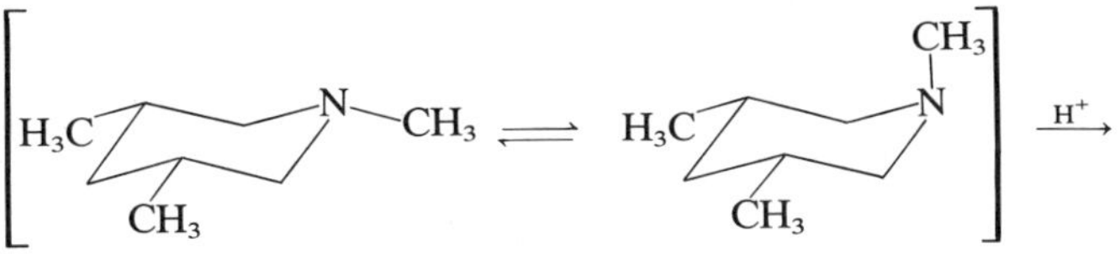

127

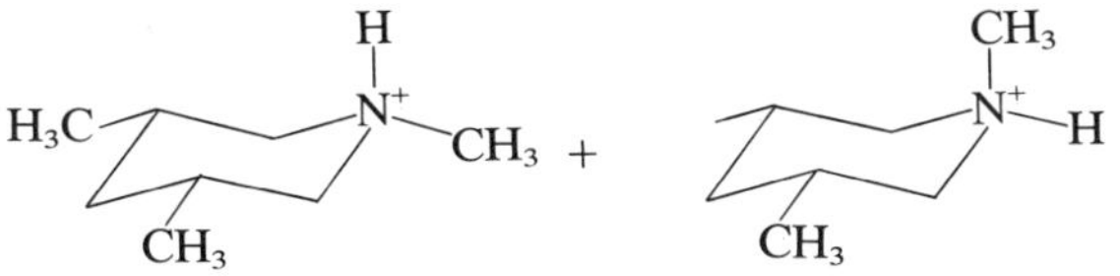

128e   **128a**

In the interpretation of nmr spectra of N-alkylpiperidines, it may thus be safely assumed that an N-alkyl group is virtually entirely equatorial and this is true even in compounds such as **129**[148] and **130**[145c,149] where the movement of an N-methyl group from the equatorial to the axial

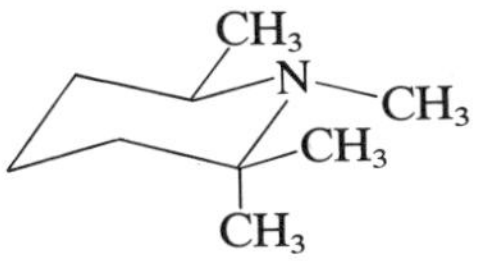

129 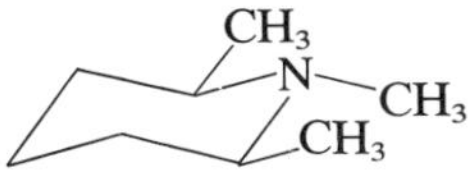   **130**

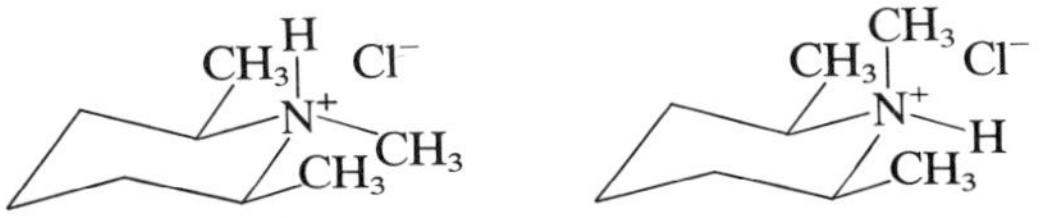

131e   **131a**

---

* The assertion in Ref. 147 that the equilibrium is only 90–95% equatorial was evidently based on unsuitable model studies.

position would lead to relief of a gauche vicinal Me/Me interaction and/or to the replacement of the more severe vicinal diequatorial to the less severe[113a] vicinal equatorial-axial Me/Me interaction.* However, the same is not true for N-alkylpiperidinium salts. Thus it has been shown by several investigators that the equilibrium of **131**, the hydrochloride of **130**, corresponds to two parts of **131e** and one part of **131a** ($K = 2$).[149-152]

The ^{13}C nmr spectra of various C-methyl substituted and disubstituted piperidines have been recorded several times.[8,113,153-156] In contrast, only partial sets of data were available for ring-substituted N-methylpiperidines; a complete set of data for these species[113b,c] is collected in Table 3.12. The shift parameters deduced from this set of data are included in Table 3.10 (column 3); given the fact that the data were calculated without computer optimization and with the inclusion of some low-temperature shifts[113a] (basically inappropriate, because of the temperature dependence of the shifts), there is no significant difference be-

Table 3.12
^{13}C Chemical Shifts in N-Methylpiperidines[a]

Methyl Substituent(s) (Position, Configuration)	C_2	C_3	C_4	C_5	C_6	Me_A[b]	Me_B[c]	N—Me
—	56.60	26.07	23.84	26.07	56.60	—	—	46.93
2	59.41	34.88	23.73	26.43	57.22	20.42	—	43.37
3	64.17	31.21	32.53	25.65	56.00	19.72	—	46.59
4	56.07	34.48	30.23	34.48	56.07	21.86	—	46.49
cis-2,6	59.59	35.17	24.79	35.17	59.59	21.62	21.62	38.07
trans-2,6	52.96	33.65	19.30	33.65	52.96	15.13	15.13	40.08
cis-3,5	63.59	31.10	41.61	31.10	63.59	19.55	19.55	46.21
trans-3,5	63.32	27.37	38.45	27.37	63.32	19.25	19.25	47.00
cis-2,3	60.75	34.82	29.37	23.55	52.91	10.89	15.71	43.42
trans-2,3	66.16	37.24	34.06	25.73	57.57	17.13	19.76	43.53
cis-2,4	59.09	43.70	31.40	34.82	57.35	20.93	22.10	42.97
trans-2,4	53.56	40.77	25.20	33.27	49.55	14.65	20.15	43.08
cis-2,5	56.48	30.86	28.68	30.18	59.29	14.12	18.91	43.42
trans-2,5	59.11	34.96	33.68	31.63	65.36	20.62	19.67	43.26

[a] Ref. 113b,c; in CDCl$_3$ with internal TMS.
[b] Lower-numbered CH$_3$ group.
[c] Higher-numbered CH$_3$ group.

* The equilibrium in compound **129** is too one-sided for study by low-temperature nmr (this is true also for **130**) and was determined[148] by a line shape analysis of the proton spectrum at low temperature, above and below coalescence. The equilibrium in **130** was determined by irreversible protonation, as explained for **127**.[145c]

tween the substitution parameters for piperidine and N-methylpiperidine and those for cyclohexane (see also Ref. 119 and 129 regarding this point).

Comparison of piperidine (125) and N-methylpiperidine (126) shows, in addition to the expected downfield β_e shift of 9.0 ppm at C(2) due to the N-methyl group, smaller upfield shifts at C(3) (γ_e, -1.3 ppm) and C(4) (δ_e, -1.8 ppm). These shifts, which recur in other piperidines upon N-methyl substitution,*[113b,c,154,155] do not appear to arise from the increased negative charge on nitrogen induced by the alkyl substituents, since similar upfield shifts are also produced (see below) when the nitrogen is protonated, that is, becomes more positive.

An interesting correlation exists[157] between ^{15}N shifts of piperidines and decahydroquinolines and the ^{13}C shifts at corresponding positions in analogous carbocycles. Protonation on nitrogen, that is, conversion of amines to their salts, has a generally well-defined effect on ring carbon shifts ($\Delta\delta$) in piperidines[155] and trans-decahydroquinolines[119]:

1. The effect on C_α [C(2) in piperidines] may be either upfield or downfield shifting with the downfield shift being produced when C_α is substituted (CHMe) and the upfield one when it is not (CH₂).

2. The effect on C_β[C(3)] is almost uniformly upfield shifting except in one case (N, 10-dimethyl-trans-decahydroquinoline[119]) where C_β was quaternary and there was also N-substitution (vide infra).

3. The effect on C_γ [C(4)] is also upfield shifting in piperidines† but generally less so than for C_β. This order ($-\Delta\delta_\beta > -\Delta\delta_\gamma$) tends to be reversed, however, when C_α is substituted [i.e., C(HMe) rather than CH₂].

4. The protonation shifts in the N—Me compounds are less upfield (or more downfield) than those in the NH compounds.

5. The gegenion effect is generally small, provided observations are made in the same solvent.[119] Thus trifluoroacetates and hydrochlorides have nearly the same shift effect (relative to the free amines) in solvent chloroform; they also have very similar shift effects in solvent trifluoroacetic acid, though the two sets of shifts are quite different from each other (cf. 6).

6. There is a marked solvent effect, protonation shifts in trifluoroacetic acid being less upfield (or more downfield) than in chloroform.

---

* However, the δ_e effect seems to be absent in trans-decahydroquinolines.[119]

† There is no palpable $\Delta\delta_\gamma$-effect in trans-decahydroquinolines.[119]

The order of the shifts, $\Delta\delta_\alpha > \Delta\delta_\beta < \Delta\delta_\gamma$, follows the alternation pattern predicted by the calculations of Pople and Gordon.[158] The fact that a positively charged nitrogen induces upfield shifts has been explained[159] by postulating that the charge induces a polarization in the ring C—H bonds in the sense C$^-$—H$^+$; thus although the ring hydrogens acquire a strong net positive charge, the induced charge on the ring carbons actually tends to be negative in most instances. This hypothesis also explains why substitution of hydrogen by carbon on a ring atom tends to convert its upfield to a downfield shift, since the positive charge can now no longer be passed off to the hydrogen substituent. The difference in $\Delta\delta_\beta$ and $\Delta\delta_\gamma$ between piperidines and N-methylpiperidines can be formally accounted for by the observation that δ_β and δ_γ in piperidinium and N-methyl-piperidinium salts[155] are nearly the same (generally within 1 ppm); the same is true for the *trans*-decahydroquinolium salts[119] (except at C-8). In other words, the difference in $\Delta\delta$ originates not in the salts but in the free amines, i.e., it is a consequence of the upfield shifting β_e and γ_e effects in N-methylpiperidines mentioned earlier. (The absence of a $\Delta\delta_\gamma$ difference between *trans*-decahydroquinolines and N-methyl-*trans*-decahydro-quinolines thus correlates with the absence of a δ_e effect in this system.) Morishima et al.[155] have tried to base the explanation on the difference in position of the lone pair in piperidine (partly axial, partly equatorial) and N-methylpiperidine (all axial), and, indeed, it appears from the spectrum of piperidine at $-172°C$[141] that C(3, 5) in the isomer with axial H (equatorial pair) resonate downfield of the corresponding carbons in the N—H-equatorial isomer. This would suggest, contrary to the N-methyl-piperidine case (cf. Scheme 3.15), that the anti-periplanar equatorial lone pair in the N—H-axial isomer shifts C(3, 5) to *lower* field!

The preferred conformation of N-(1-phenyl)cyclohexylpiperidines (**132**) and their hydrochlorides has been derived from ^{13}C nmr spectra.[160] Hagaman[161] has discussed the (non-first-order) *sford* (single-frequency

132

133

134

135

136

off-resonance decoupled) spectrum of N-methyl-4-t-butylpiperidine (**133**) notably at C(2). The structure of the alkaloids spectaline (**134**) and iso-casseine (**135**) was derived largely from their ^{13}C nmr spectra.[162] Anet[163] has measured the nitrogen inversion barrier (11.7 kcal-mol) and conformational equilibrium ($\Delta G° = 1.5$ kcal/mol in favor of the equatorial isomer) for N-chloropiperidine (**136**).[163] $\Delta G°$ here is considerably larger than in cyclohexane, just as $\Delta G°$ in N-methylpiperidine (*vide supra*) is considerably larger than in methylcyclohexane, presumably because of the proximity of the N-axial group to the syn-axial hydrogens at C(3) and C(5). Morishima and coworkers[164] have carried out an extensive investigation of the contact shifts of piperidines complexed with nickel.

A series of 4-piperidones (**137**)[138a,165] and 4-phenyl-4-piperidinols (**138**)[166] and their esters as well as salts and quaternary salts thereof have been carefully investigated; they include the analgesic promedol (the 2,5-dimethyl compound). The configuration and conformation of these compounds can be readily determined by ^{13}C nmr spectroscopy and shift parameters for the various substituents (and for protonation of the free amines to hydrochlorides) have been established.[166] The conformation of piperidine amides (**139a**)[167] and thioamides (**139b**)[168] about the N—C=X bond is readily derived from ^{13}C nmr spectroscopy and this is true also for corresponding N-acyl-4-piperidones, -morpholines, and -N'-methyl-piperazines.[167] Conformational equilibria have also been studied in N-methyl- and N-benzyl-4-methyl-4-benzoylpiperidines (**140**).[169a] The ^{13}C nmr spectrum of the alkaloid arecoline (**141**), representative of $\Delta^{3,4}$-dihydropyridines, has been reported.[169b] The spectra of enamines with nitrogen-rings of various sizes, including pyrrolidine, piperidine and morpholine have already been mentioned.[71b] The dicyanopiperidone **142** shows the usefulness of $^3J(^{13}CN/^1H)$ in configurational assignment.[170] The coupling constant for the equatorial CN is 2.1 Hz, that for the axial CN 8.8 Hz. A number of N-nitrosopiperidines (**143**) and -piperazines have been studied[171]; a series of N-nitroso-substitution parameters for the ring carbons has been established[171b] and it is easy to deduce if the nitroso group is E or Z in an unsymmetrically substituted N-nitrosopiperidine. (The N—NO bond has considerable double-bond character.) In N-nitroso-2-methylpiperidines the methyl group on the Z (syn) side is 100% axial and that on the E (anti) side 59% axial, for steric reasons.[171b]

137   138   139   140   141

142   **143**   **144**

The corresponding hydrazones (RR'C= instead of O=) have also been studied.[172] The spectra of a series of α,α- and β,β-disubstituted glutarimides have been reported.[173]

The ^{13}C nmr spectra of a series of 5-substituted[174] and 5,5-disubstituted[175-177] barbituric acids (**144**) have been investigated, including a thiourea derivative.[176] Additive parameters for higher alkyl substituent effects at C(5) and the proximal COs have been established[175] and a correlation has been made between chemical shifts, pK_as and pharmacological activity.[175]

145   **146**   **147**   **148**

The spectra of 26 1,2-oxazines (**145**) have been recorded[178] and substitution parameters established. These parameters are listed and compared with corresponding ones in cyclohexene in Table 3.13.

The substitution parameters for morpholine (**146**) and N-methyl-morpholine summarized in Table 3.14 were derived from shift data for a

Table 3.13
Shift Parameters: Comparison of Oxazines (**146**) and Cyclohexenes[178]

Parameter	α-5(Me)	α-6(Me)	α-5(Ph)	β^2-3(Me)	β^4-5(Me)
Oxazine	6.5	4.7	10.6	9.5	−4.9
Cyclohexene	7.1	7.1	9.0	9.4	−4.9

Parameter	β^6-5(Me)	β^2-3(Et)	β^4-5(Ph)	β^2-4(Ph)	β^6-5(Ph)
Oxazine	3.6	7.6	−4.3	−3.1	2.4
Cyclohexene	6.0	6.2	−3.0	−3.0	2.9

Table 3.14
Substituent Parameters in Morpholine[a]

Substituent	α_e	α_a	β_e	β_a	γ_e	γ_a
N—CH$_3$	—	—	9.1	—	−1.2	—
C(2)—CH$_3$	5.1	−2.3[a]	5.5	3.7	—	—
C(3)—CH$_3$	4.4	1.5[a]	6.8	3.5	—	−6.0

[a] See Ref. 179 for assumptions made in calculating this effect.

series of 2-methyl-, cis- and trans-2,6-dimethyl-, 2-phenyl-, cis- and trans-2-phenyl-3-methyl- and 3,3-dimethylmorpholines, and their N-methyl homologs.[179] The data are in good agreement with those for oxanes (Table 3.10), piperidines (Table 3.10) and piperazines[138a,156,167] (**147**),* in particular, the effects of N-methylation are about the same as in N-methylpiperidine (cf. **125**, **126**, p. 212), the β_e effect is attenuated† and the α_a effect is negative on the oxygen side (C-2; compare α_a for oxane in Table 3.10) but slightly positive on the nitrogen side (C-3). Morpholine salts have also been studied,[179] as have thiamorpholines (**146**, S in place of O) and derivatives thereof, including the N-benzoyl sulfoxide and N-benzoyl $\Delta^{2,3}$ dehydro compound.[180] The N—Me equatorial preference in N,N-dimethylpiperazine has been found to be 3.0 kcal/mol by the acid quenching method,[181] similar to that in N-methylpiperidine.[145] For the 1,3-oxazines **148**, see ref. 362.

The 1,2-diazane (hexahydropyridazine) system **149** has been investigated using ^{13}C nmr by Nelson and Weisman.[182] N,N-dimethyl-1,2-diazane, its 3-methyl, cis- and trans-3,6-dimethyl and 3,3,6,6-tetramethyl homologs and the bicyclic system **150** were studied, as were the N-methyl-N'-ethyl, N,N'-diethyl, N,N'-trimethylene, tetramethylene and pentamethylene analogs of **151**. Several interesting features emerged. Unlike in the piperidine system, axial-equatorial alkyl groups on the two

149   150   151

---

* The piperazine data are too sparse to warrant calculation of separate parameters.

† The β_a effect appears attenuated also in the case of morpholine.

nitrogen atoms are of the same order of stability as diequatorial ones (e.g., being only 0.3 kcal/mol less stable in the case of **149**, R = CH$_3$*). The activation barrier to ring reversal in the parent N,N'-dimethyl compound is 10.3 kcal/mol and that for nitrogen inversion $(a, e \rightleftharpoons e, a)$ 7.6 kcal/mol. However, there is a second nitrogen inversion barrier $(a, e \rightleftharpoons e, e)$ which is considerably higher (12.6 kcal/mol), and evidence was brought to bear that the high barrier is caused not by the need of the N-alkyl groups but by the need of the lone pairs to pass each other in the $a, e \rightleftharpoons e, e$ inversion. (The evidence may not be entirely conclusive, since it comes from the observation that methyl groups on N and adjacent C in N,N'-trans-3,6-tetramethylpiperazine pass each other without difficulty. However, the N—N bond (1.45 Å) is palpably shorter than N—C (1.47 Å) and therefore the substituents on adjacent nitrogens get somewhat closer to each other than those on carbon and nitrogen.) The trans-3,6-tetramethyl compound displays only the e,a combination of N-methyl groups and this combination predominates by far even in the 3,3,6,6-hexamethyl homolog, even though in this instance it requires syn-axiality of one C—Me and one N—Me group.

A series of 1,3-diazanes (hexahydropyrimidines, **151**) has been studied[184] with the findings summarized in Table 3.15. The data are fragmentary and will not be discussed here in detail. The diazane system presents two major complications that make interpretation of the data difficult. One is that the ring opens and closes rapidly on the laboratory time scale (although, fortunately, slowly on the nmr time scale) so that 2,4- and 2,5-cis and -trans isomers can be obtained only as equilibrium mixtures. The other, more serious complication is that there is an axial-equatorial N—Me equilibrium that is affected by substituents at C(2) and probably also C(4).[185] The N—Me, N'—Me diequatorial isomer suffers from an electrostatic repulsion, the generalized anomeric effect[186a] (at one time called "rabbit-ear effect"[185,186b]), which destabilizes it to the point where the equatorial-axial isomer may be of comparable stability.

1,3,5-Trimethyltriazanes (**152**) have, similarly, been shown by proton nmr to exist very predominantly in conformations in which one N-methyl group is axial and the other two are equatorial.[187]

The N,N',N"-trimethyl-1,2,4-triazane (**153**) and N,N'-dimethyl-1,3,4-thiadiazane systems (**154**) have been investigated by Katritzky and coworkers.[188a] In **154** one methyl group is apparently equatorial and the other axial; the diequatorial isomer does not contribute to the equilibrium, in contrast to the situation in **149**, and there is evidence that the compound

---

* This is probably due to a combination of high steric repulsion of adjacent equatorial groups in the puckered part of a ring[113a] and a gauche-effect[183] opposing the antiperiplanar arrangement of unshared pairs of electrons.

Table 3.15
^{13}C Chemical Shifts of 1,3-Diazanes^a

Compound	C_2	$C_{4,6}$	C_5	5—CH_3	N-Substituent	C_5-Substituent
1,3-Dimethyl	79.77	54.34	23.75	—	42.88	—
1,2,3-Trimethyl	79.19	55.20	21.84	—	39.37	17.55
1,3-Dimethyl-2-ethyl^b	84.7	55.1	21.9	—	39.4	21.4
						9.4
1,3-Dimethyl-2-iso-propyl^b	86.1	54.7	18.4	—	40.2	28.8
						19.5
1,3-Dimethyl-2-tert-butyl	89.2	51.9	17.3	—	46.1	36.4
						29.5
1,3-Dimethyl-2-phenyl	92.61	56.01	25.49	—	43.06	n.l.^{c}
1,3,5-Trimethyl	79.17	62.11	27.70	16.85	42.75	—
trans-1,2,3,5-Tetra-methyl	79.02	64.26	25.42	16.66	39.12	18.69
cis-1,2,3,5-Tetramethyl	76.07	55.20	25.92	17.18	42.26	8.38
1,3,5-Trimethyl-2-ethyl^b	84.5	64.2	25.7	16.8	39.3	23.1
						9.3
trans-1,3,5-Trimethyl-	92.02	63.87	29.61	16.81	42.89	n.l.^{c}
cis-1,3,5-Trimethyl-2-Phenyl	93.09	61.37	29.53	18.46	43.28	n.l.^{c}
1,3,5,5,-Tetramethyl	80.64	66.83	31.39	26.33	43.29	—
1,2,3,5,5-Pentamethyl	82.11	67.67	30.96	26.92	42.45	16.01
				25.88		
1,3,5,5-Tetramethyl-2-ethyl^b	86.9	68.0	30.8	26.9	42.4	21.6
				25.7		7.2
1,3,5,5-Tetramethyl-2-iso-propyl^b	89.7	67.1	31.5	27.7	44.1	29.2
				26.4		19.1
1,3,5,5-Tetramethyl-2-tert-butyl^b	89.4	63.0	32.3	29.2	48.5	39.8
				28.4		29.1
1,3-Di-iso-propyl^b	70.5	48.0	24.8	—	52.4	—
					19.0	
1,3,-Di-iso-propyl-5,5-dimethyl^b	72.7	59.5	30.4	26.6	53.0	—
					18.9	

^{a} Data from Ref. 184; in ppm relative to TMS; solutions in CCl_2=$CHCl/(CD_3)_2CO$ (1:1) unless otherwise indicated.
^{b} In $CHCl_3$.
^{c} Not listed.

exists very largely in one of the two possible *ae* conformations, the one shown as **154**. Both the other *ae* and the *ee* isomer would suffer from syn-axiality of lone pairs on sulfur and on N(3) which, as explained in conjunction with 1,3-diazane (similar data are available for N-methyl-1,3-oxazane^{188b,c}) is electrostatically unfavorable. By the same token, the major isomer of **153** is the one shown, with a sole axial N-methyl group

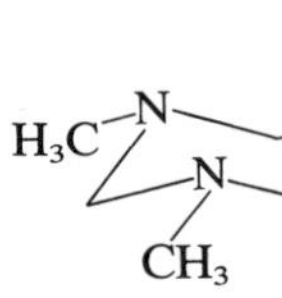

152

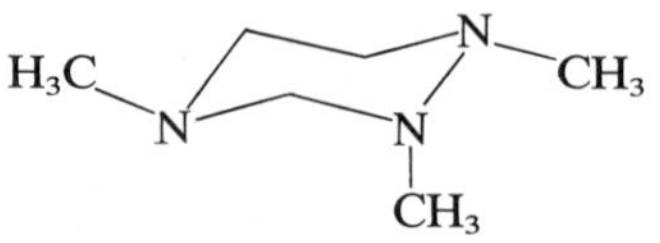

153

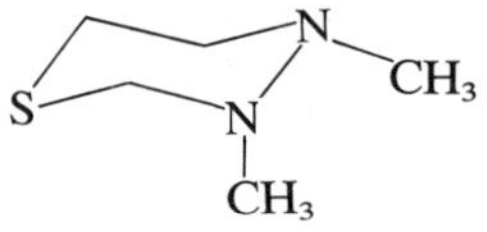

154

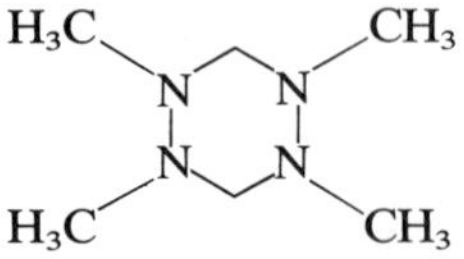

155

on N(2). An axial methyl group at N(1) is, of course, strongly disfavored (cf. the case of N-methylpiperidine) and the remaining combinations have the not particularly favorable N(1), N(2) *ee* conformation combined with either syn-axial pairs on N(2) and N(4) or an axial methyl group on N(4).

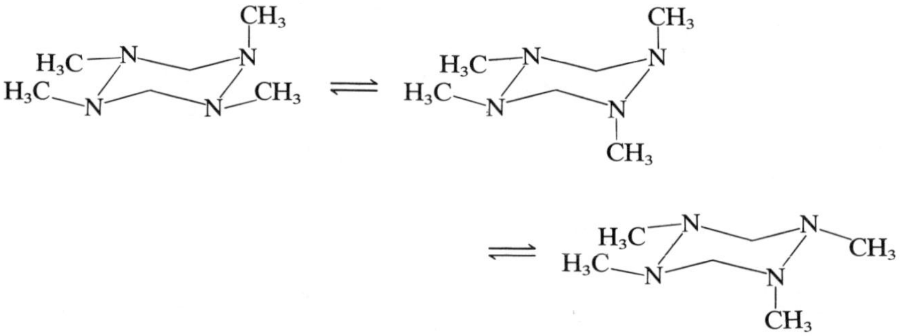

Scheme 3.18

An even more complicated situation arises in tetramethyl-1,2,4,5-tetrazane [tetramethylhexahydro-1,2,4,5-tetrazine (**155**)].[189] The preferred conformation of the system has been a subject of considerable controversy; the ^{13}C nmr evidence has been interpreted to favor the school of conformations shown in Scheme 3.18 (which can be interconverted without passage of either methyl groups or lone pairs past each other). However, it is not entirely clear why the alternate school shown in Scheme 3.19 is not more favorable since it lacks the syn-axial electron pairs present two of the structures in Scheme 3.18 and the unfavorable[182] diaxial methyl conformation on adjacent nitrogen atoms present in the third. The structure with all N—Me groups equatorial is definitively excluded.

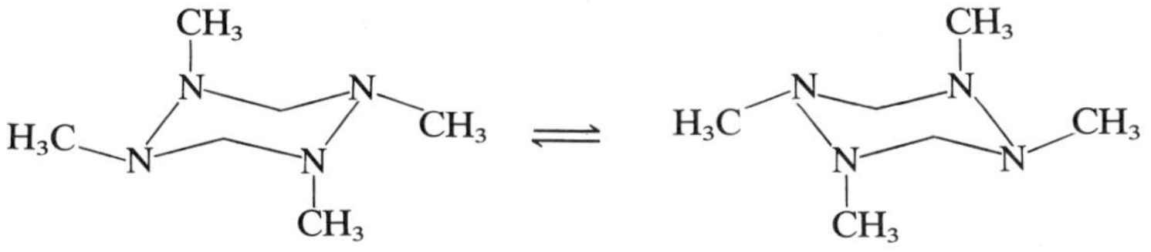

Scheme 3.19

The nmr spectra of maleic hydrazide (**156**) and N,N-dimethyluracil (**157**) are on record.[67]

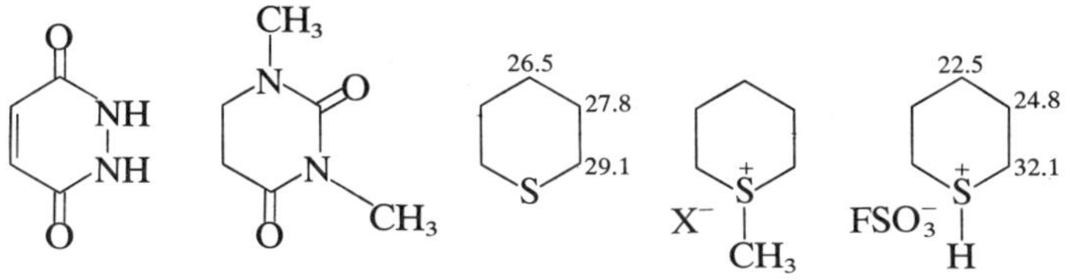

| 156 | 157 | 158 | 159 | 160 |

4. Sulfur Heterocycles

Thiane (**158**), the simplest sulfur-containing six-membered ring, has been investigated in detail.[74,114] The methyl substitution parameters are shown in Table 3.10, column 4. Of note are the large α-2 effects (explained on the basis of a high polarizability of the carbon next to sulfur) and the small β_e-2 effect compared to cyclohexane; the latter is typical of all the heterocycles shown in Table 3.10. The γ_a^4-2 effect is also somewhat enhanced, even though the steric interactions of an axial substituent at C(4) with C(2) and its axial hydrogen substituent are about the same as in cyclohexane. The γ_a^6-2 effect (acting across the sulfur) is normal. We have already pointed out (p. 207) that axial and equatorial methyl substituents at C(2) in **158** are closer together in chemical shifts than usual.

S-Methylthianium salts (**159**) have also been investigated in detail.[74,115] The α-2 effects in these compounds are even more enhanced than in the thiane precursors (cf. Table 3.10). Equatorial S-methyl (25.8 ppm) is well downfield of axial (17.3 ppm) even though $\Delta G°$ for moving the S-methyl group from the equatorial to the axial position is only 0.3 kcal/mol.[190] The shift of S—Me(e) is affected more by an axial Me(2) (-4.3 ppm) than by an equatorial one (-2.0 ppm); the reverse is true for S—Me(a), effect of Me(2e), -4.7 ppm, of Me(2a), $+2.3$ ppm. Not unexpectedly the e/a vicinal compression is greater than e/e, resulting in a greater upfield shift (if that is indeed the right interpretation). The downfield shift on the axial S—Me caused by an adjacent axial C—Me is obviously of quite different

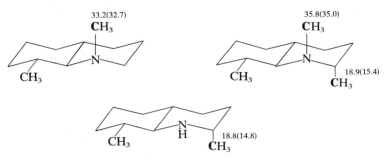

Scheme 3.20 Numbers in parentheses refer to hydrochlorides. Shifts in ppm from TMS in CDCl₃.

origin; the generality of this phenomenon is not clear at the present time. In the *trans*-decahydroquinoline series (Scheme 3.20) an axial methyl group at C(2) shifts an axial N—Me downfield in either the free amine or the hydrochloride, but there is no clear indication of a reciprocal effect of the axial *N*-methyl on the axial *C*-methyl.

The *S*-methyl group has a normal β_e effect (8.3 ppm) but the β_a effect (2.6 ppm) is somewhat low. The γ_a effect of -7.8 ppm appears enhanced, and there is a sizeable γ_e-effect of -1.8 ppm reminiscent of the *N*-methyl group in *N*-methylpiperidine (**126**). The δ_e effect is positive (1.0 ppm) contrary to that in **126**. [These effects relate to thianium fluorosulfonate (**160**) as a reference compound, so that any effect of charge is compensated.] Comparison of **160** with **158** indicates that the effect of thianium salt formation is downfield shifting at C(2, 6) and upfield shifting at C(3,5) and C(4).

The above studies include low-temperature spectra of both thianes[114] and thianium salts[115] in which ring reversal is slowed on the nmr time scale. The ¹³C nmr spectrum of thiane oxide[191] is displayed in formula **161**. C(2) is shifted substantially downfield, perhaps as a combination of charge and substituent effects plus an electric field effect, which has been calculated.[191] C(4) is shifted slightly upfield, similarly as in thianium salt **160**. There is a remarkable upfield shift at C(3) which is, however, readily explained as a γ_a effect of the sulfoxide oxygen, which is largely axial.[192] This becomes clear on inspection of the anancomeric (conformationally biassed) 4-*t*-butylthiane sulfoxides **162** and **163**. Low-temperature ¹³C nmr spectroscopy of **161**[191] leads to decoalescence of the spectra of the two conformational isomers with axial and equatorial S=O. Surprisingly, the sulfone **164**, which must have one axial oxygen, has a considerably lesser upfield shift at C(3) than the sulfoxide **161**.[138a] The 4-keto derivatives of **158** and **164** have also been studied.[138a]

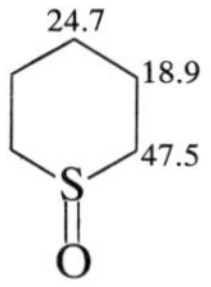

161 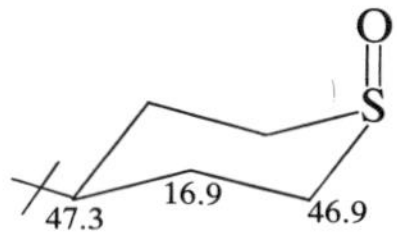   **162** 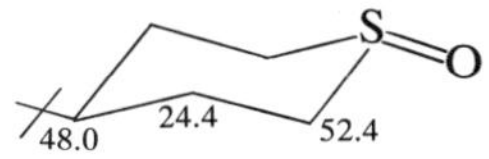   **163** 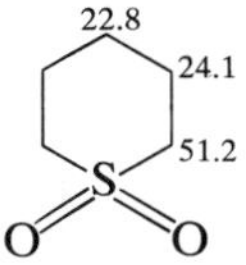   **164**

The spectrum of thiane p-chlorophenylsulfimide[193] is displayed in **165**. The p-chlorophenylsulfimide group is predominantly equatorial, as evidenced by low-temperature nmr spectroscopy;[193] formulas **165e** and **165a** show the spectra of the two conformational isomers at $-90°C$. The spectra of a large number of ring-substituted thiane p-chlorophenylsulfimides as well as p-chlorosulfimides derived from thiadecalins have also been recorded.[194] In the case of 3,3-disubstituted thiane p-tolylsulfimides (**166**, $R = CH_3$, OCH_3 or OCH_2CH_2O), some special intramolecular interactions have been postulated.[195]

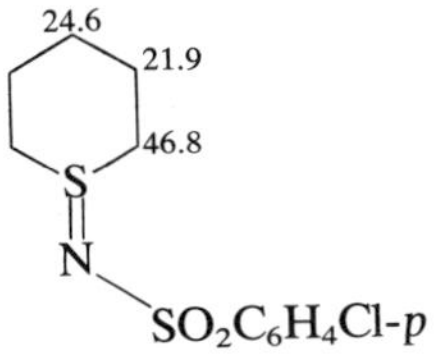

165-e

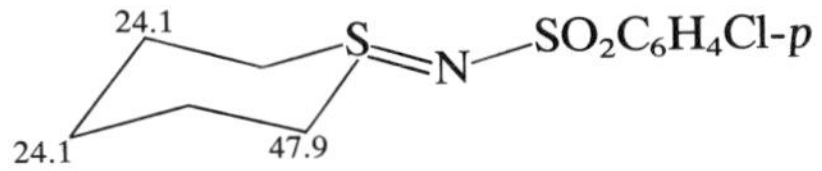

165

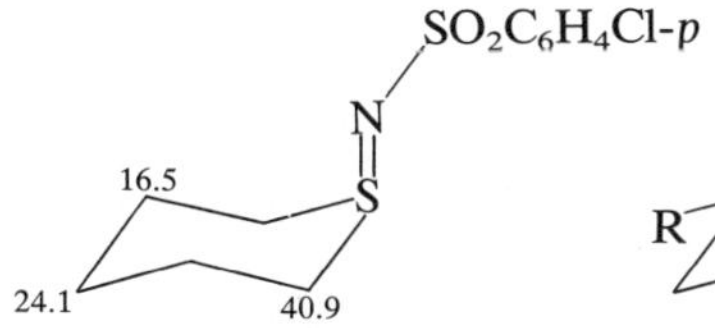

165-a 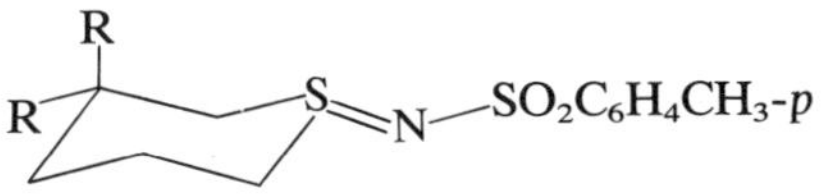   **166**

1,3-Dithiane (**167**) was an early object of study[196] and it was in this molecule that the enhanced α-2 effect of methyl next to sulfur was first seen. An extensive study of this system has now been made[117,197] and the resulting parameters are summarized in Table 3.10, column 9. Salient features in addition to the large α-2 effects and α-4 effects (it should be noted that both C-2 and C-4,6 in 1,3-diathiane are adjacent to sulfur) and the attenuated β_e-4 effect are an unusually large γ_a^4-effect and some sizeable (>1 ppm) γ_e and δ effects.

1,3-Dithiane oxide (**168**) is an equilibrium mixture of equatorial and axial sulfoxide, but, unlike in thiane oxide (*vide supra*), the equatorial

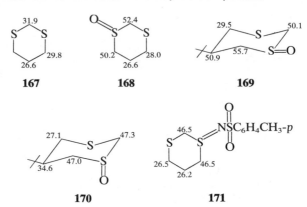

167   **168**   **169**

170   **171**

conformation predominates.[198] The resonances[199] of the parent compound and the conformationally locked (anancomeric) 5-*t*-butyl homologs are shown in **168–170**. Spectra of an additional 10 oxides derived from *cis*-4,6-dimethyl-1,3-dithiane and its 2-methyl and 2,2-dimethyl homologs as well as the 2-methyl homologs of **169** and **170** have been recorded.[199] The (major) substituent parameters are (first figure refers to equatorial oxide, figure in parentheses to axial): α_e-2, 9.6 (7.9); α_a-2, 1.6 (6.1); γ_e^2-6, −1.8 (1.0); γ_a^2-4, −7.2 (−5.6); γ_a^2-6, −10.3 (−5.6); δ_e-5, 0.8 (−1.3) ppm. The axial SO isomer holds no major surprises, but the equatorial one is characterized by an unusually large γ_a^2-6 (effect on the carbon next to sulfoxide) and an unusually small (for a sulfur compound) γ_a-2 effect.

The spectra of various 1,3-dithiane sulfimides (e.g., **171**) have been recorded[195,200,201] including low-temperature spectra.[201] The compounds were found[201] to be overwhelmingly in the conformation with equatorial imide groups.

Several 1,3-dithianes with substituents other than alkyl groups have been studied,[196,200] as have 1,3-oxathianes.[199] A systematic study of the 1,3-oxathiane (**172**) system is available[202]; substitution parameters for methyl substitution are listed in the last column of Table 3.10. The system displays the usual large α effect next to sulfur (but not next to oxygen).

The spectra of 1,4-oxathiane (**173**) and its sulfoxide (**174**) and sulfone (**175**) and some derivatives are on record[203] and have been

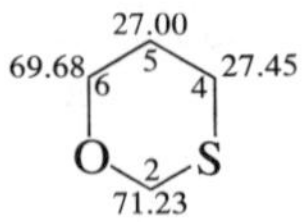

172 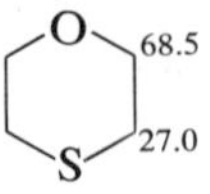   **173** 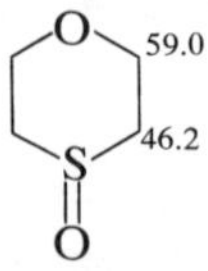   **174** 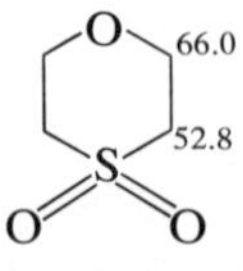   **175** 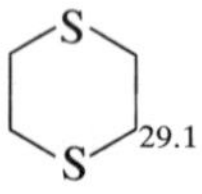   **176**

juxtaposed with the proton spectra of these compounds.[203] Low-temperature nmr[204] has shown that the sulfoxide is predominantly axial ($\Delta G° = 0.68$ kcal/mol), more so than thiane oxide ($\Delta G° = 0.17$ kcal/Mol). Accordingly, C(2,6) in the sulfoxide are shifted upfield substantially relative to the parent oxathiane. Surprisingly, this upfield shift almost disappears in the sulfone, even though the latter also has an axial oxygen. It has been suggested[203] that local anisotropy effects account for the difference between the sulfoxide and sulfone. The shift of 1,4-dithiane[138a] is shown in **176**.

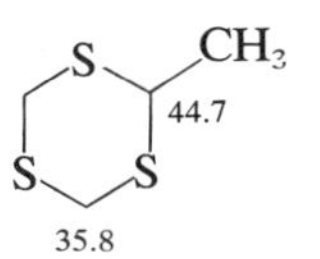

177 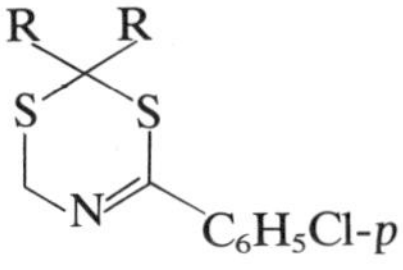   **178** 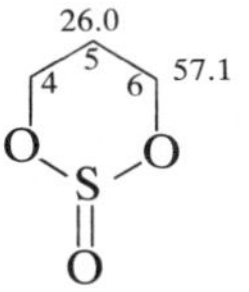   **179** 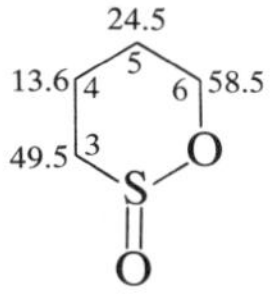   **180**

A series of methylated trithianes (e.g., **177**) has been studied by Ōki and coworkers[205]; here, as in the 1,3-dithiane system[127a] it was recognized that axial methyl groups resonate upfield of equatorial ones. The spectrum of the dithiazene **178** has been reported.[206] Both trimethylene sulfites (**179**)[207] and the structurally somewhat simpler but less accessible 1,2-oxathiane 2-oxides (**180**)[208] have been analyzed by ^{13}C nmr. In the sultines **180** each position has been substituted by methyl and, assuming that the compounds are very nearly configurationally and conformationally homogeneous with equatorial methyl and axial S=O,[208] the substitution parameters for equatorial methyl groups can be deduced; these parameters are summarized in Table 3.16.

Table 3.16
Effects of Equatorial Methyl Substitution in
180

C observed	α_e-	β_e-^{a}	γ_e-effect
C(3)	3.8	6.9	0.5
C(4)	6.5	7.2 (3)	1.2
		4.0 (5)	
C(5)	5.3	8.6 (4)	0.4
		7.5 (6)	
C(6)	6.7	5.2	0.6

^{a} Number in parentheses indicates site of origin of the effect (i.e., site of substitution).

cis-**181**   trans-**181**a   trans-**181**e

The α_e effect next to sulfur is unusually small and the β_e effects fluctuate considerably. There is the possibility that the parent **180** is not conformationally homogeneous.

In trimethylene sulfite, the oxygen on sulfur is nearly entirely axial ($\Delta G^\circ = 2.1$ kcal/mol).[209] Thus compounds in which alkyl groups are *trans* to S=O at C(4, 6) or *cis* at C(5) will be conformationally homogeneous with the alkyl group equatorial. However, when there are axial alkyl groups through *cis*- or disubstitution at C(4, 6), the molecules may be in the alternative chair or even in twist forms.* When one passes from the *cis*-5-*t*-butyl compound *cis*-**181** to *trans*-**181**, C(4, 6) shift from 59.1 to 61.5 ppm. This has been interpreted[207] in terms of a non-chair form* contributing to *trans*-**181**, but it appears to us that it is more likely that *trans*-**181** has a small contribution from the alternate chair form *trans*-**181**e, since the $-\Delta G^\circ$-value of a 5-*t*-butyl group in 1,3-dioxane (1.46 kcal/mol)[186] working against the value for S=O (2.1 kcal/mol) would lead to the prediction of about 25% *trans*-**181**e in the conformational mixture.

The coupling constants [$J(^{13}C/^{1}H)$] in trimethylene sulfites have also been examined.[210]

5.   Phosphorus Heterocycles

Although a number of different types of six-membered saturated phosphorus heterocycles have been investigated by ^{13}C nmr spectroscopy, usually in conjunction with other, more general studies, there are fewer systematic investigations of individual types of compounds than in the oxygen, nitrogen and sulfur series.

The equilibrium of *P*-methylphosphorinane (**182**) is on the side of the axial isomer at room temperature[211] because of entropy factors even though the equatorial isomer is slightly favored enthalpically. The spectrum[212] (see formula **182**) shows C(2, 6) at nearly the same shift as in cyclohexane (27 ppm), C(3, 5) shifted upfield and C(4) shifted somewhat downfield. The effect of replacing CH_2 by P should be generally upfield shifting at C_α and downfield shifting at C_β and C_γ (cf. p.198); that the

---

* For evidence against twist forms see H. Nikander, V.-M. Mukkala, T. Nurmi, and K. Pihlaja, *Org. Mag. Res.*, **8**, 375 (1976).

28.3

23.4(3)

26.7(13)

CH₃

10.9(19)

182

74.6

P 13.6(12)

CH₃

29.4

OH (7.5)

23.6(10)

183

4.4(16)

CH₃

P

74.4

23.8

OH (0)

17.8(12)

184

HO   CH₃

P

CH₃

185

HO   H

P

R

186

R = CH₃,
C₂H₅

actual effect is nil at C_α (C-2,6) and upfield shifting at C_β (C-3,5) is, no doubt, due to the β effect and γ_a effect of the P-alkyl substituent, respectively. Of note, also, are the $^{13}C/^{31}P$ couplings shown in parentheses in **182**. The α coupling of 13 Hz and the exocyclic coupling of 19 Hz are enhanced to 18 Hz and 28 Hz, respectively, in the P-t-butyl homolog of **182**; this is believed to reflect an increase in the C—P—C bond angle.[213] There is also an increase in β coupling from 3 to 7 Hz in the t-butyl compound which is ascribed[212] to greater predominance of the equatorial P—R conformation for reasons to be discussed presently. Accordingly, C(3,5) also moves downfield, to 25.0 ppm.

The epimeric 4-t-butyl-4-phosphorinanols **183** and **184**[214] differ in configuration at phosphorus and thus exemplify structures with equatorial and axial methyl groups at P. It is clear that the axial isomer (**184**) has the higher field shift at C(2,6) (lesser α_a effect) as well as at C(3,5) (greater negative γ_a effect). C(4) has nearly the same resonance in the two isomers, in accord with the assigned conformations, and the axial P-methyl in **184** resonates substantially upfield of the equatorial one in **183**. The coupling constants also differ, the ^1Js being larger in **184** than in **183**; this has been ascribed[214] to hybridization differences (cf. the situation in phospholenes, p. 189). A characteristic difference is seen in ^2J, which is considerably larger in **183** than in **184**. This difference reflects the general proposition[214] that a carbon anti-periplanar to a lone pair on phosphorus (as C-3 in **184**) couples less than one positioned syn-clinal to the pair (as in **183**). Shift and coupling information thus gathered from **183** and **184** has been applied to assigning configuration to cis- and trans-**185** and -**186**.[214]

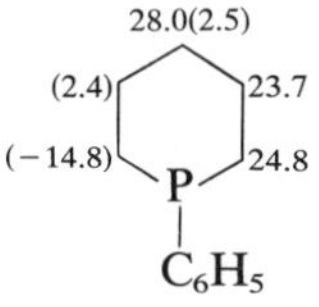

28.0(2.5)
(2.4)   23.7
(−14.8)   24.8
C$_6$H$_5$
187

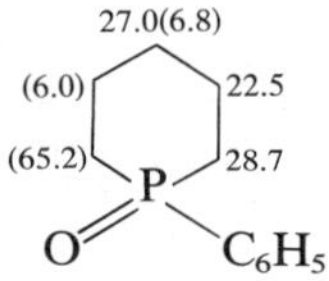

27.0(6.8)
(6.0)   22.5
(65.2)   28.7
O
C$_6$H$_5$
188

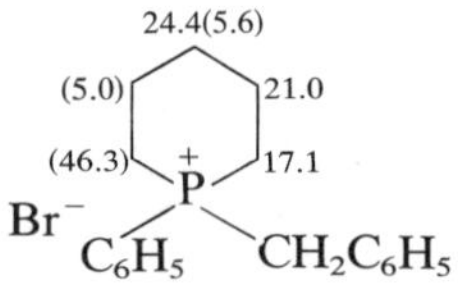

24.4(5.6)
(5.0)   21.0
(46.3)   17.1
Br$^-$ C$_6$H$_5$ CH$_2$C$_6$H$_5$
189

O
CH$_3$
190

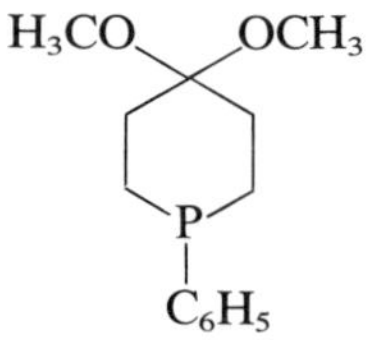

H$_3$CO   OCH$_3$
C$_6$H$_5$
191

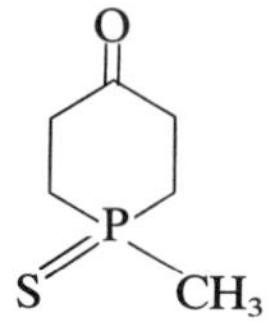

O
S   CH$_3$
192

The phenylphosphorinanes **187** and their oxides **188** and phenyl and benzyl (**189**) phosphonium salts have been investigated[33] as have methylphosphorinan-4-one[215] (**190**) and the dimethyl ketal of the corresponding P-phenyl compound (**191**).[216]

Compound **190** and its P-ethyl homolog have been converted to the corresponding P-methylphosphonium salts, phosphine oxides, and phosphine sulfides.[212,215] The trends of the shifts in the salts and oxides are similar as in **187–189**. The shielding of C(4) both in these compounds and the corresponding sulfide **192** has been ascribed[215] to an electric field effect.

The effect of phosphine sulfide formation is seen most clearly[212,217] in the conformationally homogeneous compounds **193** and **194**, which should be compared to their phosphorinane precursors **183** and **184**. Sulfide formation shifts the P-methyl as well as C(2) downfield (geminal or β effect of sulfur); as expected, the β effect is greater for equatorial sulfur in **194** than for axial sulfur in **193**. The γ effect of the axial sulfur in **193** is shielding (relative to **183**) but in **194** the equatorial sulfur produces a remarkably large (4.2 ppm) downfield shift, which has been ascribed[212] to bond angle changes at phosphorus resulting in lessened

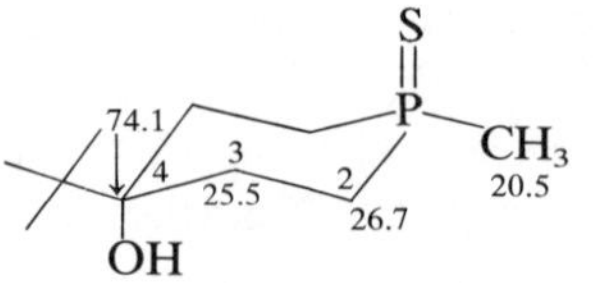

S
P   CH$_3$
74.1   20.5
4   25.5   3   2
26.7
OH
193

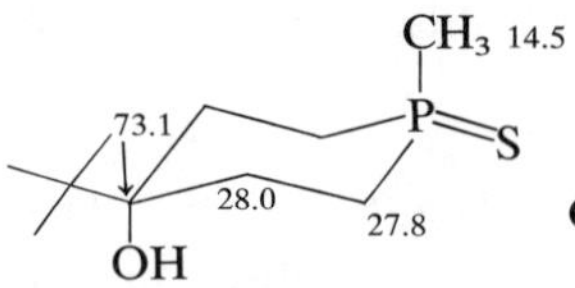

CH$_3$ 14.5
P=S
73.1
28.0   27.8
OH
194

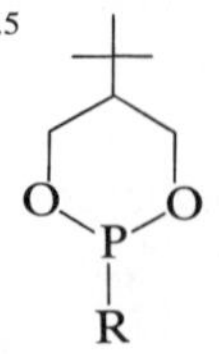

O   O
P
R

O   O
P
Cl

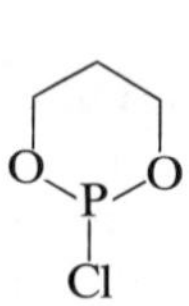

R = CH$_3$, NHMe, NMe$_2$ **196**
195

steric compression of the axial methyl group in the sulfide. Attention has also been drawn[212,217] to the fact that the shielding of C(3,5) by the axial sulfur in **193** is greater than that by the axial methyl in **194**, even though methyl has the greater steric requirement. Clearly factors other than strictly steric ones affect the γ upfield shift.

Bentrude and coworkers have studied the phosphonites **195**.[218] When R = CH$_3$, the axial configuration is preferred, but when R = NHMe or NMe$_2$ the equatorial isomer predominates. The ^{13}C nmr spectra were very helpful in configurational assignments in this series. NMe$_2$-substituted compounds as well as various chloro (**196**) and alkoxy analogs were also studied by Nifant'ev et al.[219] and a systematic study of alkoxy derivatives (cyclic phosphites) was carried out by Reisse and coworkers.[220] Of particular interest are the conformationally homogeneous compounds **197** and **198**. The usual axial "compression shift" is seen at C(4,6) in the axial isomer **197**. The $^3J(^{31}P/^{13}C)$ coupling constants for **197** and **198** are also different. The difference is small (3.2 Hz in **197** vs 1.6 Hz in **198**) for the anti-periplanar methyl substituents but much larger and opposite (4.2 Hz in **197** vs 13.5 Hz in **198**) for the syn-clinal ring carbon C(5). The difference may reside in the different orientation of the pair or the methoxyl group, but its rationale is not clear, and it may seem surprising that the anti-periplanar coupling is smaller than the syn-clinal regardless of the methoxy orientation (see also below). However, the facts are borne out in the parallel study,[219] and are useful for configurational assignments,[221] e.g., in the di-*t*-butyl-phosphino compound **199** in which the small ^3J between P and C(5) (4.1 Hz) leads to the axial (trans) assignment.[222] Differential conformational effects are also seen in the 5,5-dimethyl compound **200**[220] in which the $^4J(P/CH_3)$ is 1.2 Hz for the equatorial methyl group but zero for the axial.

197

198

199

200

cis-**201**　　　　　trans-**201**

trans-**202**　　　　　cis-**202**

Two diastereomers of the oxide analogous to **199**, namely cis- and trans- **201** have been obtained[223] and it was observed that the exocyclic phosphorus nucleus couples with C(4) and C(6) only in the (P-axial) trans isomer. This bears out the "anti-Karplus" relationship alluded to above, that is, it is the syn-clinal but not the anti-periplanar phosphorus nucleus that couples, in contrast to what is commonly observed and, in proton spectroscopy, is called the Karplus relationship.

The $^3J(^{31}P/^{13}C)$ coupling relationship in the 1,3,2-dioxaphosphorinan-2-ones (**202**)[224] resembles that in the phosphites mentioned earlier in that the cis isomer has the larger coupling to the methyl group and the smaller to C(5) relative to the trans. However, for the analogy to hold (see formulas **197**, **198**, and **202**) axial P=O must be equated with equatorial P—OCH₃; that is, the doubly bonded oxygen in **202** takes the place of the lone pair in **197** and **198** and the H(-P) in **202** the place of the OMe in **197**, **198**. The $^3J(P/C-5)$ coupling constants can be used to infer configuration and conformation in six-membered cyclic phosphates.[225] As far as chemical shifts in **202** are concerned, we note a surprisingly small difference in the C(4) and C(6) shifts as between the cis and trans isomers, even though the trans isomer has an axial oxygen at phosphorus.

A series of 1,3,2-dithiaphosphorinanes (**203**) with the substituent at C(2) Me, Ph, t-Bu, OMe, Cl, N(CH₂)₂, NH-t-bu, piperidino, N(i-Pr)₂, and N(i-Pr)(t-Bu) has been investigated.[226] Conclusions as to the preferred conformation of the substituent on phosphorus could be reached on the basis of the chemical shift of C(4,6), since axial P-substituents shield these carbons, and on the basis of the $^{31}P/^{13}C$ coupling constants for C(5). The latter, in analogy to those in 1,3,2-dioxaphosphorinanes (vide supra), are smaller (near zero) when the phosphorus substituent is axial and larger (up to 12.1 Hz) when it is equatorial. Thus it was found that methyl, phenyl, methoxy, chlorine, and aziridino were largely axial, the bulky diisopropyl- and isopropyl-t-butylamino groups largely equatorial and the remaining amino groups intermediate (i.e., with both axial and equatorial conformations populated).[226] In the case of the 4-methyl homologs (**204**)[227,228] [X = H, Cl, OC$_2$H$_5$, O = P(OCH$_3$)$_2$] both stereoisomers were obtained in some cases; chemical shift (high-field C-4 and C-6) and coupling criteria (large ^3J for C-5) suggest that both exist predominantly with the substituent on phosphorus axial.

The trithia compounds **205** have also been studied[229] and it was concluded that axial X predominates if X = Cl or CH$_3$, equatorial X for the bulky N(i-Pr)$_2$, NMe(t-Bu) and N(i-Pr)(t-Bu) substituents whereas the remainder (X = Me, Ph, t-Bu, NMe$_2$, NHt-Bu and dimethylaziridino) were conformational mixtures.

Cyclophosphamide (**206**) has attracted attention as an antitumor agent and its spectrum has been studied[230] as has that of the diastereomeric 4-methyl homologs.[231] The parent compound is a rapidly inverting conformational mixture;[230] of the two methyl compounds, the one with the bischloroethylamino group *cis* to methyl appears to be conformationally homogeneous as suggested by a large $^3J(^{31}P/^{13}C$-5) of 12.1 Hz and a small $^3J(^{31}P/^{13}CH_3)$ of 3.2 Hz. In contrast, the *trans* isomer has intermediate values for both coupling constants (7.0 Hz to C-5 and 7.6 Hz to CH$_3$) suggesting that it is conformationally heterogeneous. Evidently the syn-axial CH$_3$/O=P interaction is not very severe.

The 1,3,2-diazaphosphorinanes **207** (X = H, Cl, OC$_2$H$_5$) appear to have axial X groups, as judged from the high-field position of C(4,6) and the large $^3J(P/C$-5).[227]

6. Miscellaneous Heterocycles

The spectra of the cyclic siloxanes **208** (R = CH$_3$ and R = C$_6$H$_5$)[232] are on record. The shifts indicated in **208** are for R = CH$_3$. The cyclic silane analog is listed in Table 3.9. In formula **209** are displayed the chemical shifts and ^{119}Sn/^{13}C coupling constants for a six-membered dimethylstannane.[233] Compound **210** has also been studied[232] as has the five-membered analog.[233] Arsonium salts **211** and **212** have been investigated[234]; additional arsenic derivatives are listed in Table 3.9. It has been

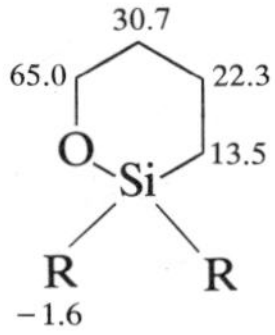

208 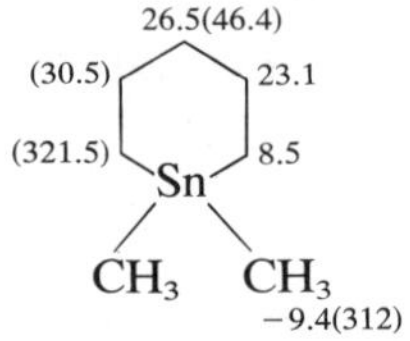   **209** 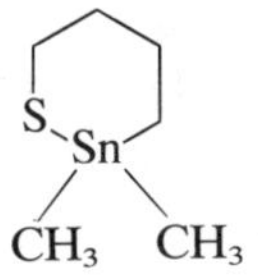   **210**   **211**

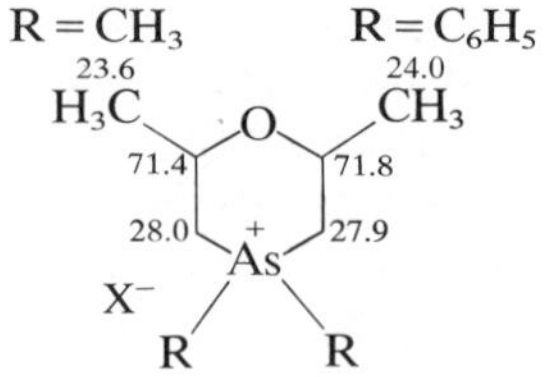

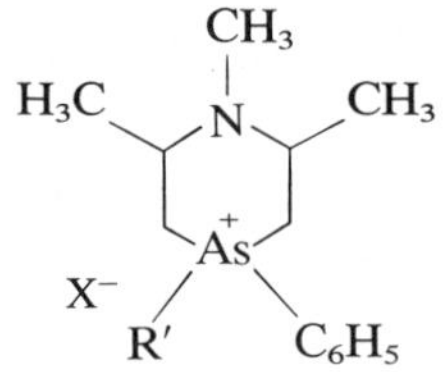

R′ = C₆H₅ or CH₂C₆H₅

212

claimed[234] that reversal of the ring is slow on the nmr time scale in **211** and in **212**, R′ = C₆H₅ on the basis of the finding that there are *two* sets of signals for the two substituents on As. However, this claim is almost certainly erroneous—the ring systems are anancomeric and therefore the diastereotopic phenyl or methyl groups on As (*cis* and *trans*, respectively, to the C-methyls) are strongly anisotropic, regardless of the barrier to ring reversal. That no doubling occurs in **212** (R′ = CH₂C₆H₅) does *not* mean that ring reversal is any faster in this compound than in **211** or **212** (R = C₆H₅), but probably indicates that a single diastereomer of the quaternary salt was obtained in this case.

In the study of 1,3-dithianes (p. 226) it was shown that there are twist contributions in some cases and that a multiple linear regression analysis indicates that the chemical shifts in the parent twist form are upfield of those in the corresponding chair. This finding has been generalized[235] through an investigation of structures of type **213** (X = NMe, NMe₂⁺, NHCH₂Ph⁺, S, S—Me⁺, SO and SO₂). These compounds are believed to exist mainly or entirely in the twist form. It was pointed out,[235] however, that high-field shifts (relative to equatorially substituted chair isomers) cannot, by themselves, taken as proof of the existence of twist forms, since axial substitution in a chair will also shift α, β, and γ carbons upfield relative to equatorial substitution. A good criterion of the contribution of twist forms is the strong shielding observed in conformational mixtures containing such forms when the temperature is raised.[235] This is due to

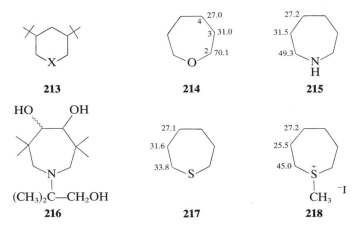

213   214   215

216   217   218

the relatively high entropy of the twist form and the resulting influence of the $T\Delta S$ term on the percent contribution of that form.

E. Seven-Membered Rings

The conformation of seven-membered rings is often ill-defined (several easily interconvertible conformations may be of nearly equal energy) and the interpretation of ^{13}C nmr spectra of such systems thus tends to be less incisive than that for corresponding six-membered rings.

A recent study of oxepane[236] is summarized in formula **214**; the 4-ketone and its ethylene ketal were also studied. Taking the shift in cycloheptane as 28.4 ppm[236] the effects of oxygen at C(2), C(3), and C(4) may be computed and compared with corresponding values in oxane (Table 3.9), indicated in parentheses, as follows: C(2), 41.7 (40.3); C(3), 2.6 (−1.1); C(4), −1.4 (−4.1) ppm. The oxygen effect is thus more deshielding at all positions in the seven-membered ring than in the six-membered one. Azacycloheptane (hexahydro-1H-azepine) **(215)** and its 4-ketone and ethylene ketal have similarly been investigated.[236] The shifts relative to cycloheptane with the corresponding comparison for piperidine in parentheses is as follows: C(2), 20.9 (19.8); C(3), 3.1 (−0.5); C(4), −1.2 (−2.2) ppm. Again the heteroatom is more deshielding in the seven-membered ring; the effect is particularly notable at C_β. A similar situation is seen in the N—Me homolog.[2] The spectra of the substituted azacycloheptanes *cis-* and *trans-***216** are on record.[237] Unpublished spectra[54] of thiacycloheptane **217** and its methiodide **218** are shown in the formulations. The shifts in **217** relative to cycloheptane differ substantially from corresponding shifts in the six-membered analog (in parentheses): C(2), 5.4 (1.6); C(3), 3.2 (0.5); C(4), −1.3 (−0.8) ppm. A similar

219   **220**   **221**

222   **223**

comparison may be made for the phosphine **219**[33]: C(2), 1.5 (−2.9); C(3), −3.2 (−4.0); C(4), −0.1 (+0.3) ppm. The sulfur compounds differ more at C_α in their differential relative to the carbocycle than the other series and the phosphorus compounds less at C_β. Formulas **220** and **221** show seven-membered phosphine oxides and phosphonium salts, respectively.

The 1,3-dioxacycloheptane system **222** is easy to synthesize and has been studied by several groups of investigators.[238,239] Gianni and coworkers[238] have recorded the ^{13}C nmr spectrum of the parent and the 2-*t*-Bu, 4-, and 5-Me, 5,5-di-Me, *cis*- and *trans*-4,7-di-Me, *cis*- and *trans*-2-*t*-Bu-4-Me, *cis*- and *trans*-2-*t*-Bu-5-Me and r-2-*t*-Bu-*cis,trans*-4,7-di-Me and the corresponding *cis,cis* epimer. The similarity to analogous 1,3-dioxane spectra was noted (compare **222** to **117**, p. 209; the principal difference is at C(4) which is 3.3 ppm more downfield in the seven-membered than in the six-membered one.) The nmr spectra proved useful in assigning configuration in the *cis/trans*-4,7-di-Me pair making reasonable conformational assumptions but, in general, left the conformation of the 1,3-dioxacycloheptanes open. In another, rather extensive study of the same system[239] the ^{13}C nmr spectra of *cis*- and *trans*-2-methyl-5-methoxy-1,3-dioxacycloheptane were recorded; the interpretation is in terms of a "pseudochair" conformation but in some cases a second conformation was manifested by low-temperature nmr. The spectra of 1,3-dioxo-5-cycloheptane (**223**) and its 2-*t*-Bu, 2,2-di-Me, *cis*- and *trans*-4,7-di-Me and r-2-Bu-*cis*, *cis*- and -*cis,trans*-4,7-di-Me homologs have been recorded[240]; that of the parent system was interpreted in terms of a twist–boat conformation, but this is not necessarily the same for the derivatives; for example, the r-2-*t*-Bu-*cis*-4,*cis*-7-di-Me compound appears to be a chair.[240]

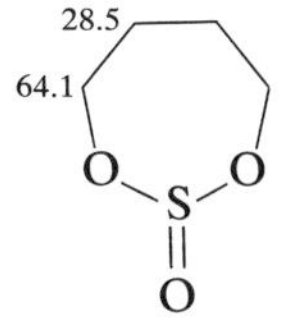

224 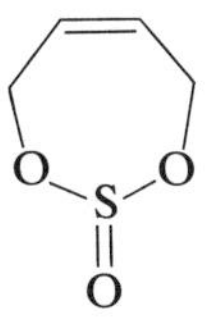   **225** 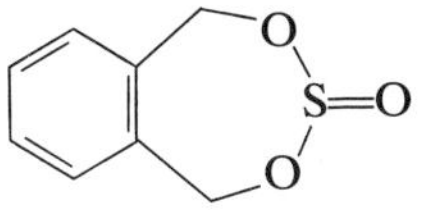   **226** 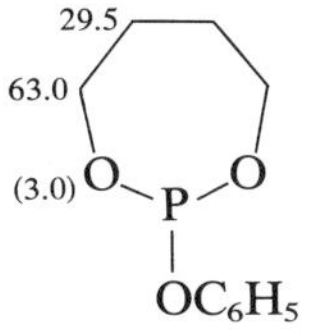   **227**

The cyclic sulfite of 1,4-butanediol (**224**) and the related unsaturated compounds **225** and **226** have been studied[241]; **224** appears to be a chair, **225** a twist and **226** a mixture of chair and twist conformations.[241] The corresponding phosphites (P—OC_6H_5 instead of S=O) have also been studied[242]; the spectrum for the saturated species (**227**) is shown. Large changes of coupling constants with temperature suggest that these compounds are conformationally heterogeneous.[242]

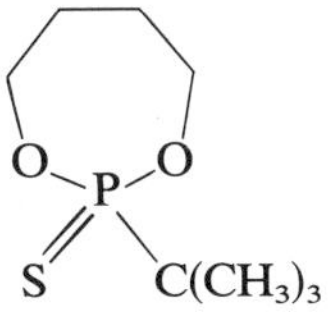

228 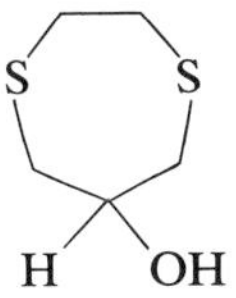   **229**   **231**

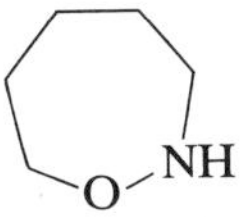

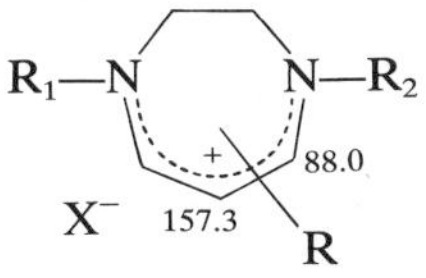

(shifts for R = R_1 = R_2 = H)

Data are also available for the thiaphosphonate **228**,[243] for 6-hydroxy-1,4-dithiacycloheptane (**229**)[244] and for substituted 1,2-oxazepanes (**230**).[236] The 2,3-dihydro-1,4-diazepinium salts **231** have been studied by two groups of investigators[245,246] one of which[245] also looked at a number of methyl and phenyl (and a few other) homologs. As shown in the formulation, there is a remarkable shift alternation in the conjugated system. The saturated part appears to be a rapidly inverting half-chair[245]; methyl substituents in this part assume pseudoaxial and pseudoequatorial conformations about equally easily.[245]

F. Eight-Membered Rings

The ^{13}C nmr spectra of eight-membered rings have been investigated almost exclusively as a tool in conformational study.[247] The simplest compounds to have been studied are the phosphine **232** and the corresponding phosphine oxide and phenyl and benzyl phosphonium salts.[33] The shifts might be compared to that in cyclooctane (27.8 ppm^2), which is virtually the same as the cyclohexane shift; the shifts in the eight- and six-membered phosphorus analogs differ somewhat.[33] 5-Oxocanone[248] (**233**) is a boat-chair as are 1,3-dioxocane and its 2,2- and 6,6-dimethyl

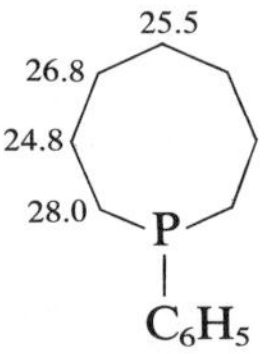

232 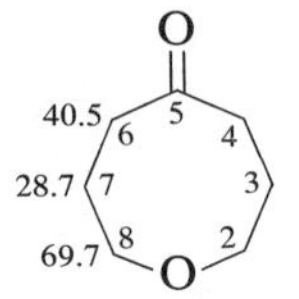   **233** 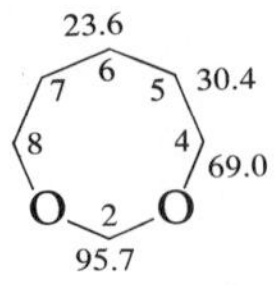   **234** 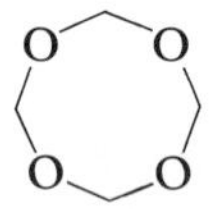   **235**

and 2,2,6,6-tetramethyl derivatives.[249] The shifts shown in **234** were obtained at −85°C; for the conformational analysis the spectrum was also recorded at −165°C.

The tetroxacane **235** and its 2,6-bistrichloromethyl homolog have been studied[250]; **235** is predominantly in the crown form with some boat-chair whereas the bis-CCl$_3$ derivative is a pseudorotating boat-chair. An attempt to study transannular interaction in 5-heterocyclooctanones (**236**, X = S, O, **237**)[251] was partially successful; a clear-cut result was obtained only with **237** where the carbonyl resonance was reduced from 184.7 ppm (position in cyclooctanone) to 171.9 ppm in solvent cyclohexane and, much more dramatically, from 190.5 ppm to 102.0 ppm in a chloroform-cyclohexane mixture (9:1). Hydrogen bonding presumably favors the transannularly bridged tautomer with its negatively charged oxygen. Transannular bonding (N → P) has also been found in the phosphatranes **238** when X = H or CPh$_3$[252]; this is suggested *inter alia* by a large ^{31}P/^{13}C coupling constant (approximately 13 Hz) at the carbons marked in **238b**. No transannular interactions have been seen by ^{13}C nmr in six-[138a] or seven-membered[236] heteracyclanones. Low temperature ^{13}C nmr data on

236 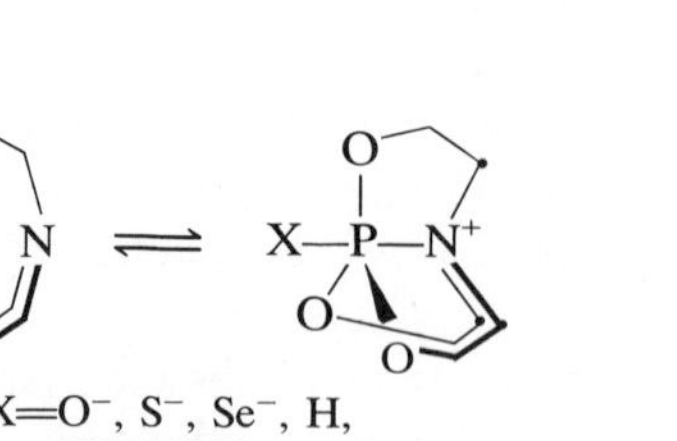   **237**

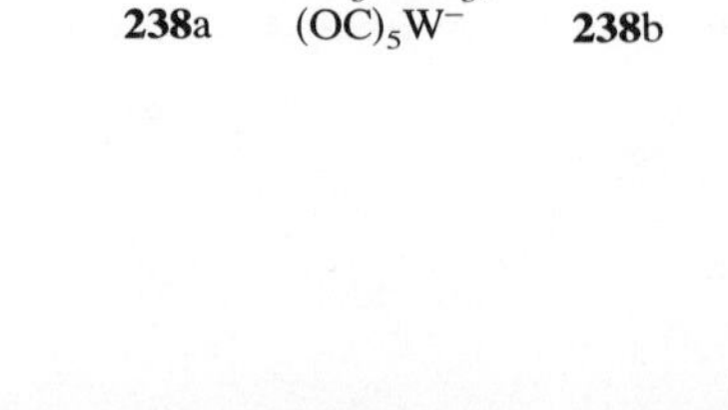

X = O⁻, S⁻, Se⁻, H,
CPh$_3$, BH$_3^-$,
238a   (OC)$_5$W⁻   **238b**

the eight-membered azo compound **239** are on record.[253a] The carbon spectra of azacyclooctane (azocane) and its N-methyl and N-chloro derivatives are temperature invariant.[253b]

G. Rings Larger than Eight-Membered

The studies of medium-sized and large rings containing heteroatoms by ^{13}C nmr are widely dispersed. One subject which has attracted interest is the effect on metal complexing on the conformation of crown ethers.[254–256] Dibenzo-18-crown-6,[254,255a] 2,3-naphtho-18-crown-6,[255b] dicyclohexyl-18-crown-6,[254] nonactin,[254] valinomycin,[254] benzo-18-crown-6,[255] and dibenzo-30-crown-10 were studied, both through determination of chemical shifts in the uncomplexed and complexed state[254,255] and through determination of T_1 relaxation times.[254b,255] The conclusion was reached that conformational changes take place on complexing with the metal. 1,4,5,11-Tetrathiacyclotetradecane has been investigated both in the free state[244] and as the nickel complex[257]; there is evidence for a dynamic conformational equilibrium in the complex.[257] Higher polythia rings with up to eight sulfur atoms and up to 48 ring members have also been studied.[244]

The problem of *cis-trans* isomerism in lactams (**240**) has been studied by ^{13}C nmr spectroscopy.[258,259] A survey with R = H and n = 5–9 and 13[258] confirmed earlier findings that the nine-membered lactam is the smallest one to have the *trans* isomer present at equilibrium. Amide rotational isomerism in large-ring exocyclic amides (**241, 242**) has also been investigated.[260]

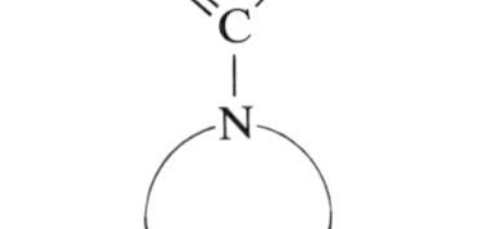

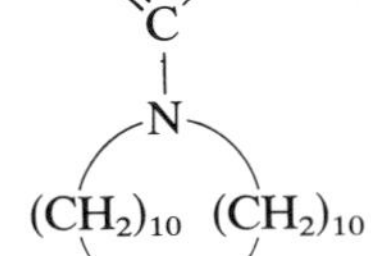

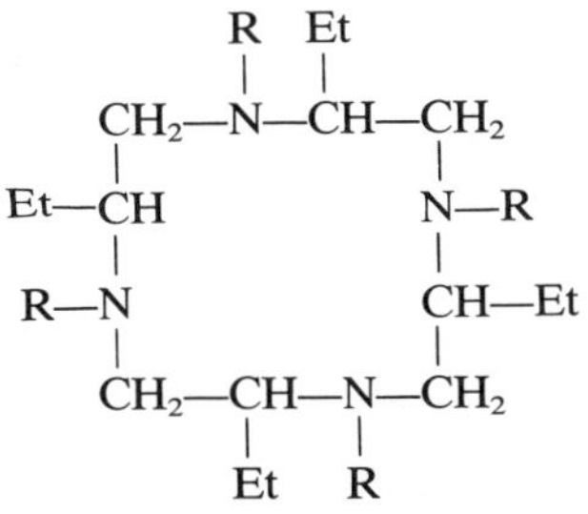

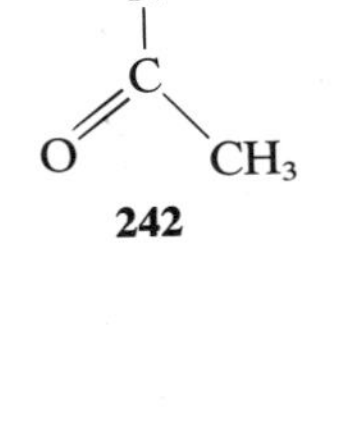

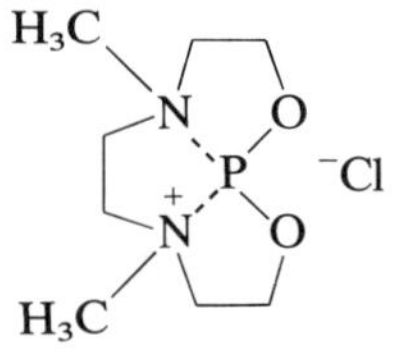

The stereoisomerism of the 1,4,7,10-tetraazadodecane **243** derived from tetramerization of 2-ethylazirine has been determined by ^{13}C nmr spectroscopy on the basis of symmetry principles.[261]

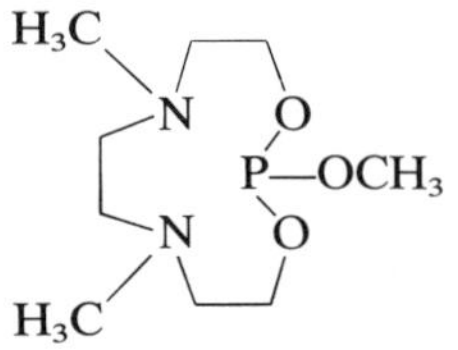

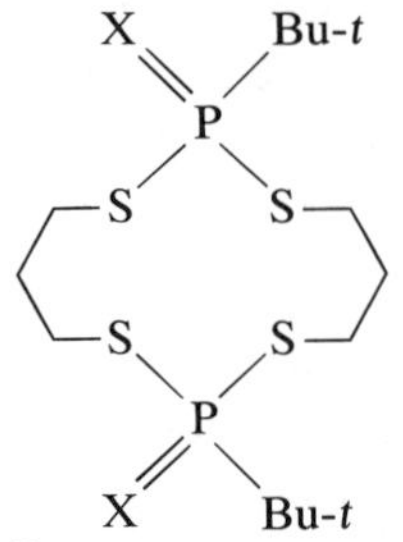

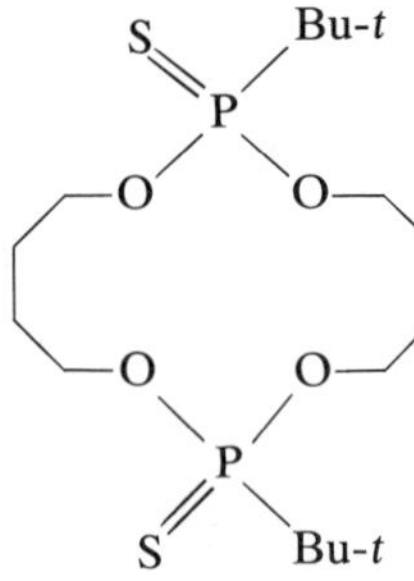

The large-ring phosphorus heterocycles **244**,[262] **245**,[262] **246**,[263] and **247**[243] have been studied.

III. POLYCYCLIC SYSTEMS

A. General

The literature on ^{13}C nmr spectra of polycyclic systems: spiro, fused, and bridged is both quite extensive and quite diffuse. Moreover most alkaloids and other natural products (other than monosaccharides), which are excluded from this review, fall into this category and the borderline between naturally occurring materials and artifacts is not always sharp. Finally, unlike in the monocyclic series where many spectra were studied for their own sake, in the polycyclic compounds the recording of spectra is often incidental to other work. For these reasons and because of space limitations, we have kept this section brief. In many cases only references to pertinent work are given, in others, formulas are shown with references but with no substantive discussion. Only the more interesting systems (in our subjective view) are discussed in any detail.

B. Fused Ring Systems

1. Annelated Three- and Four-Membered Rings

The spectrum of 2-methyl-3-phenyl-1-azabicyclobutane[53] is shown in formula **248**. The spectra of cyclohexene oxide (**249**, X=O), cyclohexene episulfide (**249**, X = S) and their *cis*- and *trans*-3-methyl homologs have been recorded[264]; the epimeric methyl compounds differ considerably at C(5) which is about 3 ppm upfield in the *trans* isomer. ^{13}C nmr analysis at

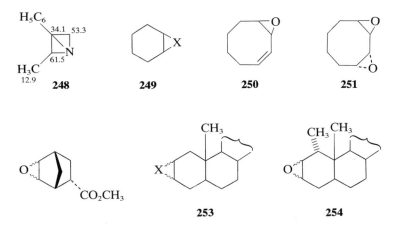

low temperatures[265] has disclosed the existence of two conformations in cycloheptene oxide. The earlier cited[20] extensive study of epoxide spectra includes epoxides of cyclopentene, cyclohexene, cycloheptene, cyclooctene, and other cyclic systems. The mono- and bis-epoxides of 1,3-cyclooctadiene (**250, 251**) have been subjected to conformational analysis,[266] as have the epoxy-5-cyclooctenes, both E and Z at the double bond.[20] Epoxides derived from norbornanes,[20,267,268] for example the steroisomers of **252**, and of more complex polycyclic structures[20,269] have been studied. In the stereoidal 2,3-epoxides and -episulfides **253** the earlier mentioned (Scheme 3.11) delta-compression shift (deshielding) is observed,[270] even though the A-ring is a half-chair (in which syn-axial interactions are less than in a true chair). The shift of the 19-methyl group goes from 12.9 ppm in the α epoxide to 14.0 ppm in the β and from 12.6 ppm in the α episulfide to 14.9 ppm in the β. Methyl substituents at C(1α) in 2,3-epoxysteroids (**254**) may cause shielding, instead of the usual deshielding, α_a-effects[271] (see also Table 3.10 for case of upfield shifting α_a effects). The same group of investigators who studied monocyclic aziridines (p. 174) also made an extensive study of substituted 7-azabicyclo[4.1.0]heptanes (**255**).[272] Sulfones (**256**)[273] and phosphine oxides (**257**)[59a,274] annelated with cyclopropane rings as well as epoxyphospholanes (**258**) and their oxides[275] have had their ^{13}C nmr spectra

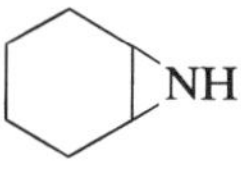

255

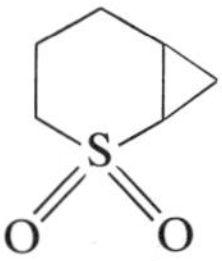

256

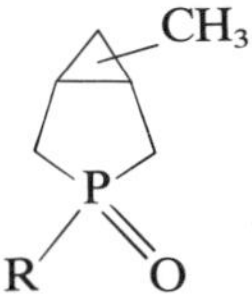

R = Ph or Me
257

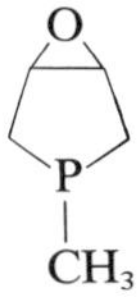

258

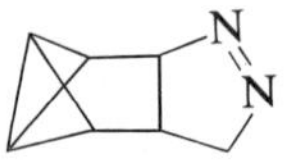

CF$_3$

259

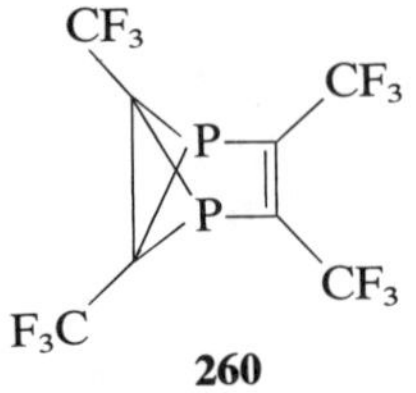

260

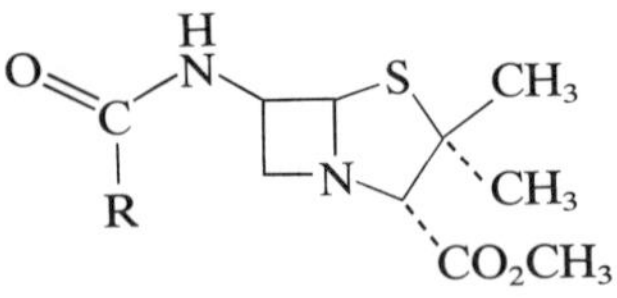

261

reported, as have the aza compound **259**,[276] and the trifluoromethyl-triphospha compound **260**.

Penicillins (**261**) and their sulfoxides[278] have been studied as such as well as complexed to lanthanides.[279] Pseudocontact shifts with Eu(thd)$_3$ have also been determined in **262**.[280]

2. Bicyclo[3.3.0]octanes

1-Aza-[3.3.0]bicyclooctane (pyrrolizidine, **263**, R = R' = H) has been investigated;[281] the shift of C(8) in the 3-t-butyl and cis-3,5-dimethyl derivatives (**263**, R = t-Bu, R' = H or R = R' = Me) is downfield by 7–8 ppm of that in corresponding homologs and this has been ascribed[281] to these two derivatives being in the form in which the rings are $trans$-fused,* whereas generally the ring fusion is cis.[281] The downfield shift may be a consequence of the angle distortion, which probably occurs in the strained $trans$-fused 5-5 ring system. Other examples of hetero analogs of the 5-5 systems are the lignans **264**,[282] the disulfonium salt **265**[283] and the disulfone **266**.[284]

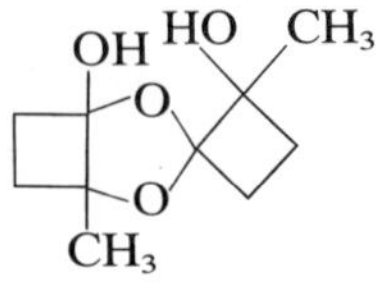

262

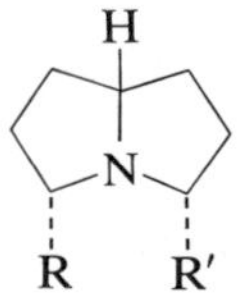

263

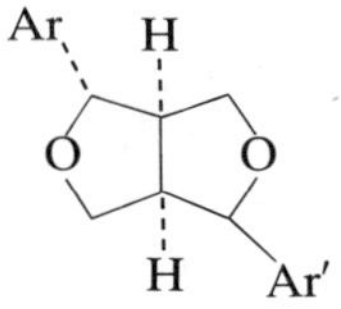

264

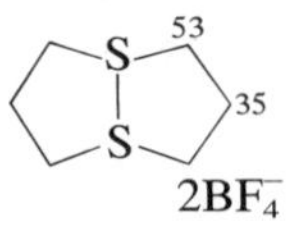

265

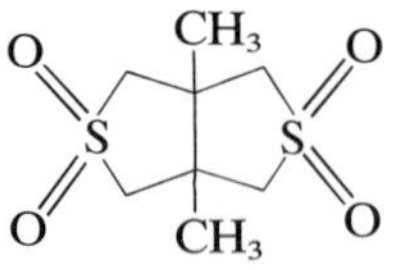

266

---

* cis Fusion with two endo-methyl groups or one endo-t-butyl group would give rise to excessive strain.

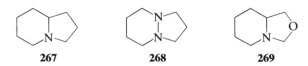

267   268   269

3. Bicyclo[4.3.0]nonanes

The spectra of indolizidine (**267**),[8] and its diaza (**268**)[182b] and oxaza (**269**) analogs[285] are on record. The latter is predominantly *trans*-fused, according to comparison of chemical shifts and $^{13}C/^1H$ coupling constants with model compounds.[285] Other compounds investigated by ^{13}C nmr are the annelated lactones and anhydrides **270**[286] and annelated phospholene oxides **271**[287] and **272**[87] and corresponding phospholine sulfides (S in lieu of O in **271**). In **270** and **272** ^{13}C nmr serves for configurational assignment. Other studies in this series[288-290] deal with rather highly aromatic systems; we mention here only compounds of type **273** and **274** (X = O, NH, S, SO$_2$ as well as C=O and CF$_2$)[290] in which γ-anti- and γ-gauche-effects on aromatic carbon atoms were studied.

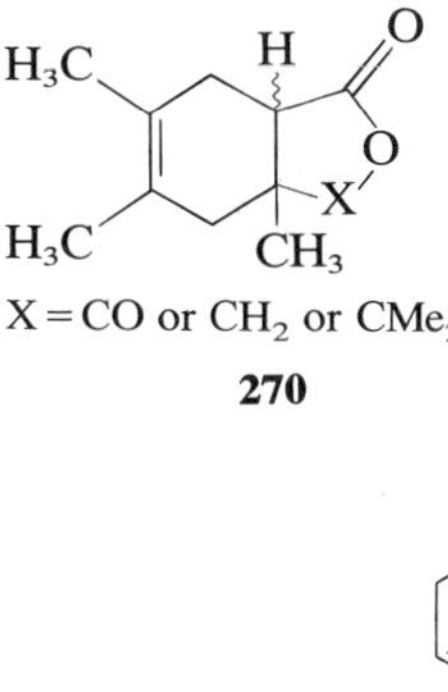

X = CO or CH$_2$ or CMe$_2$

270   271 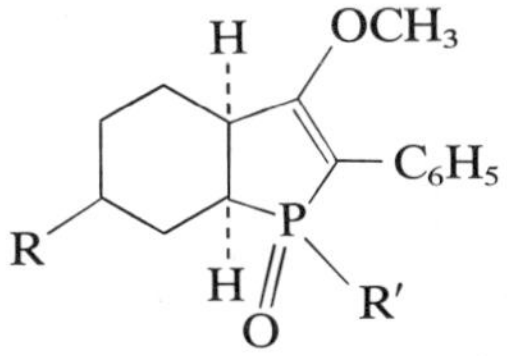   272

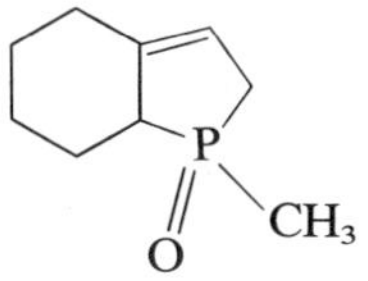

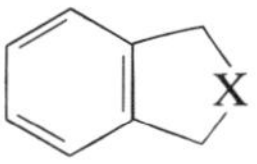

273 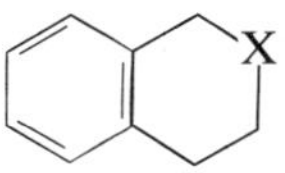   274

4. Bicyclo[4.4.0]decanes

This system is probably the most interesting of the fused ring systems. It comprises the hetero-analogs of decalin, including the conformationally rigid 1- and 2-hetero-*trans*-decalins as well as conformationally mobile species in which the heteroatom occupies a ring fusion position, notably the perhydroquinolizidines.

Among the fused oxanes, system **275** and the $\Delta^{2,3}$ unsaturated analogs (flavanols, chroman-4-ols) have been studied as to their ^{13}C nmr spectra.[140] Several groups have investigated the spectra of decahydroquinolines.[119,126,129,147,154,291,292] *trans*-Decahydroquinoline (**276**, R = H)

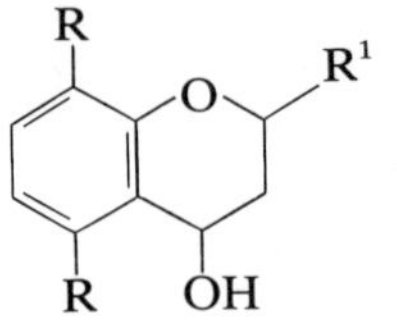

275

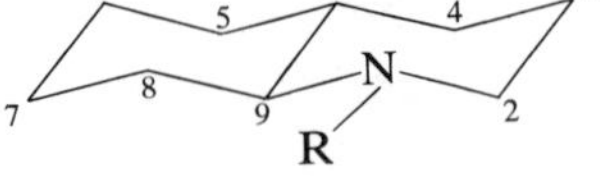

276

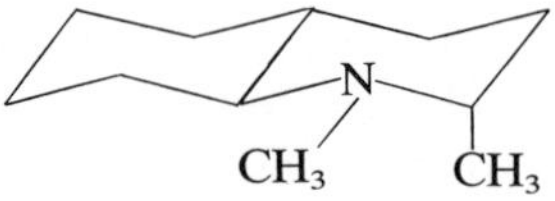

277

is conformationally rigid as is its *N*-methyl homolog (**276**, R = CH$_3$). However, in the latter the *N*-methyl group alternates (rapidly on the nmr time scale) between equatorial and axial positions (Scheme 3.21). Comparison of the ^{13}C signal of the (mobile) *N*-methyl group in *N*-methyl-*trans*-decahydroquinoline with those of the (nonmobile) equatorial and axial *N*-methyl groups in the methyl homologs shown in Scheme 3.21 has led to the conclusion[147] that equatorial *N*-methyl is preferred by 1.8 kcal/mol in chloroform and 2.3 kcal/mol in benzene.* However, there is some question regarding the accuracy of this value, since it was computed directly from the *N*-methyl shifts indicated in Scheme 3.21 and

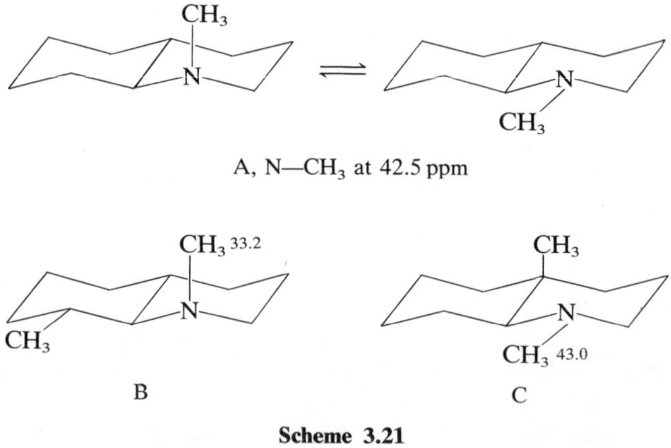

A, N—CH$_3$ at 42.5 ppm

B

C

Scheme 3.21

---

* We had earlier (p. 213) indicated that the corresponding value in *N*-methylpiperidine is larger, of the order of 3.0 kcal/mol. However, since *N*-methyl-*trans*-decahydroquinoline corresponds to an *N*-methylpiperidine with an equatorial substituent at C(2) and since it is known[149] that such a substituent will tend to shift the N—Me equilibrium slightly toward the axial side, the two results may be not incompatible.

does not take into account possible δ-effects of the ring-methyl groups on the N-methyl group in the model compounds B and C. Such δ-effects would slightly falsify the reference shifts of purely axial and purely equatorial N—Me in the two isomers of A, which is of particular concern in the case of the equatorial N—Me shift, where the difference between the mobile system A and the rigid reference compound C is quite small.

The chemical shifts in *trans*-decahydroquinolines (**276** and its ring-methyl homologs)[119,154] have already been discussed in connection with piperidine, since **276** provides a good model for a rigid nitrogen-containing six-membered ring. It was noted[119] that methyl-substitution parameters in **276** are very similar to those in cyclohexane. The earlier discussed (cf. Scheme 3.12) shielding γ-anti-effect of the nitrogen atom is clearly in evidence. An equatorial N-methyl group in **276** also has a sizeable (-1.5 ppm) γ_e shielding effect at C(3) and C(10). The effect disappears upon conversion of the free amines to their hydrochlorides (whereas the γ-anti-effect of the nitrogen is actually enhanced upon protonation[43]). A large upfield shift in the axial C(2)-methyl resonance in **277** has already been discussed; this shift is more upfield than in the corresponding carbocycle or in the corresponding amine hydrochloride and thus seems to be caused in part by the *anti*-periplanar pair[119] (cf. discussion on p. 206). The deshielding effect of an axial methyl at C(2) on axial N—Me in either the free amines or their hydrochlorides has also been mentioned earlier (Scheme 3.20). Finally system **276** has provided a good vehicle for studying[119] protonation shifts in conformationally rigid systems; it is of note that these shifts are strongly solvent dependent, so much so that the α carbon in a given amine may be shielded upon protonation in chloroform but deshielded in trifluoroacetic acid.*

The ^{13}C nmr spectra of *cis*-decahydroquinoline (**278**), its N-methyl and various ring-methylated homologs have been studied in detail both at room temperature or somewhat above where the conformational equilibrium is fast on the nmr time scale[129,291] and at low temperature where it is slow so that both conformations can be seen in the spectrum and their equilibrium studied.[129,292] The results are of interest mainly in the context of the conformational analysis of **278**.

Coupling constants $^1J(^{13}C/^1H)$ in decahydroquinolines can be used as indicators of conformation.[293]

---

* As a result, the protonation shifts are not the same when, on the one hand, an amine hydrochloride is compared with the free amine in chloroform-*d* and, on the other, the effect of addition of trifluoroacetic acid to an amine is examined. The difference is largely due to the solvent rather than the gegenion; the latter has only a minor (although sometimes palpable[127b]) effect when the solvent is chloroform-*d* throughout.[119]

278   **279**   **280**

Spectra of $\Delta^{1,9}$-octahydroquinolines have been recorded.[129] There appears to be no spectral information (^{13}C nmr) on the decahydroisoquinolines but tetrahydroisoquinolines (**279**)[290,294] and dihydroisoquinoline (**280**),[294] its salt, N-oxide and N-oxide salt[294b] have been examined both as to chemical shifts and $^1J(^{15}N/^{13}C)$. It was noted that the coupling constant of ^{15}N to C(1) increases greatly when the amine is protonated or converted to its N-oxide or N-oxide salt.[294b] Hexa- and octahydrophenanthroline (**281, 282**) spectra are known.[295]

cis, trans
281

cis, trans; R = H or Me
282

283

Quinolizidine (**283**) has been investigated by several groups.[155,296–299] The spectra of the parent compound and eight of its ring-methyl homologs were interpreted either by analogy with corresponding trans-decalins using a nitrogen-substitution parameter for the α, β, and γ positions or by the Grant-Dalling procedure[34] using methyl-substitution parameters in conjunction with the spectrum of the parent compound. It is of interest that, although the quinolizidine system can readily alternate between cis and trans ring fusion by mere nitrogen inversion, all the compounds investigated, including those bearing axial methyl groups, were exclusively (or nearly exclusively) in the trans configuration. This may be seen very clearly in the comparison of the shift of the carbons indicated in **284** and **285**[298]; if there were any contribution from the alternative conformation of **285** indicated in brackets, the carbon in question should be strongly shifted upfield by the syn-axial ring juncture on nitrogen. $^1J(^{13}C/^1H)$ in quinolizidines have been studied[298] as have protonation shifts[155,298]; it is of interest that the axial 2-methyl group in **285**, which is at an unusually high field, is the only carbon beta to nitrogen, which is shifted downfield upon protonation; all other β carbons experience the usual shielding effect. Here, again, one sees the earlier-mentioned anti-periplanar upfield shift of the lone pair on nitrogen on C_β, which disappears upon protonation.

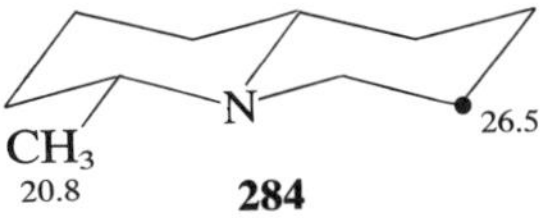

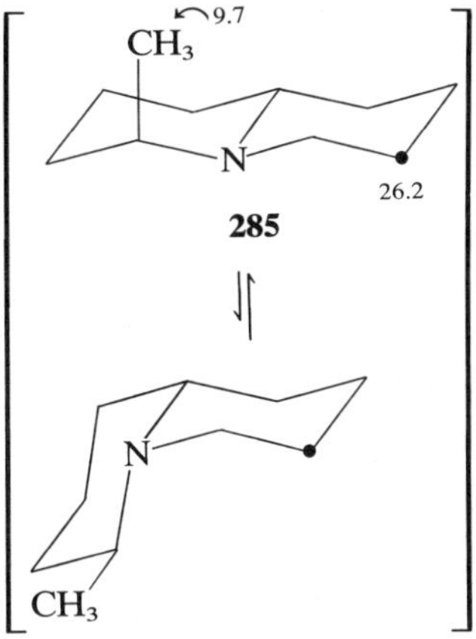

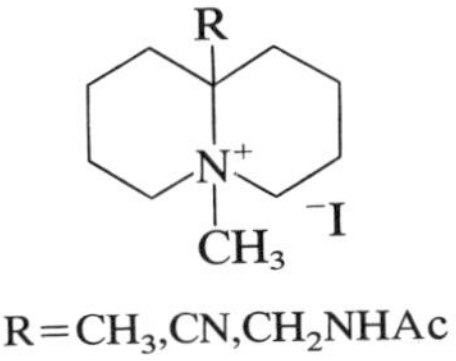

As in other systems $^1J(^{13}C/^1H)$ is affected by the orientation of the proton under consideration relative to the lone pair on nitrogen: An anti-periplanar lone pair leads to a decrease in ^1J, e.g., from 140–141 Hz in cis-quinolizidine systems to 123–130 in trans-quinolizidines.[293,298]

The methiodides of 283[298,299] and some angular substitution products (286)[297] have been investigated spectrally; since quaternization of the nitrogen atom stops inversion, cis and trans isomers can be isolated in the methiodide series and the ^{13}C spectra of both types have been recorded. The ring inversion (enantiomerization) of the cis methiodide (287) has been studied by low-temperature ^{13}C nmr; the barrier to inversion is 13.7 kcal/mol.[297]

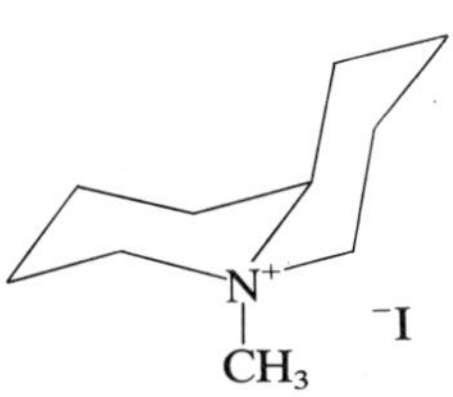

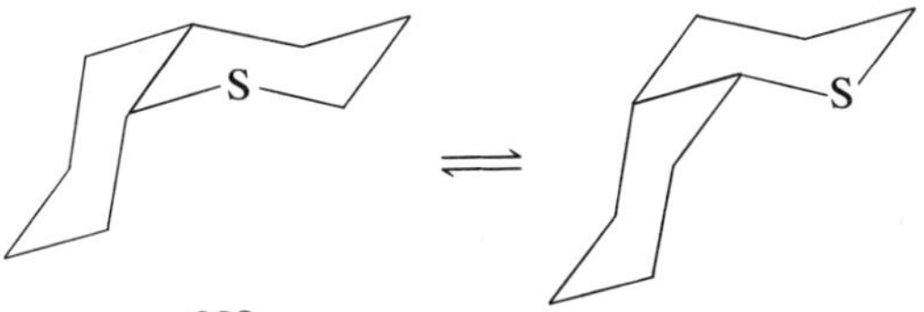

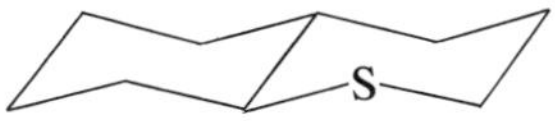

The trans-quinolizidine system is found in the Nuphar alkaloids.[75] It is also found in such alkaloids as yohimbine and reserpine and the ^{13}C nmr spectra of benzo- and indolo-fused quinolizines have been studied in this context.[293]

The ^{13}C nmr spectra of the *cis* (**288**) and *trans* (**289**) isomers of 1-thiadecalin and a number of their methyl substitution products have been studied in detail.[300] Assignments were made on the basis of the known spectra of *cis*- and *trans*-decalin[301] using additive increments for the replacement of C(1) by sulfur. These increments are greatest (2.4–6.4 ppm) in the α positions (C-2, C-9), but are palpable at other positions as well; it was noted[300] that they vary appreciably from one compound to another, perhaps because varying geometric changes (in going from decalins to thiadecalins) are superimposed upon electronic factors. The shift increments for methyl substitution, although similar to those in thianes (Table 3.10), also vary appreciably from compound to compound. Conformational equilibria in the *cis*-1-thiadecalins (e.g., **288**) were assessed from spectra recorded at low temperature (−70°C).

The spectra of the *cis*- and *trans*-9-thiadecalinium salts[302] are recorded in formulations **290** and **291**, respectively. Inversion at sulfur in this system is slow, the barrier being $\Delta H^{\ddagger} = 28$ kcal/mol. The barrier is reduced to 20.5 kcal/mol in the corresponding ylide.[302] The spectrum of the dithianium salt **292** has been mentioned.[283]

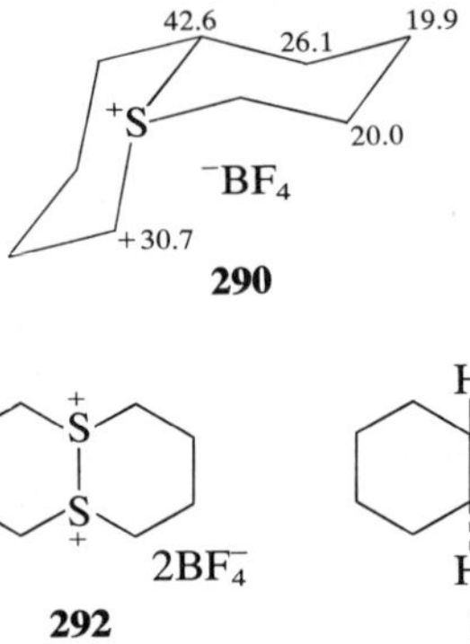

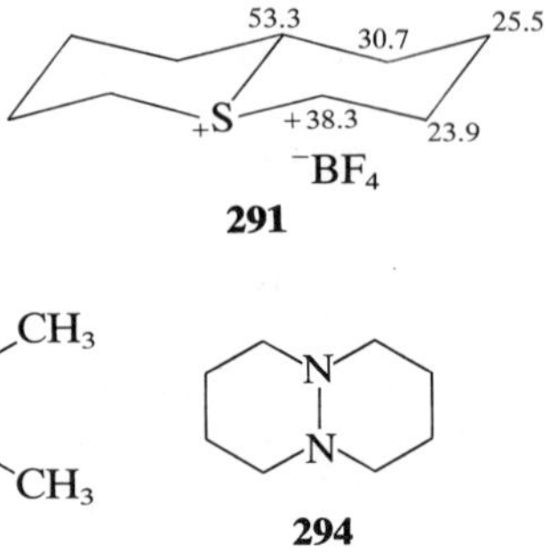

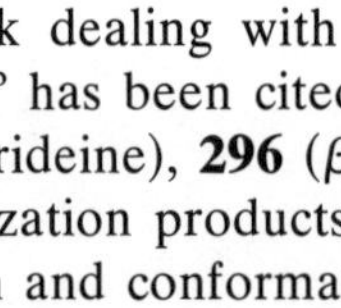

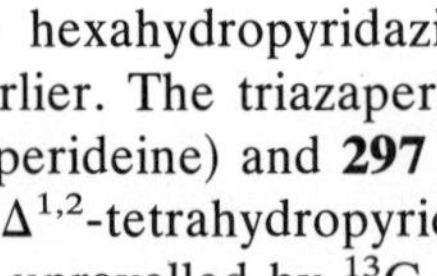

Work dealing with the hexahydropyridazine derivatives **293**[182a] and **294**[182b] has been cited earlier. The triazaperhydrotriphenylenes **295** (α-tripiperideine), **296** (β-triperideine) and **297** (isotripiperideine), all three trimerization products of $\Delta^{1,2}$-tetrahydropyridine, have had their configuration and conformation unravelled by ^{13}C nmr.[303] The alpha and beta isomers are readily identified in that the former, for reasons of symmetry (C₃) displays only five signals at room temperature whereas the latter shows fifteen. At −67.5°, however, the spectrum of the alpha isomer also

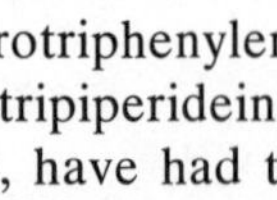

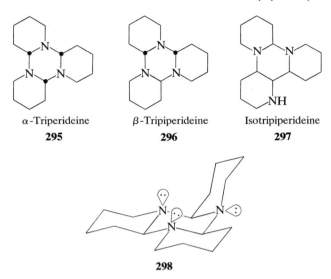

α-Triperideine
295

β-Tripiperideine
296

Isotripiperideine
297

298

decoalesces into a total of 15 resonances, presumably because ring inversion of the quinolizidine systems is slowed* and the most stable conformation is one with one axial N-alkyl group (**298**) and devoid of the C_3 axis. (This is to be expected by analogy with N,N',N''-trimethyl-1,3,5-triazane (**152**) in which one methyl group is axial[187] so as to avoid the triple "rabbit-ear" interaction of the lone pairs on the three nitrogen atoms; see the discussion on p. 220). There are, of course, three equivalent conformations of this type and their rapid interconversion by nitrogen inversion leads to the degenerate five-line spectrum at room temperature.

The spectra of the bicyclic phosphites **299** and **300** display the same systematic differences in chemical shifts and $^3J(^{31}P/^{13}C)$ between stereoisomers that had been earlier discussed for their monocyclic analogs

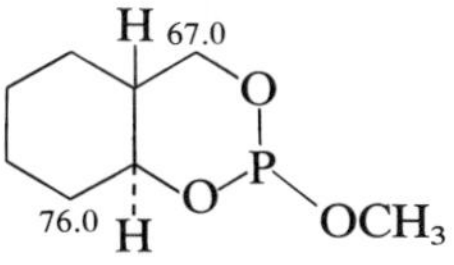

H 67.0

O
P
O   OCH₃
76.0 H

299

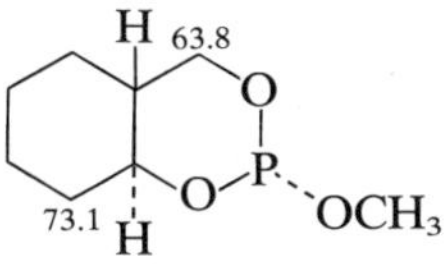

H 63.8

O
P
O   OCH₃
73.1 H

300

O

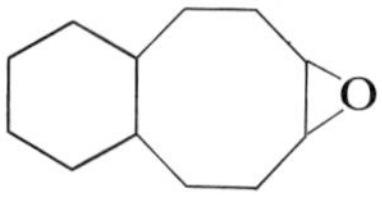

301

---

* It had been previously noted that nitrogen inversion, while normally fast on the nmr time scale even at −100°, is slowed greatly when the N-alkyl moiety is flanked by equatorial alkyl groups in the adjacent (alpha) positions.[148] Presumably there is considerable steric hindrance to the N-alkyl group passing through the conformation in which it is eclipsed with both adjacent alkyl groups (which it must do for nitrogen inversion to occur).

(p. 231).[220] It had earlier been implied that replacement of a CH_2 group in a carbocyclic compound by a heteroatom X produces a shift increment that can be used in the assignment of resonances. Indeed such a procedure has been used for the decahydroquinolines[119,129] and 1-thiadecalins[300] whose signals were assigned on the basis of those of corresponding decalins and methyldecalins. Other investigators,[296,303] however, have used a two-parameter equation: $\delta_{het} = A\delta_{carb} + B$ where δ_{het} is the chemical shift of a given carbon atom in the heterocycle, δ_{carb} is the chemical shift in the analogous carbocycle and A and B are empirical constants.* However, if experimental shifts for the carbocycle are used for reference, the value of A in the cases studied[296,303] is between 0.97 and 1.03 (i.e., close to unity), and, since one is dealing with reference shifts of 20–50 ppm, the difference incurred in setting $A = 1$ (which amounts to using the simple additive relationship referred to above) is less than 1.5 ppm, in most cases less than 1 ppm. Since the standard deviations in such calculations range from 0.3 to 0.8 ppm, and since shifts which are closer together than 1 ppm cannot be safely assigned by parametric methods anyway, the advantage of the two-parameter equation over the single-parameter one is probably ephemeral.[303]

5. Higher Annelated Fused Ring Systems

Very little seems to be known about heteroatom containing fused ring systems with two rings larger than six-membered. Various stereoisomers and an unsaturated analog of **301** have been studied as to their 13C nmr spectra; these compounds are closely related to epoxides listed in Section III.B.1.

B. Bridged Ring Systems

1. Bicyclo[2.2.1]heptanes (Heteranorbornanes)

13C nmr spectra of a number of substituted norbornanes have been investigated[2] and found useful in the assignment of configuration. This usefulness extends to heteranorbornanes and their unsaturated analogs, such as 1-methyl-7-oxanorbornenes (**302**).[304] A semitheoretical study of 2,3-dimethylenenorbornenes has been extended to the 7-oxa analogs (e.g., **303**).[305] Orbital interactions have also been studied[306] in 7-azanorbornene (**304**), its N-methyl homolog, salts, and lanthanide complexes. Nickel contact shifts of the 1-azanorbornane derivative, **305**, have been reported.[164b] Low-temperature 13C nmr study of N,N'-dimethyl-2,3-diazanorbornane (**306**) indicates a slow inversion (on the nmr time scale) interconverting the two enantiomeric trans isomers; there is no

---

* The two-parameter equation was first reported in Ref. 40 for cyclohexanols; here A for C_α is 0.83 and cannot be set equal to unity.

302

303

304

305

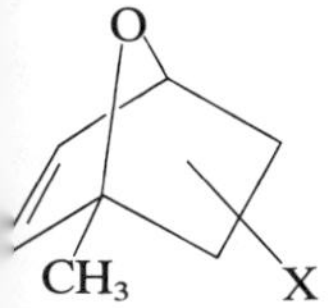

indication for the presence of either of the *cis* isomers (which would have eclipsed lone pairs) at equilibrium.[307]

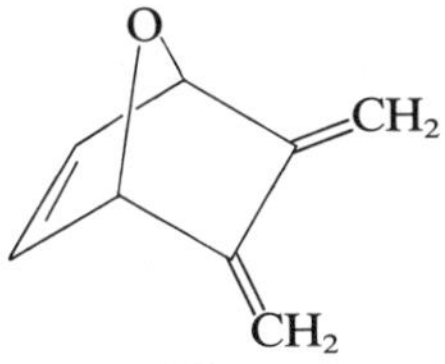

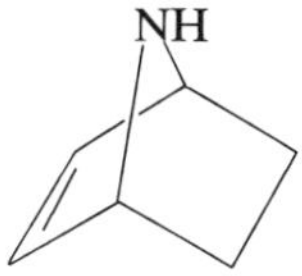

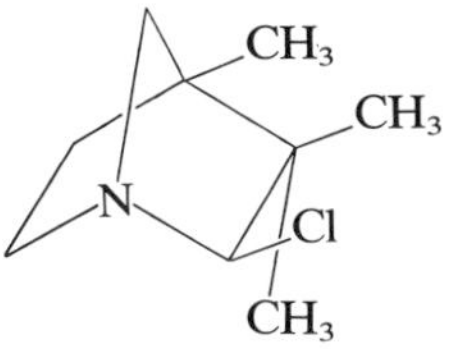

The stereoisomeric 2-phenyl-1,4-dimethylphosphanorbornane oxides and corresponding *P*-methyl analogs (**307**) differ in both chemical shifts and $^2J(^{31}P/^{13}C)$.[308] In the oxide of 1-phosphanorbornane (**308**) a rather large $^3J(^{31}P/^{13}C)$ is evident, perhaps partly due to multiple coupling paths and partly due to the syn-periplanar arrangement of the phosphorus and the bridgehead carbon, suggesting a Karplus relationship with a large coupling constant for a torsional angle near zero.[308] The spectrum of the phosphine sulfide **309** has been reported.[82]

306

R = Me, Ph and stereoisomer

307

308

309

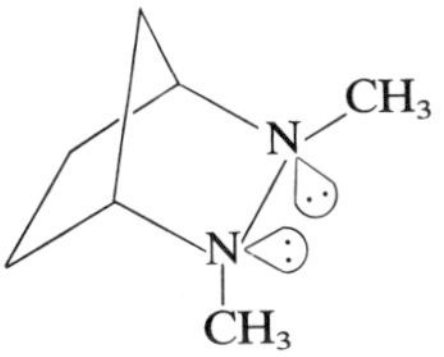

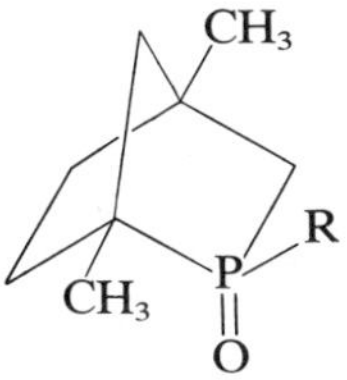

2. *Bicyclo[2.2.2]octanes*

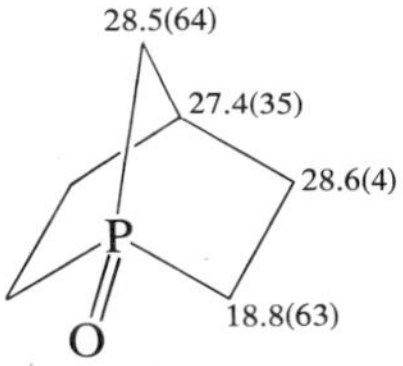

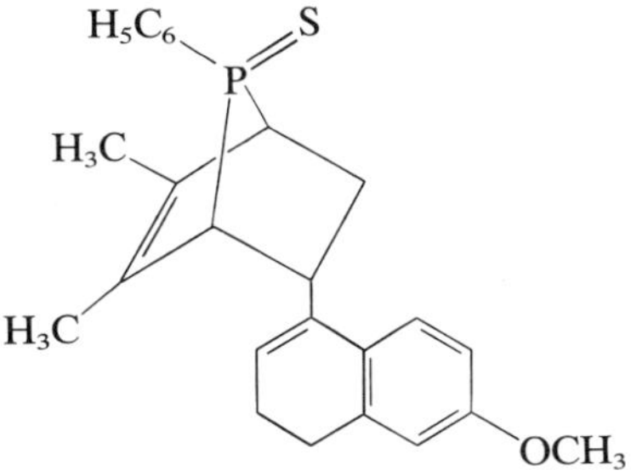

The spectrum of the 1-aza compound (quinuclidine, **310**) has been studied by several groups[8,39,164a]; the studies include the effect of protonation[164a] and *N*-oxide formation.[8] Comparison[39] with the carbocyclic

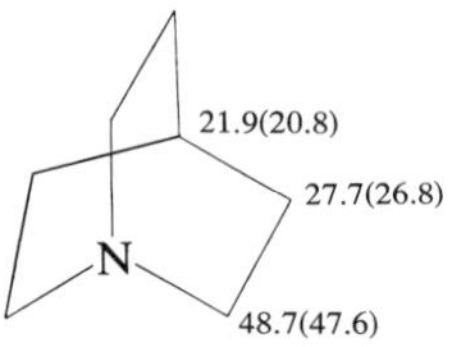

310* 21.9(20.8) 27.7(26.8) 48.7(47.6)

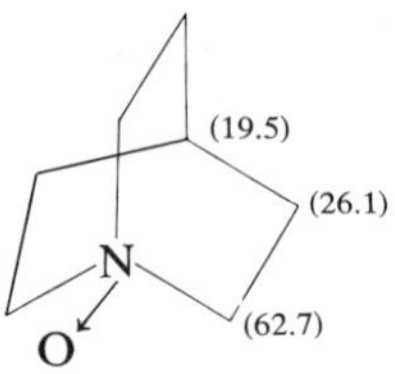

311* (19.5) (26.1) (62.7)

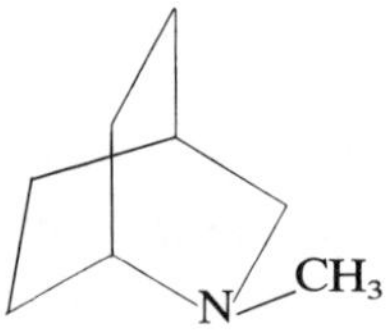

312

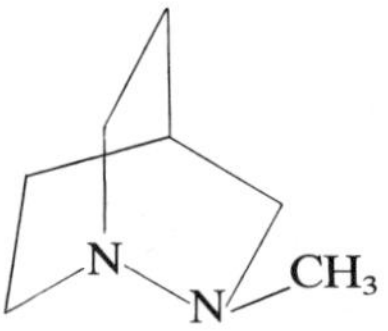

313

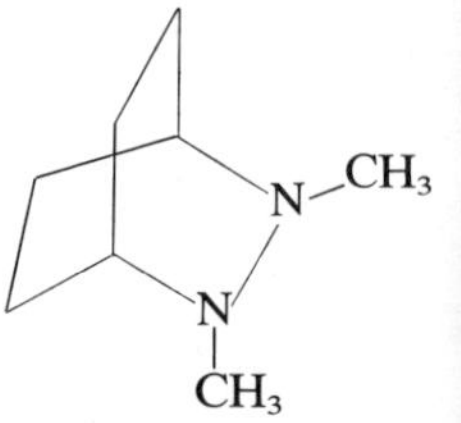

314

parent shows nearly normal changes at C_α (22.5 ppm, the normal is approximately 20 ppm, cf. Table 3.9) and C_γ (-2.2 ppm), but the shift at C_β, normally zero or slightly upfield, is downfield by 1.5 ppm in this case as the result of the introduction of nitrogen. This appears to be typical of bridgehead nitrogen compounds.[39] The further shifts[8] upon conversion to the N-oxide **311** are also typical as are the protonation shifts.[164a] The 2-azabicyclo[2.2.2]octane system **312** and the 1,2- (**313**) and 2,3-diaza analogs (**314**) have also been studied as to their spectra, as has the bicyclooctene corresponding to the latter.[309] The stable configuration of the N,N'-dimethyl-2,3-diaza compound has the methyl substituents *trans* to each other, similar as in the corresponding norbornane **306**, and the barrier to the double inversion at nitrogen is 12.2 kcal/mol.

Spectra of 2-thiabicyclo[2.2.2]octane (**315**), the corresponding unsaturated system **316** and one of its sulfoxides, its sultine (O instead of CH_2 next to SO), and its sulfone have been recorded.[310] The spectrum of the phosphine oxide **317** [$J(^{31}P/^{13}C)$ in parentheses] has been studied[308] together with that of its lower homolog **308** mentioned earlier.

3. Bicyclo[3.2.1]octanes

Although the ^{13}C nmr spectrum of the 8-oxa analog itself appears not to have been recorded, the spectrum of the *trans*-2,*trans*-4-dibromo analog **318** is on record[311] along with spectra of the corresponding [4.2.1], [5.2.1] and two diastereomeric [8.2.1] homologs. The spectrum of the lactone **319** is on record[312] and the 6,8-dioxa system **320** is found in

* Data from Ref. 39; data in parenthesis from Ref. 8.

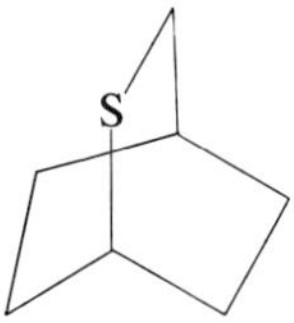

315

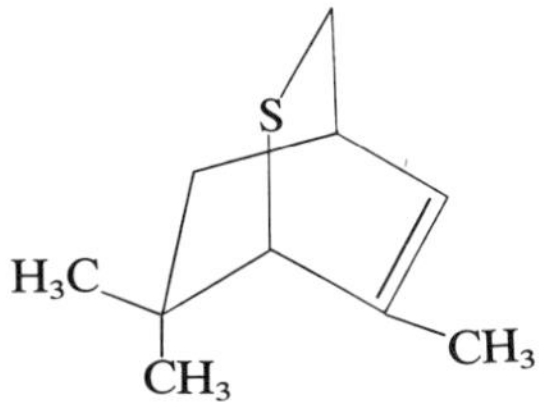

316

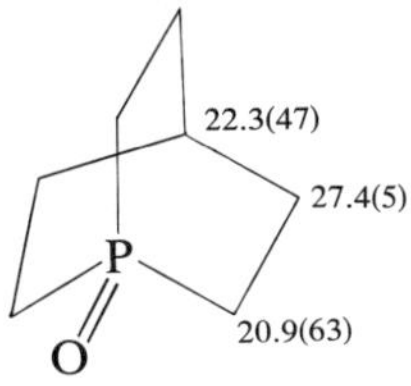

22.3(47)

27.4(5)

20.9(63)

317

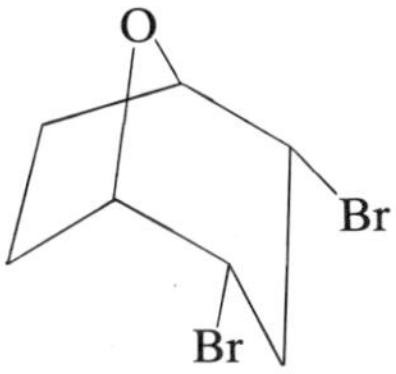

318

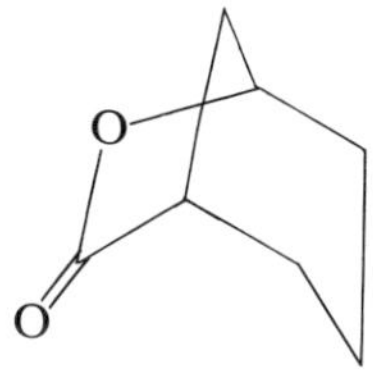

319

several beetle pheromones (frontalin, brevicomin, multistratin) whose spectra* have been reported.[313]

8-Azabicyclo[3.2.1]octane (nortropane) and its N-methyl derivative, tropane (321) are the parent compounds of the tropine alkaloids and much information is available regarding ^{13}C nmr spectra of these compounds and their derivatives.[8,314–319] The assignments for the parent compound are shown in the formula; some of the early[314] assignments in some of the derivatives were later[317,318] changed. Nitrogen inversion in the strained bridge is slower than in N-methylpiperidine and can be "frozen" on the nmr time scale at $-100°C$[316]; the preference of N—Me for the equatorial position in the six-membered chair so determined[316] is

---

* In invoking a possible γ-anti-effect,[43] the authors[313] claim that C(3) is anti-periplanar to O(6) which is, of course, not correct. However, in α-multistratin (i) the two ring methyl substituents are anti-periplanar to O(6), and it is interesting that these two methyls are within 0.4 ppm of each other, even though one is gauche to the ethyl group (and thus should be shifted upfield) and the other is not. This is most likely a manifestation of the effect discussed earlier (p. 205): Me(2) (authors' numbering[313]) is shifted upfield by the γ-anti-effect of O(6) but Me(4) is not (or may be even shifted downfield by that effect acting by itself) because the interposed C(5) is quaternary.

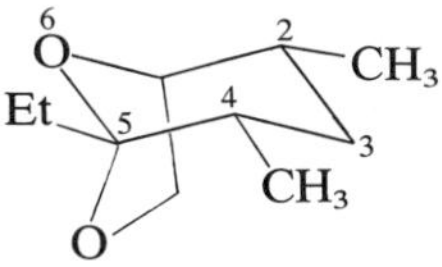

(i)

320

321

322

323

0.9 kcal/mol. An alternative determination based on line shape analysis (see Ref. 148) yielded a value[319] of 1.1 ± 0.15 kcal/mol.* The salts[315,318] and quaternary salts[315] of **321** and related compounds have been investigated as have solvent effects (CDCl$_3$ vs CD$_3$OD) on the chemical shifts.[318] From the sometimes anomalous substitutent effects in tropanes (e.g., a very small γ_a effect of what should be an axial hydroxyl group in the cyclohexane chair), it was concluded[318] that some tropane derivatives (notably those in which an attractive transannular interaction between the nitrogen and a substituent at C(3), such as a hydroxyl group, is possible) exist predominantly in the boat form.

Nickel contact shifts in the ^{13}C nmr spectrum of the bridgehead aza compound **322** have been reported.[164b] The spectrum of one sulfur analog, **323**, is on record.[310]

4. Bicyclo[3.3.1]nonanes

This system has long been known to exist in a (slightly distorted) double-chair form, a fact which is confirmed by ^{13}C nmr study[39,124a320–322] of the parent system **324**. C(1), C(2), and C(9) seem to have normal positions considering they are located in a cyclohexane system with α_a and single or double β_a effects. C(3) is not at as high a field as one would anticipate on the basis of a double γ_a effect; there is no doubt (see below) a partially compensating downfield δ-compression effect as a result of the proximity of C(3) and C(7). *endo*-Substituents at C(3) and/or C(7) may change the system to boat-boat or chair-boat.[321,323,324] No line broadening is seen in **324** itself at 160°K, suggesting that the chair-chair is at least 1.4 kcal/mol more stable than the chair-boat.[320]

324

325

---

* The values for *N*-methylmorpholine and *N,N'*-dimethylpiperazine reported in Ref.[319] were later withdrawn.[188a] See also Ref. 181 and p. 219.

The heteroanalogs have been subject to systematic study.[124a,323-325] The spectrum of the 9-oxa compound[39,124a] is shown in **325**; the shifts relative to **324** are normal for the change from a cyclohexane to an oxane. Derivatives of **324** and **325** (as well as **326–328**; see below) with electronegative substituents, Cl and/or OH, at C(1) have also been investigated[39,124a]; the bicyclo[3.3.1]nonane system and its 9-heteroanalogs are among the compounds where bridgehead substitution of an electronegative group causes a downfield shifting rather than the normal upfield shifting γ-anti-effect because of the quaternary nature of the bridgehead (cf. p. 205). It has also been pointed out[124a] that this deshielding γ-anti-effect is greater at the other bridgehead carbon (C-5) than at the nonbridgehead carbons C(3,7).*

326†   **327**   **328**

The 9-methyl-9-aza compound **326** has been studied by two groups.[323,326] If one compares **326** with **324** and takes into account the normal change in going from cyclohexane to N-methylpiperidine (see Table 3.9), one is struck that both C(1,5) and C(2,4,6,8) are approximately 5 ppm higher in field than one might have anticipated, whereas the shift of C(3,7) is normal. This is presumably because, whereas in N-methylpiperidine the N—Me is entirely equatorial (see earlier discussion: $\delta_{Me} = 47.0$ ppm), in **326** it must necessarily be axial in each ring half the time (and hence $\delta_{Me} = 41.4$ ppm). As a result, the β_e effect (at C-1,5) is replaced by the average of half a β_e and half a β_a effect and the γ_e effect (at C-2,4,6,8) is replaced by the mean of a γ_e and a γ_a effect. Both of these replacements produce upfield shifts since the β_a effect is less deshielding than β_e and γ_a is more shielding than γ_e. That this explanation is most likely correct is indicated by the low-temperature spectrum[326] of **326**. The C(2,4,6,8) signal decoalesces into one at 21.2 ppm (in the

* Reference 124a contains a summary of a number of bridgehead-substituted systems with electronegative substituents; the γ-anti-effects are sometimes shielding and sometimes deshielding. Most of the systems in question are not heterocyclic and are not considered elsewhere in this review.

† Unparenthesized data from ref. 326, parenthesized from Ref. 323.

ring with axial N—Me) and one at 33.2 ppm (in the ring with equatorial N—Me); the latter signal is at nearly the same position as C(2,4,6,8) in **324** just as C_β in N-methylpiperidine is nearly at the same frequency as the carbon in cyclohexane. Incidentally, there is only a trivial split between C(3) and C(7) at low temperature (21.3 vs 21.7 ppm) indicating that, contrary to assertion,[326] there is no major effect on C_γ caused by the position (axial or equatorial) of the lone pair on nitrogen. (What effect there is may just as well be considered as arising from the difference of δ_e and δ_a effects of the N-methyl.)

Substitution of a hydroxyl group at C(3) has little effect at the distal C(7) when the group is *exo* but has a large shielding effect when it is *endo*[323] (compare **328** to **327**). This suggests[323] that the *endo* hydroxyl group in fact does not occupy this highly crowded position to any extent but forces the substituted ring into the boat form, as shown in formula **328**. The shielding of C(7) then results from the disappearance of the δ-compression shift in the chair-boat **328**. A similar though lesser upfield shift is also seen at C(3) (compare **328** to **327**). The upfield shifts at C(1,5) and C(6,8) are believed[326] to result from the fact that the N-methyl group is now entirely axial in the chair-shaped ring, rather than half-axial-half-equatorial. This is confirmed by the near-identity of N— Me shifts in **327** and **328**: If the N—Me were appreciably chair-equatorial in **328**, it should experience a strong δ-compression shift arising from C(3), at the opposite tip of the boat.

The N-methyl inversion in **326** can be seen on the nmr time scale at $-118°C$[326] and the barrier to inversion is $\Delta G^\ddagger = 7.1$ kcal/mol. Spectra of the 1-hydroxy and 3-keto analogs of **326** and some higher N-alkyl analogs of the latter have also been recorded[323]; of particular interest is the N-t-Bu homolog in which the t-butyl group must necessarily be axial in one or other of the piperidine chairs. The analog of **326** with NMe$_2$ in lieu of CH$_3$ has also been looked at.[326] Surprisingly, the barrier to nitrogen inversion in this hydrazine could not be established. The ^{13}C nmr spectrum of the methochloride of **326** is unexceptional.[326]

The 9-thia compound **329** has been studied along with its sulfoxide, sulfone, the *trans*-2,*trans*-6-dichloro analog and its sulfone, the $\Delta^{2,3}$ unsaturated analog and its sulfone, and the corresponding *trans*-6-chloro homologs and the diastereomeric 2-hydroxy analogs of the sulfone of **329**.[124a,325] The deshielding effect at C(1,5) of converting the sulfide to a sulfoxide or sulfone is considerably less than in thiane (Table 3.9) but similar to that in the bicyclo[2.2.2]octene analog **316**; the effects of oxidation of sulfur at the more distant carbons are normal. Comparison of the parent compound **329** with **324** would, at first sight, suggest a somewhat unusual downfield shift at C(1,5) due to sulfur. (The shifts at

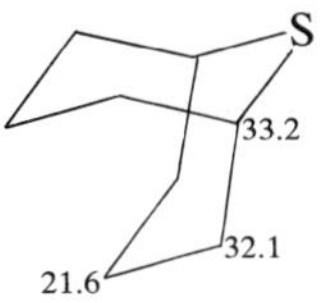

S
33.2
32.1
21.6
329

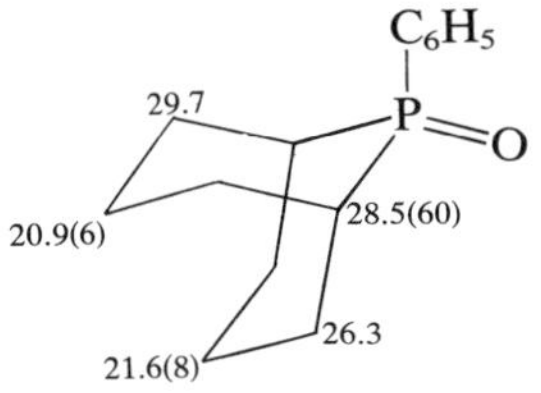

C_6H_5
P=O
29.7
20.9(6)
28.5(60)
21.6(8)
26.3
330

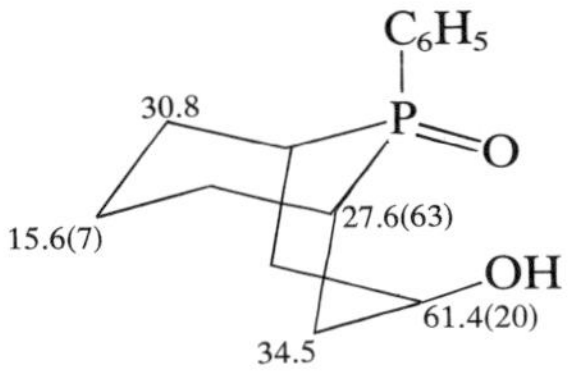

C_6H_5
P=O
30.8
15.6(7)
27.6(63)
OH
61.4(20)
34.5
331

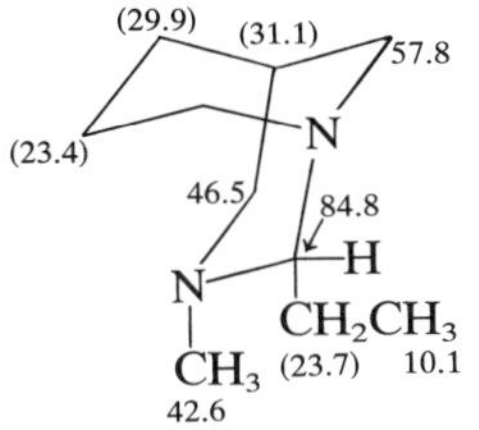

(29.9) (31.1)
57.8
(23.4)
N
46.5
84.8
N
H
CH₂CH₃
CH₃ (23.7) 10.1
42.6
332

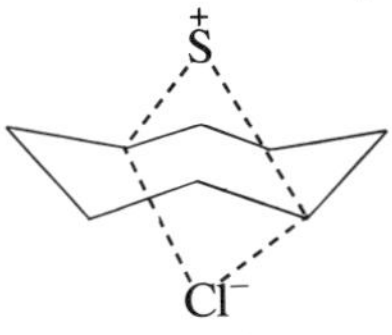

$\overset{+}{S}$
Cl⁻
333

the other positions are normal, see Table 3.9.) However, it must be taken into account that the deshielding substituent α effects at C(2,6) in a thiane are considerably larger than in cyclohexane; this no doubt explains why the deshielding at C(1,5) caused by the introduction of sulfur at C(9) in **324** is about twice as large as the deshielding at C_α in thiane relative to cyclohexane.

The spectrum of the phosphine oxide **330** is shown in the formula. In analogy to **328**, one of the 3-hydroxy analogs (**331**) is in the chair-boat form; this not only leads to substantial shielding at C(7) but also to a greatly enhanced coupling constant $^3J(^{31}P/^{13}C\text{-}3)$. The reason for this is that a Karplus relationship holds for $^3J(^{31}P/^{13}C)$ and C(3) in **331** is syn-periplanar to the phosphorus whereas in **330** (and the diastereomer of **331**) it is syn-clinal; the decrease in torsional angle from 60 to 0° leads to a considerable increase in ^3J.

Data of the bicyclic diazane **332** and its 2-phenyl (in lieu of ethyl) analog are as yet unpublished.[327] There are ^{13}C nmr data on several chloro- and amino-derivatives of **329** and mono- and di-unsaturated analogs.[328] The sulfonium salt **333** has been detected as a reaction intermediate by ^{13}C nmr.[329]

5. Bicyclo[4.2.1]nonanes

In contrast to the [3.3.1]-system, the [4.2.1] has received little study. The spectrum of the 9-oxa analog **334** is on record.[330] The compound appears to be in a conformation in which the cyclooctane part is a chair. The

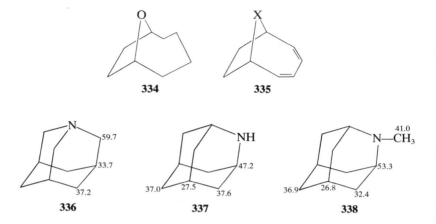

2,5-dibromo derivative has already been mentioned earlier.[311] Spectra of several 9-heterobicyclo[4.2.1]nona-2,4,7-trienes (**335**, X=O, NH, NCN, S) have been reported.[331]

6. Adamantanes

A systematic ^{13}C nmr study of 1- and 2-heteraadamantanes and some of their ring substitution products has been made.[39] 1-Azaadamantane (**336**) and 2-aza- (**337**), 2-methyl-2-aza- (**338**), 2-oxa- (**339**), and 2-thia-adamantane (**340**) are included in this study; the 1-aza-[164a] and 2-methyl-2-aza-[326] compounds have also been studied by others. The chemical shifts should be compared to those of adamantane, CH$_2$, 37.8, CH, 28.5 ppm.[39] With respect to **336**, the unusually marked deshielding of the β (bridgehead) carbon has been noted.[39] The γ effect is less upfield shifting than usual; both C$_\beta$ and C$_\gamma$ are more downfield than one would expect. The spectrum of the 2-aza compound (**337**) is unexceptional. In the N-methyl derivative **338** the bridgehead carbon next to nitrogen and the methylene carbon next to it appear, at first sight, to be at too high a field relative to the NH compound **337** if one uses piperidine and N-methylpiperidine (Table 3.9) for comparison. However, as in the 9-methyl-9-azabicyclo[3.3.1]nonane system **326** (*vide supra*), the N-methyl group is now half equatorial and half axial, rather than all-equatorial as in N-methylpiperidine, and the diminished downfield β shift and the upfield γ shift can thus be readily accounted for.

The 2-oxa- (**339**) and 2-thiaadamantanes (**340**) are unexceptional; with respect to the upfield shift next to sulfur the same point regarding sulfur next to a tertiary as compared to next to a secondary (as in thiane) carbon should be made which had earlier been brought up in connection with the bicyclo[3.3.1]nonane system (**329**, *vide supra*).

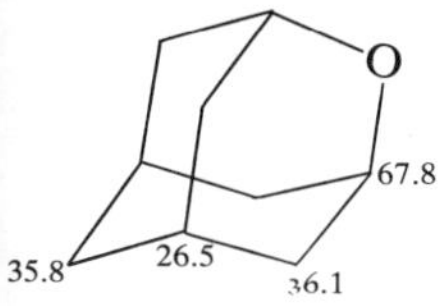

339

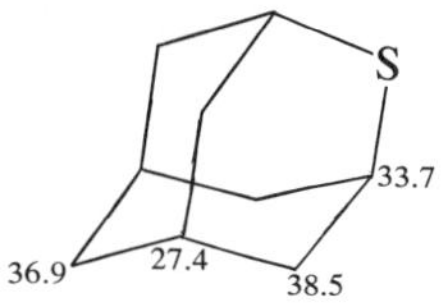

340

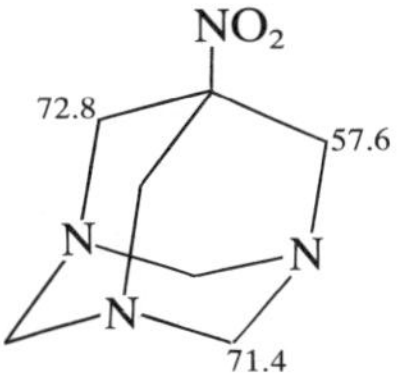

341

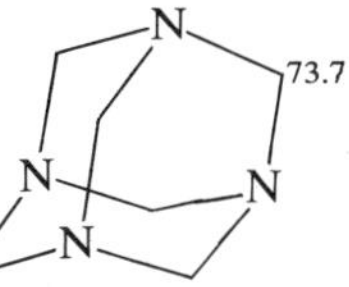

342

The γ-anti-effect of electronegative substituents at C(1) in adamantane and its heteroanalogs is deshielding[39] as had been noted earlier.[43] Derivatives of 1,3,5-triazadamantane (e.g., **341**) and the tetraaza compound **342** (hexamethylene tetramine) have been studied[332]; spectral investigations were supplemented with INDO calculations. A large difference of T_1s in **342** compared to the methylene group in adamantane in solvents $CDCl_3$ (1.89 vs 13.0 sec) and $CHBr_3$ (1.04 vs 9.2 sec) was ascribed to a diminished tumbling rate of the nitrogen compound due to intermolecular hydrogen bonding to solvent.[333]

The spectra of 2,4,6,8-tetraoxa-, tetrathia-, and tetraselena-adamantanes (**343**) have been recorded[334] and interpreted in terms of contributions of diamagnetic and paramagnetic effects. The shifts are 97.5 (CH) and 40.3 ppm (CH_2) for X = O; 51.4 and 52.0 ppm for X = S; and 40.0 and 57.0 ppm for X = Se. Other heteraadamantanes whose spectra have been recorded[328,335,336] are shown in formulas **344**–**347**.

X = O,
S, Se

343

X = OH or = O
R_1, R_2 = CH_3 or
R_1 = CH_3, R_2 = OH, OMe
or R_1 = CH_2Br, R_2 = Br

344[328]

R_1 = R_2 = O
or —CN

345[335]

346[336]

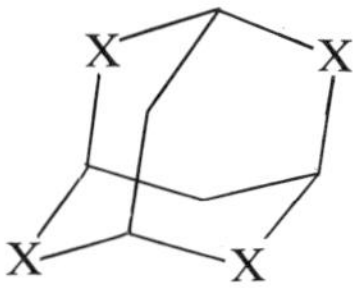

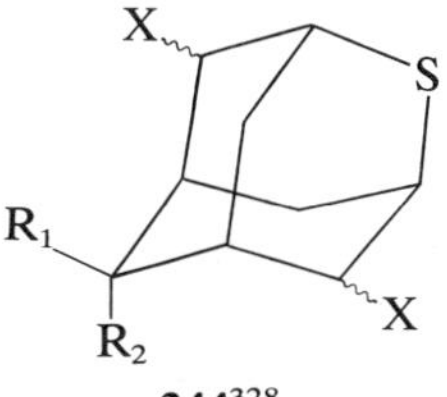

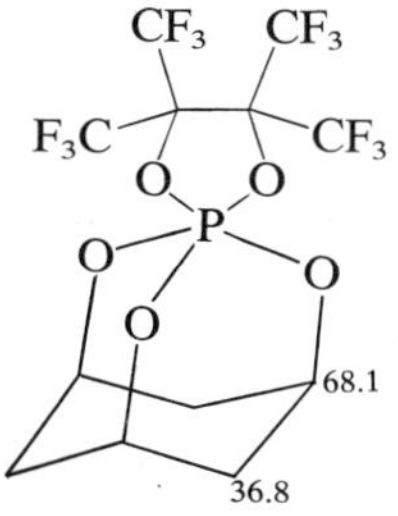

347[336]

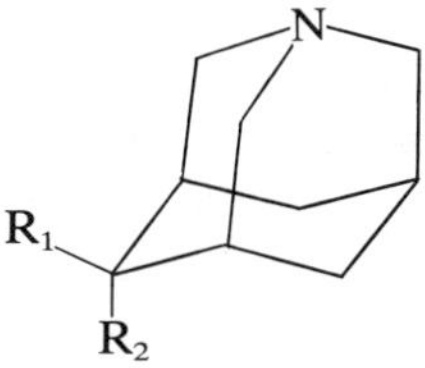

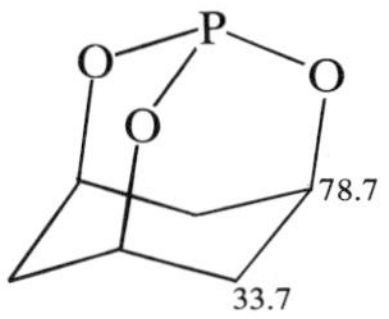

7. Miscellaneous Polycyclic Systems

Under this rubric we deal with polycyclic systems for which there are only isolated reports.

Miscellaneous Fused Ring Systems. A spectral study of the *Buxus* alkaloids[337] provides data for the otherwise unexplored 1,3-oxazane system **348**. Studies on dihydrolysergic acid and 10-methoxydihydrolysergic acid derivatives (**349**) exemplify a case where the preferred conformation of a natural product was derived from ¹³C nmr data.[338] The *Lycopodium* alkaloids include examples for axial *N*-methyl groups in piperidine rings.[339] (Applications of model studies to alkaloid structure in general have been reviewed in Ref. 8.) Gated decoupling of protons has proved useful in studies of pentalenolactones **350**.[340]

R = H or OCH₃
R¹ = H or CH₃
X = OCH₃ or NH₂

348

349

350

Miscellaneous Bridged Ring Systems. The *Aconitium* and *Delphinium* alkaloids provide an application of ¹³C nmr studies to quite complex structures, including conformational study, for example identification of a six-membered twist.[341] The orientation of the *N*-methyl group in cannivonine (**351**) was inferred from a ¹³C nmr study that included investigation of nickel contact shifts.[342] The bridged systems **352**,[343] **353**,[344] **354** [including Eu(fod)₃ studies],[345] **355**,[346] and **356** and congeners[344] have been investigated by ¹³C nmr and, in the case of **356**, configurational assignments are based on the spectral study. In **357** the effect of the proximity and orientation of the lone pairs on phosphorus on ¹³C–³¹P coupling constants was seen; the study includes investigation of the ³¹P(¹³C) satellite spectra.[348] The nitrogen containing twistane homolog **358** has been studied spectrally.[349] (Twistane itself has been investigated

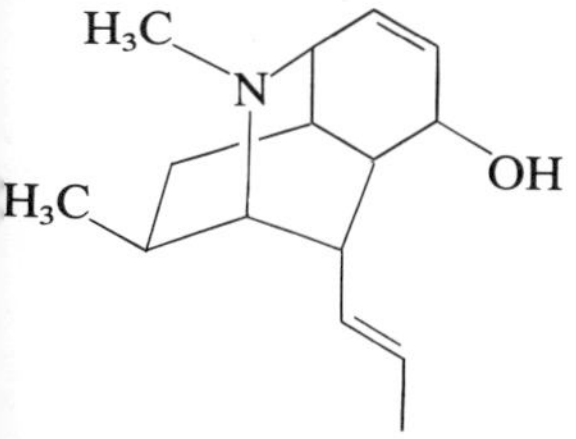

351

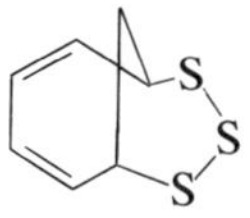

352

2 stereoisomers
R = H or Me
R^1 = Ac or CO$_2$Me

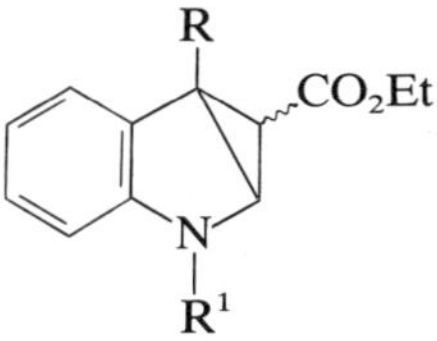

353

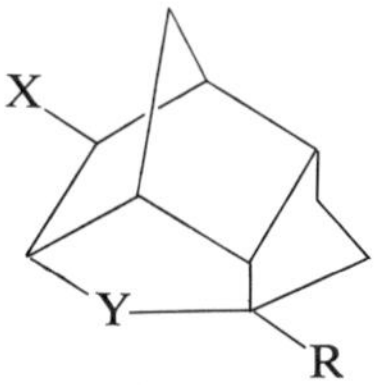

354

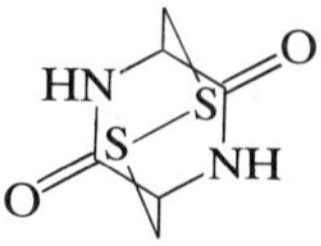

355

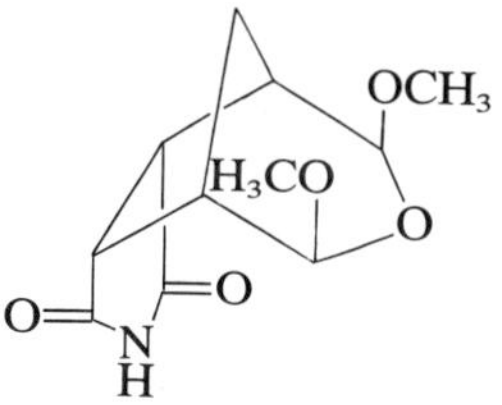

356

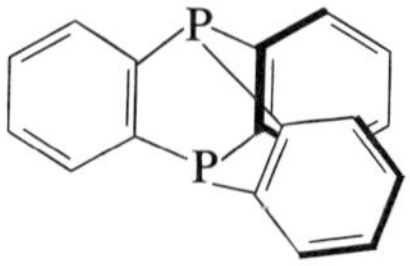

357

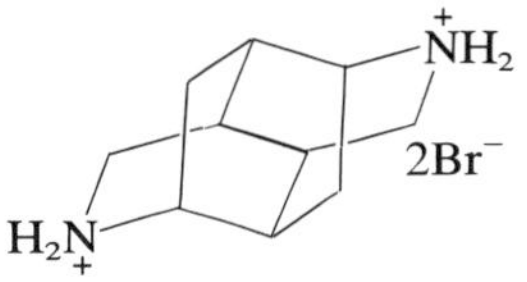

358

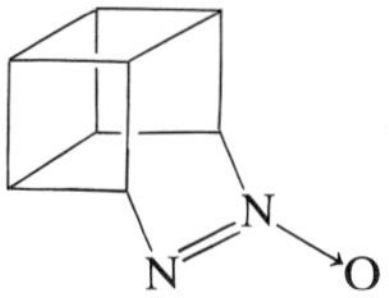

359

as to the γ shift produced by electronegative substituents, which has been compared with the corresponding shift in adamantane.)

Cage Compounds. The asymmetry of compound **359** was confirmed[351] by ^{13}C nmr. The spectrum of the endrin derivative **360** has been reported.[352] Several cage compounds of the type exemplified by **361** have been studied by a South African group[353–355]; the assignment of the sometimes complex ^{13}C nmr spectra was aided by the use of the selective population inversion technique[353] and deuterium labeling.[354]

Propellanes. Propellanes of type **362** and derivatives[356] and **363** (and corresponding sulfones and chlorosulfones)[357,358] have been investigated by ^{13}C nmr.

Spiranes. Hiemstra and Wynberg[359] have studied the enantiomeric purity of cyclohexanones by ^{13}C nmr study of their ketals or dithioketals with

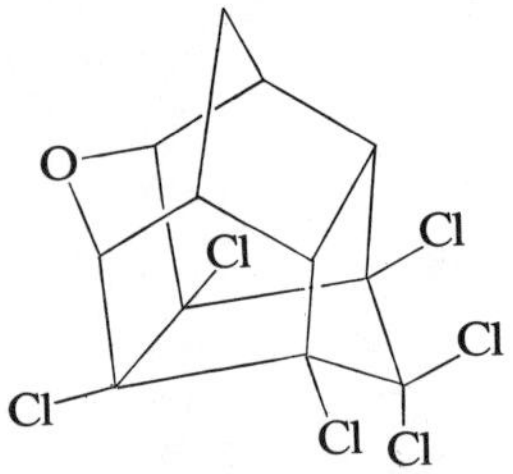

360

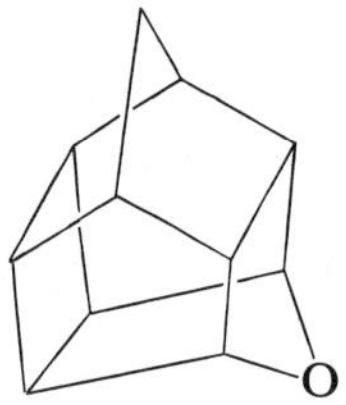

361

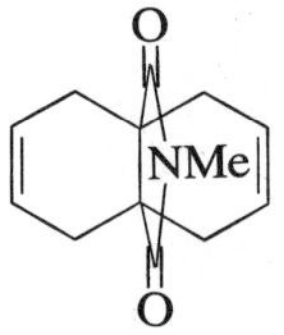

362

$n = 1-5$
$X = O$ or S

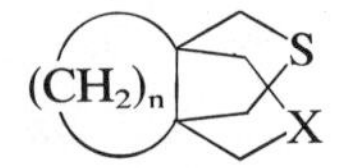

363

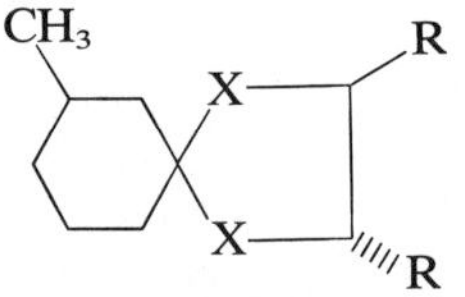

364

$X = O$, $R = CH_3$ or CO_2Et
$X = S$, $R = CH_3$

chiral diethyl tartrate, 2,3-butanediol or 2,3-butanedithiol (**364**). The spectra of several oxa, thia, and oxathia-spira[4.5]decanes, some conformationally locked, have been recorded[360]; sample shifts are shown in formulas **365–369**. Other cyclic ketal, mono-, and dithioketal spectra are undoubtedly in the literature.

Catenanes. The spectrum of catenane **370** has been reported[361] and

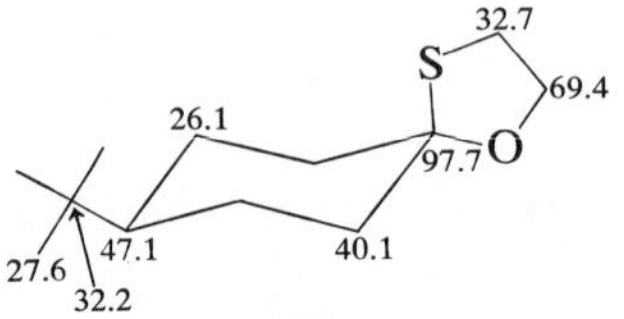

365

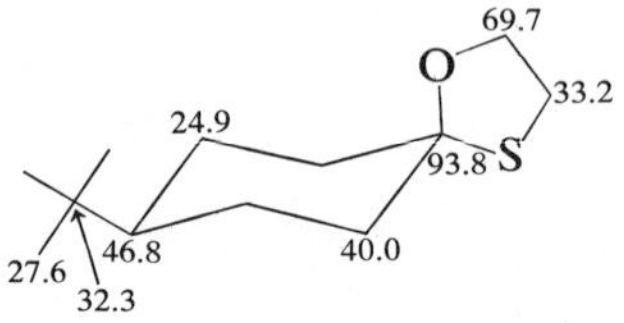

366

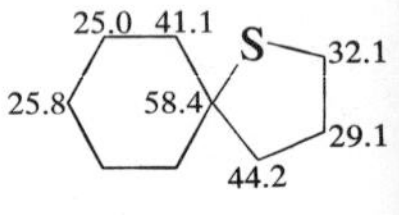

367

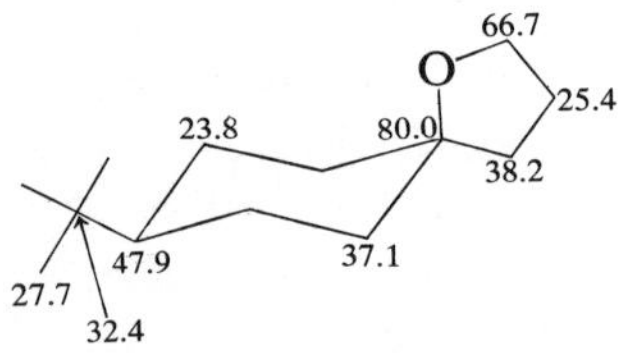

368

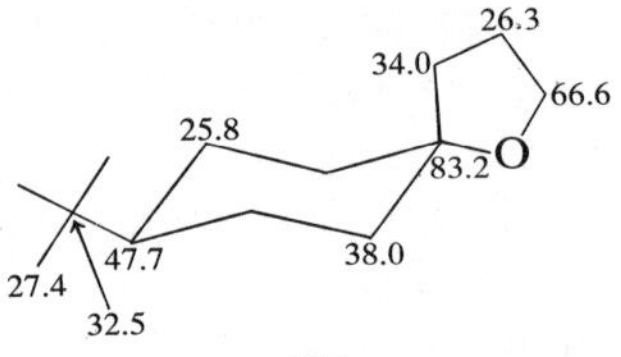

369

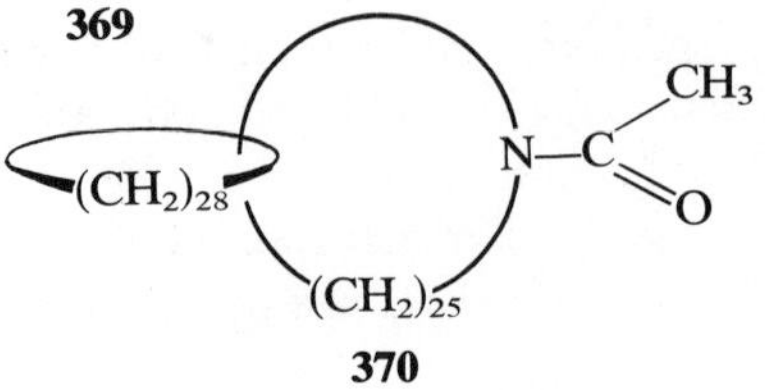

370

compared with that of an equimolar mixture of the two isolated constituent rings; there is a small but palpable difference.

IV. CONCLUSION

Although ^{13}C nmr spectroscopy has been a readily accessible tool for no more than five or six years, an enormous amount of experimental material has been accumulated by the end of 1977. By a combination of substituent parametrization, basically following Grant and collaborators,[34] and parametrization using increments for the replacement of carbon by heteroatoms, first introduced by Roberts and co-workers,[40] it becomes, in many cases, quite easy to assign signals in the spectra of nonaromatic heterocycles. A number of interesting effects, such as the γ-anti-effect, the effect of an anti-periplanar lone pair, the constancy of most substituent parameters, save α effects next to sulfur and β_e effects next to many heteroatoms are noteworthy in their own right and present challenges to the theoretician. Unfortunately the theoretical understanding of all these effects is still extremely rudimentary and major advances in this area are to be hoped for, since the empirical interpretation of spectra may have gone about as far as it can go.

V. ACKNOWLEDGMENT

This work was supported by NSF Grant CHE-75-20052 and by a grant-in-aid from Allied Chemical Corporation. We are grateful to Mrs. Lynn V. Buckley for typing the chapter at very short notice.

REFERENCES

1. G. C. Levy and G. L. Nelson, *Carbon-13 Nuclear Magnetic Resonance for Organic Chemists*, Wiley-Interscience, New York, 1972.

2. J. B. Stothers, *Carbon-13 NMR Spectroscopy*, Academic Press, New York, 1972.

3. E. Breitmaier and W. Voelter, ^{13}C *NMR Spectroscopy*, Verlag Chemie, Weinheim, Federal Republic of Germany, 1974.

4. S. N. Rosenthal and J. H. Fendler, *Adv. Phys. Org. Chem.* **13**, 279 (1976).

5. O. Oster, E. Breitmaier, and W. Voelter, *Nuclear Magnetic Resonance Spectroscopy of Nuclei Other than Protons*, T. Axenrod and G. A. Webb, Eds., Wiley-Interscience, New York, 1974, p. 233.

6. R. A. Komoroski, I. R. Peat, and G. C. Levy, *Topics in Carbon-13 NMR Spectroscopy* **2**, 179 (1976).

7. A. S. Shashkov and O. S. Chizhov, *Bioorg. Khim.* **2**, 437 (1976).

8. E. Wenkert, J. S. Bindra, C. J. Chang, D. W. Cochran, and F. M. Schell, *Acc. Chem. Res.* **7**, 46 (1974).

9. E. Wenkert, B. L. Buckwalter, I. R. Burfitt, M. J. Gasic, H. E. Gottlieb, E. W. Hagaman, F. M. Schell, P. M. Wovkulich, and A. Zheleva, *Topics in C-13 NMR Spectroscopy* **2**, 81 (1976).

10. A. Allerhand, *Pure Appl. Chem.* **41**, 247 (1975).

11. J. G. Nourse and J. D. Roberts, *J. Am. Chem. Soc.* **97**, 4584 (1975).

12. Y. Terui, K. Tori, K. Nagashima, and N. Tsuji, *Tetrahedron Lett.* 2583 (1975).

13. S. Omura, A. Nakagawa, A. Neszmelyi, S. D. Gero, A.-M. Sepulchre, F. Piriou, and G. Lukacs, *J. Am. Chem. Soc.* **97**, 4001 (1975).

14. H. Minato, T. Katayama, and K. Tori, *Tetrahedron Lett.* 2579 (1975).

15. S. Omura, A. Nakagawa, H. Takeshima, J. Miyazawa, C. Kitao, F. Piriou, and G. Lukacs, *Tetrahedron Lett.* 4503 (1975).

16. W. Breitenstein and C. Tamm, *Helv. Chim. Acta* **58**, 1172 (1975).

17. (a) A. Neszmelyi, S. Omura, and G. Lucacs, *J. Chem. Soc. Chem. Comm.* 97 (1976); (b) A. Neszmelyi, S. Omura, T. T. Thang, and G. Lukacs, *Tetrahedron Lett.* 725 (1977).

18. P. S. Pregosin and E. W. Randall, in *Nuclear Magnetic Resonance Spectroscopy of Nuclei Other than Protons*, T. Axenrod and G. A. Webb, Eds., Wiley-Interscience, New York, 1974, p. 243.

19. G. A. Olah, *Halonium Ions* Wiley-Interscience, New York, 1975.

20. S. G. Davies and G. H. Whitham, *J. Chem. Soc. Perkin 2*, 861 (1975).

21. D. R. Paulson, F. Y. N. Tang, G. F. Moran, A. S. Murray, B. P. Pelka, and E. M. Vasquez, *J. Org. Chem.* **40**, 184 (1975).

22. M. J. Shapiro, *J. Org. Chem.* **42**, 1434 (1977).

23. B. Everatt, A. H. Haines, and B. P. Stark, *Org. Mag. Res.*, **8**, 275 (1976).

24. R. Martino, P. Mison, F. W. Wehrli, and T. Wirthlin, *Org. Mag. Res.* **7**, 175 (1975).

25. P. Mison, R. Chaabouni, Y. Diab, R. Martino, A. Lopez, A. Lattes, F. W. Wehrli, and T. Wirthlin, *Org. Mag. Res.* **8**, 79 (1976).

26. T. Yonezawa, I. Morishima, K. Fukuta, and Y. Ohmori, *J. Mol. Spectrosc.* **31**, 341 (1969); T. Yonezawa and I. Morishima, *J. Mol. Spectrosc.* **27**, 210 (1968).

27. W. B. Jennings, D. R. Boyd, C. G. Watson, E. D. Becker, R. B. Bradley, and D. M. Jerina, *J. Am. Chem. Soc.* **94**, 8501 (1972).

28. (a) G. J. Jordan and D. R. Crist, *Org. Mag. Res.* **9**, 322 (1977). (b) D. R. Crist, G. J. Jordan, and J. A. Hashmall, *J. Am. Chem. Soc.* **96**, 4927 (1974).

29. R. Radeglia, *J. Prakt. Chem.* **318**, 871 (1976).

30. G. E. Maciel and G. B. Savitsky, *J. Phys. Chem.* **69**, 3925 (1965).

31. (a) V. Nair, *Org. Mag. Res.* **6**, 483 (1974); (b) K. Isomura, H. Taniguchi, M. Mishima, M. Fujio, and Y. Tsuno, *Org. Mag. Res.* **9**, 559 (1977). The data in the two references are in very poor agreement.

32. G. Capozzi, O. De Lucchi, V. Lucchini, and G. Modena, *J. Chem. Soc. Chem. Comm.*, 248 (1975).

33. G. A. Gray, S. E. Cremer, and K. L. Marsi, *J. Am. Chem. Soc.* **98**, 2109 (1976).

34. (a) D. K. Dalling and D. M. Grant, *J. Am. Chem. Soc.* **89**, 6612 (1967); (b) D. K. Dalling and D. M. Grant, *J. Am. Chem. Soc.* **94**, 5318 (1972); (c) see also Ref. 1, p. 44.

35. J. P. Monti, R. Faure, and E. J. Vincent, *Org. Mag. Res.* **7**, 637 (1975).

36. K. M. Crecely, R. W. Crecely, and J. H. Goldstein, *J. Phys. Chem.* **74**, 2680 (1970); see also Ref. 2.

37. J. J. Burke and P. C. Lauterbur, *J. Am. Chem. Soc.* **86**, 1870 (1964); see also Ref. 1.

38. For exceptions in adamantanoid systems containing O, see Ref. 39. We are aware of no other upfield β shifts when CH_2 is replaced by NH.

39. H. Duddeck and P. Wolff, *Org. Mag. Res.* **8**, 593 (1976).

40. J. D. Roberts, F. J. Weigert, J. I. Kroschwitz, and H. J. Reich, *J. Am. Chem. Soc.* **92**, 1338 (1970).

41. H. Eggert and C. Djerassi, *J. Am. Chem. Soc.* **95**, 3710 (1973).

42. J. E. Sarneski, H. L. Surprenant, F. K. Molen, and C. N. Reilley,

Anal. Chem. **47,** 2116 (1975); see also J. G. Batchelor, *J. Mag. Res.* **28,** 123 (1977).

43. E. L. Eliel, W. F. Bailey, L. D. Kopp, R. L. Willer, D. M. Grant, R. Bertrand, K. A. Christensen, D. K. Dalling, M. W. Duch, E. Wenkert, F. M. Schell, and D. W. Cochran, *J. Am. Chem. Soc.* **97,** 322 (1975).

44. A. Mannschreck, R. Radeglia, E. Grundemann, and R. Ohme, *Chem. Ber.* **100,** 1778 (1967).

45. R. G. Kostyanovskii, A. A. Fomichev, G. K. Kadorkina, and Z. E. Samoilova, *Proc. Acad. Sci. USSR, Phys. Chem. Soc.* **195,** 875 (1970).

46. M. E. Bishop, C. D. Schaeffer, Jr., and J. J. Zuckerman, *J. Organometallic Chem.*, **101,** C19 (1975).

47. G. Levy and P. de Loth, *Compt. Rend.* **279,** 331 (1974).

48. S. Berger, F. R. Kreissl, D. M. Grant, and J. D. Roberts, *J. Am. Chem. Soc.*, **97,** 1805 (1975).

49. D. F. Ewing, K. A. Holbrook, and R. A. Scott, *Org. Mag. Res.* **7,** 554 (1975).

50. Y. Y. Samitov, A. V. Bogatskii, and G. A. Filip, *Proc. Acad. Sci. USSR, Phys. Chem. Sec.* **192,** 372 (1970), also *J. Org. Chem. USSR* **7,** 591 (1971).

51. C. Konno and H. Hikino, *Tetrahedron* **32,** 325 (1976).

52. A. K. Bose, M. Sugiura, and P. R. Srinivasan, *Tetrahedron Lett.* 1251 (1975).

53. A. Hassner, J. O. Currie, A. S. Steinfeld, and R. F. Atkinson, *J. Am. Chem. Soc.* **95,** 2982 (1973).

54. R. L. Willer and R. O. Hutchins, unpublished observation.

55. (a) G. C. Levy and D. C. Dittmer, *Org. Mag. Res.* **4,** 107 (1972); (b) G. Fronza, A. Gamba, R. Mondelli, and G. Pagani, *J. Mag. Res.* **12,** 231 (1973).

56. A. M. Bui, A. Cave, M. M. Janot, J. Parello, P. Potier, and U. Scheidegger, *Tetrahedron* **30,** 1327 (1974).

57. G. A. Gray and S. E. Cremer, *J. Org. Chem.* **37,** 3458, 3470 (1972).

58. G. A. Gray and S. E. Cremer, *Tetrahedron Lett.* 3061 (1971); also *Chem. Comm.*, 367 (1972).

59. (a) M. Rotem and Y. Kashman, *Tetrahedron Lett.* 63 (1978); (b) J. A. Gibson, G.-V. Roschenthaler, and V. Wray, *J. Chem. Soc. Dalton* 1492 (1977).

60. (a) B. Fuchs, *Topics Stereochem.* **10,** 1 (1978); (b) R. M. Moriarty *Topics Stereochem.* **8,** 271 (1974); (c) K. Pihlaja, J. Jokisaari, P.

Olavi, I. Virtanen, H. Ruotsalainen, and M. Anteunis, *Org. Mag. Res.*, **7**, 286 (1975).

61. M. Christl, H. J. Reich, and J. D. Roberts, *J. Am. Chem, Soc.* **93**, 3463 (1971).

62. E. L. Eliel, V. S. Rao, and K. M. Pietrusiewicz, *Org. Mag. Res.*, in press; for 1,3-dioxolanes see also Ref. 63.

63. (a) K. Pihlaja and T. Nurmi, *Finn. Chem. Lett.* 141 (1977); (b) Y. Senda, J.-i. Ishiyama, and S. Imaizumi, *Bull. Chem. Soc. Japan* **50**, 2813 (1977).

64. W. E. Willy, G. Binsch, and E. L. Eliel, *J. Am. Chem. Soc.* **92**, 5394 (1970).

65. K. M. Pietrusiewicz and E. L. Eliel, unpublished observations.

66. F. W. Vierhapper and R. L. Willer, *Org. Mag. Res.* **9**, 13 (1977).

67. (a) L. F. Johnson and W. C. Jankowski, *Carbon-13 NMR Spectra*, Wiley-Interscience, New York, 1972; (b) see also J. W. Triplett, G. A. Digenis, W. J. Layton, and S. L. Smith, *Spectrosc. Lett.* **10**, 141 (1977).

68. (a) H. Pyysalo and j. Enqvist, *Finn. Chem. Lett.* 136 (1975); H. Pyysalo, J. Enqvist, E. Honkanen, and A. Pippuri, *Finn. Chem. Lett.*, 129 (1975); (b) A. H. Andrist and M. J. Kovelan, *Spectrosc. Lett.* **8**, 719 (1975); (c) G. W. Buchanan and C. Benezra, *Can. J. Chem.* **54**, 231 (1976); (*d*) see also J. Altman, H. Gilboa, and D. Ben-Ishai, *Tetrahedron* **33**, 3173 (1977).

69. See also J. A. Deyrup and H. L. Gingrich, *J. Org. Chem.* **42**, 1015 (1977).

70. (a) W. Voelter, S. Fuchs, R. H. Seuffer, and K. Zech., *Monats. Chem.* **105**, 1110 (1974); (b) R. Deslauriers and I. C. P. Smith, *Biopolymers* **16**, 1245 (1977); (c) K. S. Bose, T. H. Witherup, and E. H. Abbott, *J. Mag. Res.* **27**, 385 (1977).

71. M. G. Ahmed and P. W. Hickmott, *J. Chem. Soc. Perkin 2*, 838 (1977); see also D. Tourwé, G. Van Binst, S. A. G. De Graaf, and U. K. Pandit, *Org. Mag. Res.* **7**, 433 (1975).

72. (a) D. J. Chadwick and D. H. Williams, *J. Chem. Soc. Perkin 2*, 1202 (1974); (b) G. Montaudo, A. Recca, J. Verhoeven, and C. Kruk, *Gazz. Chim. Ital.* **105**, 443 (1975).

73. A. Solladié-Cavallo and G. Solladié, *Org. Mag. Res.* **7**, 18 (1975).

74. G. Barbarella, P. Dembech, A. Garbesi, and A. Fava, *Org. Mag. Res.* **8**, 108, 469 (1976).

75. R. T. LaLonde, T. N. Donvito, and A. I-M. Tsai, *Can. J. Chem.* **53**, 1714 (1975).

76. K. Kabzińska, *Bull. Acad. Polon. Ser. Sci. Chim.* **24,** 363 (1976).

77. See Ref. 2, p. 71 ff.

78. D. E. Dorman, M. Jautelat, and J. D. Roberts, *J. Org. Chem.* **36,** 2757 (1971).

79. G. W. Buchanan and D. G. Hellier, *Can. J. Chem.* **54,** 1428 (1976).

80. H.-W. Tan and W. G. Bentrude, *Tetrahedron Lett.* 619 (1975); W. G. Bentrude and H.-W. Tan, *J. Am. Chem. Soc.* **98,** 1850 (1976).

81. G. Ponchoulin, J. R. Llinas, G. Buono, and E. J. Vincent, *Org. Mag. Res.* **8,** 518 (1976).

82. Y. Kashman, I. Wagenstein, and A. Rudi, *Tetrahedron* **32,** 2427 (1976).

83. J. J. Breen, S. I. Featherman, L. D. Quin, and R. C. Stocks, *J. Chem. Soc. Chem. Comm.* 657 (1972).

84. (a) L. D. Quin, S. G. Borleske, and R. C. Stocks, *Org. Mag. Res.* **5,** 161 (1973); (b) T. Bungaard and H. J. Jakobsett, *Tetrahedron Lett.* 3353 (1972).

85. S. G. Borleske and L. D. Quin, *Phosphorus* **5,** 173 (1965).

86. C. Symmes, Jr. and L. D. Quin, *J. Org. Chem.* **41,** 1548 (1976).

87. R. Bodalski and K. M. Pietrusiewicz, unpublished observations.

88. J. A. Gibson and G.-V. Röschenthaler, *J. Chem. Soc. Dalton* 1440 (1976).

89. G. Buono and J. R. Llinas, *Tetrahedron Lett.* 749 (1976).

90. D. M. McKinnon and T. Schaefer, *Can. J. Chem.* **49,** 89 (1971); T. Schaefer, K. Chum, D. McKinnon, and M. S. Chauhan, *Can. J. Chem.* **53,** 2734 (1975).

91. G. A. Kalabin, M. V. Sigalov, D. F. Kushnarev, A. N. Mirskova, and T. S. Proskurina, *Khim. Geterosikl. Soedin* 1176 (1976); *Chem. Abstr.* **86,** 54894p (1977).

92. S. H. Pines and A. W. Douglas, *J. Am. Chem. Soc.* **98,** 8119 (1976).

93. C. K. Sauers and H. M. Relles, *J. Am. Chem. Soc.* **95,** 7731 (1973).

94. J. A. Deyrup and H. L. Gingrich. *J. Org. Chem.* **42,** 1015 (1977).

95. (a) J. Palmer, J. L. Roberts, P. S. Rutledge, and P. D. Woodgate, *Heterocycles* **5,** 109 (1976); (b) J. M. Riordan, T. L. McLean, and C. H. Stammer, *J. Org. Chem.* **40,** 3219 (1975); (c) D. M. Rackham, S. E. Cowdrey, N. J. A. Gutteridge, and D. J. Osborne, *Org. Mag. Res.* **9,** 160 (1977).

96. (a) H.-O. Kalinowski and H. Kessler, *Org. Mag. Res.* **6,** 305 (1974); (b) D. Leibfritz, *Chem. Ber.* **108,** 3014 (1975).

97. H.-O. Kalinowski and H. Kessler, *Org. Mag. Res.* **7,** 128 (1975).

98. G. Assef, J. Kister, J. Metzger, R. Faure, and E. J. Vincent, *Tetrahedron Lett.*, 3313 (1976); R. Faure, E.-J. Vincent, G. Assef, J. Kister, and J. Metzger, *Org. Mag. Res.* **9**, 688 (1977).

99. N. Motohashi, I. Mori, and Y. Sugiura, *Chem. Pharm. Bull.* **24**, 1737 (1976).

100. J. P. Allorand, A. Cogue, D. Gagnaire, and J. B. Robert, *Tetrahedron* **27**, 2453 (1971).

101. K.-H. Pook, H. Stähle, and H. Daniel, *Chem. Ber.* **107**, 2644 (1974).

102. R. Faure, J.-R. Llinas, E.-J. Vincent, and J.-L. Larice, *Compt. Rend.*, **279**, 717 (1974).

103. S. Toppet, P. Claes, and J. Hoogmartens, *Org. Mag. Res.* **6**, 48 (1974).

104. (a) J. V. Paukstelis, E. F. Byrne, T. P. O'Connor, and T. E. Roche, *J. Org. Chem.* **42**, 3941 (1977); (b) K. Sakamoto, N. Nakamura, M. Ōki, and J. Nakayama, *Chem. Lett.* 1133 (1977).

105. (a) C. T. Pedersen and K. Schaumburg, *Org. Mag. Res.* **6**, 586 (1974); (b) C. T. Pedersen, E. G. Frandsen, and K. Schaumburg, *Org. Mag. Res.* **9**, 546 (1977).

106. J. H. Hargis, S. D. Worley, W. B. Jennings, and M. S. Tolley, *J. Am. Chem. Soc.* **99**, 8090 (1977).

107. (a) R. Radeglia, *J. Prakt. Chem.* **316**, 344 (1974); (b) M. L. Filleux-Blanchard, N.-D. An, and G. Manuel, *J. Organomet. Chem.*, **137**, 11 (1977).

108. C. H. Bushweller, G. Bhat, L. J. Letendre, J. A. Brunelle, H. S. Bilofsky, H. Ruben, D. H. Templeton, and A. Zalkin, *J. Am. Chem. Soc.*, **97**, 65 (1975).

109. J. B. Lambert, D. A. Netzel, H-N. Sun, and K. K. Lilianstrom, *J. Am. Chem. Soc.* **98**, 3778 (1976).

110. J. B. Lambert and D. A. Netzel, *J. Am. Chem. Soc.* **98**, 3783 (1976).

111. cf. G. C. Levy, *Acc. Chem. Res.* **6**, 161 (1973).

112. D. G. Gorenstein, *J. Am. Chem. Soc.* **99**, 2254 (1977); H.-J. Schneider and E. F. Weigand, *J. Am. Chem. Soc.* **99**, 8362 (1977).

113. (a) E. L. Eliel and D. Kandasamy, *Tetrahedron Lett.*, 3765 (1976); (b) C.-Y. Yen, M.A. Dissertation, University of North Carolina, Chapel Hill, N.C. (1977); (c) C.-Y. Yen and E. L. Eliel, unpublished observations.

114. R. L. Willer and E. L. Eliel, *J. Am. Chem. Soc.* **99**, 1925 (1977); see also Ref. 74.

115. R. L. Willer and E. L. Eliel, *Org. Mag. Res.* **9**, 285 (1977).

116. (a) G. M. Kellie and F. G. Riddell, *J. Chem. Soc. B*, 1030 (1971); (b) F. G. Riddell, *J. Chem. Soc. B*, 331 (1970); (c) A. J. Jones, E. L. Eliel, D. M. Grant, M. C. Knoeber, and W. F. Bailey, *J. Am. Chem. Soc.* **93**, 4772 (1971).

117. E. L. Eliel, V. S. Rao, and F. G. Riddell, *J. Am. Chem. Soc.* **98**, 3583 (1976).

118. S. H. Grover, J. P. Guthrie, J. B. Stothers, and C. T. Tan, *J. Mag. Res.* **10**, 227 (1973); S. H. Grover and J. B. Stothers, *Can. J. Chem.* **52**, 870 (1974); J. B. Stothers and C. T. Tan, *Can. J. Chem.* **54**, 917 (1976); J. B. Stothers, C. T. Tan, and K. C. Teo, *Can. J. Chem.* **54**, 1211 (1976). See also H. Duddeck, *Org. Mag. Res.* **7**, 151 (1975).

119. E. L. Eliel and F. W. Vierhapper, *J. Org. Chem.* **41**, 199 (1976).

120. For a different view, see G. Engelhardt, H. Jancke, and D. Zeigan, *Org. Mag. Res.* **8**, 655 (1976).

121. See also Ref. 40 and J. B. Grutzner, M. Jautelat, J. B. Dence, R. A. Smith, and J. D. Roberts, *J. Am. Chem. Soc.* **92**, 7107 (1970).

122. W. Kitching, M. Marriott, W. Adcock, and D. Doddrell, *J. Org. Chem.* **41**, 1671 (1976).

123. W. A. Ayer, L. M. Browne, S. Fung, and J. B. Stothers, *Can. J. Chem.* **54**, 3272 (1976); also *Org. Mag. Res.*, **11**, 73 (1978).

124. There are now a number of other examples; see (a) J. R. Wiseman and H. O. Krabbenhoft, *J. Org. Chem.* **42**, 2240 (1977); (b) the data of Y. Senda, J. Ishima, and S. Imaizumi, *Tetrahedeon* **31**, 1601 (1975) there cited.

125. G. E. Maciel and H. C. Dorn, *J. Am. Chem. Soc.* **93**, 1268 (1971); G. E. Maciel, H. C. Dorn, R. L. Greene, W. A. Kleschick, M. R. Peterson, and G. H. Wahl, *Org. Mag. Res.* **6**, 178 (1974); see also Ref. 39.

126. H. Beierbeck, J. K. Saunders, and J. W. ApSimon, *Can. J. Chem.* **55**, 2813 (1977).

127. (a) E. L. Eliel, V. S. Rao, F. W. Vierhapper, and G. Z. Juaristi, *Tetrahedron Lett.*, 4339 (1975). (b) G. Z. Juaristi and E. L. Eliel, unpublished observations.

128. E. W. Della and E. L. Eliel, unpublished observations.

129. F. W. Vierhapper and E. L. Eliel, *J. Org. Chem.* **42**, 51 (1977).

130. (a) H. P. Hamlow, S. Okuda, and N. Nakagawa, *Tetrahedron Lett.*, 2553 (1964). (b) J. B. Lambert, D. S. Bailey, and B. F. Michel, *J. Am. Chem. Soc.* **94,** 3812 (1972).

131. A. J. de Hoog, *Org. Mag. Res.* **6,** 233 (1974).

132. (a) J. Thiem and B. Meyer, *Tetrahedron Lett.* 3573 (1977); (b) J. P. Zahra, B. Waegell, J. Reisse, G. Pouzard, and J. Fournier, *Bull. Soc. Chim. Fr.*, 1896 (1976); (c) G. W. Buchanan and J. H. Bowen, *Can. J. Chem.* **55,** 604 (1977).

133. O. Achmatowicz, Jr., M. Chmielewski and J. Jurczak, *Roczniki Chem.* **48,** 481 (1974).

134. V. P. Lezina, A. U. Stepanyants, F. A. Alimirzoev, S. S. Zlotski, and D. L. Rakhmankulov, *Bull. Acad. Sci. USSR Div. Chem. Sci,* **25,** 772 (1976).

135. E. L. Eliel, D. Kandasamy, and R. C. Sechrest, *J. Org. Chem.* **42,** 1533 (1977).

136. F. R. Jensen and R. A. Neese, *J. Am. Chem. Soc.* **97,** 4345 (1975).

137. F. I. Carroll, G. N. Mitchell, J. T. Blackwell, A. Soloti, and R. Meck, *J. Org. Chem.* **39,** 3890 (1974).

138. (a) J. A. Hirsch and E. Havinga, *J. Org. Chem.* **41,** 455 (1976); (b) F. J. Koer, A. J. de Hoog, and C. Altona, *Rec. Trav. Chim.* **94,** 75 (1975).

139. P. Äyräs, *Acta Chem. Scand.* **B30,** 957 (1976).

140. Y. Senda, J.-i. Ishiyama, S. Imaizumi, and K. Hanaya, *J. Chem. Soc. Perkin 1,* 217 (1977); Y. Senda, A. Kasahara, T. Izumi, and T. Takeda, *Bull. Chem. Soc. Japan* **50,** 2789 (1977).

141. (a) F. A. L. Anet and I. Yavari, *J. Am. Chem. Soc.* **99,** 2794 (1977); (b) F. W. Vierhapper and E. L. Eliel, *J. Org. Chem.*, in press.

142. cf. G. Binsch. *Topics Stereochem.* **3,** 97 (1968).

143. E. L. Eliel, *Israel J. Chem.* **15,** 7 (1976/77).

144. (a) I. D. Blackburne, A. R. Katritzky, and Y. Takeuchi, *Acc. Chem. Res.* **8,** 300 (1975); (b) J. B. Lambert and S. I. Featherman, *Chem. Rev.* **75,** 611 (1975).

145. (a) P. J. Crowley, M. J. T. Robinson, and M. G. Ward, *J. Chem. Soc. Chem. Comm.* 825 (1974); (b) M. J. T. Robinson, *J. Chem. Soc. Chem. Comm.* 844 (1975); (c) P. J. Crowley, M. J. T. Robinson, and M. G. Ward, *Tetrahedron* **33,** 915 (1977).

146. D. C. Appleton, J. McKenna, J. M. McKenna, L. B. Sims, and A. R. Walley, *J. Am. Chem. Soc.* **98,** 292 (1976).

147. E. L. Eliel and F. W. Vierhapper, J. Am. Chem. Soc. **97**, 2424 (1975).

148. F. A. L. Anet, I. Yavari, I. J. Ferguson, A. R. Katritzky, M. Moreno-Mañas, and M. J. T. Robinson, J. Chem. Soc. Chem. Comm. 399 (1976).

149. E. L. Eliel, C.-Y. Yen, and G. Z. Juaristi, Tetrahedron Lett. 2931 (1977).

150. Y. Kawazoe, M. Tsuda, and M. Ohnishi, Chem. Pharm. Bull. **15,** 51 (1967); Y. Kawazoe and M. Tsuda, Chem. Pharm. Bull. **15,** 1405 (1967).

151. J. C. N. Ma and E. W. Warnhoff, Can. J. Chem. **43,** 1849 (1965).

152. B. Bianchin and J. J. Delpuech, Tetrahedron **30,** 2859 (1974).

153. D. Wendisch, H. Feltkamp, and U. Scheidegger, Org. Mag. Res. **5,** 129 (1973).

154. H. Booth and D. V. Griffiths, J. Chem. Soc. Perkin 2, 842 (1973); see also H. Booth and M. L. Jozefowicz, J. Chem. Soc. Perkin 2, 895 (1976).

155. I. Morishima, K. Yoshikawa, K. Okada, T. Yonezawa, and K. Goto, J. Am. Chem. Soc. **95,** 165 (1973).

156. (a) G. Ellis and R. G. Jones, J. Chem. Soc. Perkin 2, 437 (1972); (b) see also M. W. Duch, Ph.D. Dissertation, University of Utah, Salt Lake City, Utah, 1970; cf. Ref. 2.

157. R. O. Duthaler, K. L. Williams, D. D. Giannini, W. H. Bearden, and J. D. Roberts, J. Am. Chem. Soc. **99,** 8406 (1977).

158. J. A. Pople and M. Gordon, J. Am. Chem. Soc. **89,** 4253 (1967).

159. W. Horsley and H. Sternlicht, J. Am. Chem. Soc. **90,** 3738 (1968); W. Horsley, H. Sternlicht, and J. S. Cohen, ibid., **92,** 680 (1970).

160. P. Geneste and J. M. Kamenka, Org. Mag. Res. **7,** 579 (1975).

161. E. W. Hagaman, Org. Mag. Res. **8,** 389 (1976).

162. I. Christofidis, A. Welter, and J. Jadot, Tetrahedron **33,** 977 (1977).

163. F. A. L. Anet and I. Yavari, Tetrahedron Lett. 3207 (1977); see also K. W. Baldry and M. J. T. Robinson, Tetrahedron **31,** 2621 (1975).

164. (a) I. Morishima, K. Okada, T. Yonezawa, and K. Goto, J. Am. Chem. Soc. **93,** 3922 (1971); (b) I. Morishima, K. Yoshikawa, and K. Okada, ibid., **98,** 3787 (1976); (c) K. Yoshikawa, M. Hashimoto, H. Masuda, and I. Morishima, J. Chem. Soc. Perkin 2, 809 (1977).

165. A. J. Jones and M. M. A. Hassan, J. Org. Chem. **37,** 2332 (1972).

166. A. J. Jones, A. F. Casy, and K. M. J. McErlane, *Tetrahedron Lett.* 1727 (1972); also *Can. J. Chem.*, **51,** 1782 (1973); A. J. Jones, C. P. Beeman, A. F. Casy, and K. M. J. McErlane, *Can. J. Chem.* **51,** 1790 (1973); P. Hanisch and A. J. Jones, *Can. J. Chem.* **54,** 2432 (1976).

167. J. A. Hirsch, R. L. Augustine, G. Koletar, and H. G. Wolf, *J. Org. Chem.* **40,** 3547 (1975).

168. C. Piccinni-Leopardi, O. Fabre, D. Zimmermann, J. Reisse, F. Cornea, and C. Fulea, *Org. Mag. Res.* **8,** 536 (1976); also *Can. J. Chem.*, **55,** 2649 (1977); see also U. Berg, *Can. J. Chem.* **55,** 2297 (1977).

169. (a) P. J. Wagner and B. J. Scheve, *J. Am. Chem. Soc.* **99,** 1858 (1977); (b) P. R. Srinivasan and R. L. Lichter, *Org. Mag. Res.* **8,** 193 (1976).

170. C. A. Kingsbury and M. E. Jordan, *J. Chem. Soc. Perkin 2,* 364 (1977).

171. (a) G. E. Ellis, R. G. Jones, and M. G. Papadopoulos, *J. Chem. Soc. Perkin 2,* 1381 (1974); (b) R. R. Fraser and T. B. Grindley, *Can. J. Chem.* **53,** 2465 (1975).

172. L. Lunazzi, G. Placucci, and G. Cerioni, *J. Chem. Soc. Perkin 2,* 1666 (1977).

173. C. Dorlet and G. Van Binst, *Anal. Lett.* **6,** 785 (1973).

174. (a) J. Okado and T. Esaki, *Yakugaku Zasshi* **93,** 1014 (1973); *Chem. Abstr.,* **79,** 104414k (1973); also *Yakugaku Zasshi* **95,** 56 (1975); *Chem. Abstr.,* **82,** 139051d (1975).

175. J. Okada and T. Esaki, *Chem. Pharm. Bull.* **22,** 1580 (1974).

176. A. Fratiello, M. Mardirossian, and E. Chavez, *J. Mag. Res.* **12,** 221 (1973).

177. F. I. Carroll and C. G. Moreland, *J. Chem. Soc. Perkin 2,* 374 (1974).

178. H. Labaziewicz, F. G. Riddell, and B. G. Sayer, *J. Chem. Soc. Perkin 2,* 619 (1977).

179. A. J. Jones, C. P. Beeman, M. U. Hasan, A. F. Casy, and M. M. A. Hassan, *Can. J. Chem.* **54,** 126 (1976); see also Ref. 180.

180. B. M. Pinto, D. M. Vyas, and W. A. Szarek, *Can. J. Chem.* **55,** 937 (1977).

181. F. A. L. Anet and I. Yavari, *Tetrahedron Lett.* 2093 (1976).

182. (a) S. F. Nelsen and G. R. Weisman, *J. Am. Chem. Soc.* **98,** 3281

(1976). (b) G. R. Weisman and S. F. Nelsen, *J. Am. Chem. Soc.* **98,** 7007 (1976).

183. Compare S. Wolfe, *Acc. Chem. Res.* **5,** 102 (1972).

184. L. D. Kopp, Ph.D. Dissertation, University of Notre Dame, Notre Dame, IN, 1972; E. W. Della, unpublished observations.

185. R. O. Hutchins, L. D. Kopp, and E. L. Eliel, *J. Am. Chem. Soc.* **90,** 7174 (1968); E. L. Eliel, L. D. Kopp, J. E. Dennis, and S. A. Evans, Jr., *Tetrahedron Lett.* 3409 (1971).

186. (a) E. L. Eliel, *Acc. Chem. Res.* **3,** 1 (1970); (b) also *Kemisk Tidskrift* **22,** 6/7 (1969).

187. C. H. Bushweller, M. Z. Lourandos, and J. A. Brunelle, *J. Am. Chem. Soc.* **96,** 1591 (1974).

188. (a) A. R. Katritzky, R. C. Patel, and D. M. Read, *Tetrahedron Lett.* 3803 (1977); (b) F. G. Riddell and J. M. Lehn, *J. Chem. Soc. B* 1224 (1968); see also I. J. Ferguson, A. R. Katritzky, and D. M. Read, *J. Chem. Soc. Perkin 2,* 818 (1977).

189. V. J. Baker, A. R. Katritzky, J.-P. Majoral, A. R. Martin, and J. M. Sullivan, *J. Am. Chem. Soc.* **98,** 5748 (1976).

190. E. L. Eliel and R. L. Willer, *J. Am. Chem. Soc.* **99,** 1936 (1977).

191. G. W. Buchanan and T. Durst, *Tetrahedron Lett.,* 1683 (1975).

192. R. R. Fraser, T. Durst, M. R. McClory, R. Viau, and Y. Y. Wigfield, *Int. J. Sulfur Chem. A,* **1,** 133 (1971).

193. P. K. Claus, W. Rieder, F. W. Vierhapper, and R. L. Willer, *Tetrahedron Lett.* 119 (1976); see also Refs. 195, 201.

194. P. K. Claus, W. Rieder, and F. W. Vierhapper, *Monatsh. Chem.* **109,** 631 (1978).

195. J. R. DeMember, R. B. Greenwald, and D. H. Evans, *J. Org. Chem.* **42,** 3518 (1977).

196. A. M. Sepulchre, B. Septe, G. Lukacs, S. D. Gero, W. Voelter, and E. Breitmaier, *Tetrahedron* **30,** 905 (1974).

197. See also K. Pihlaja and B. Björkqvist, *Org. Mag. Res.* **9,** 533 (1977).

198. For example, M. J. Cook and A. P. Tonge, *Tetrahedron Lett.* 849 (1973); also *J. Chem. Soc. Perkin 2,* 767 (1974).

199. J. K. Koskimies, Ph.D. Dissertation, University of North Carolina, Chapel Hill, NC, 1976.

200. R. B. Greenwald, D. H. Evans, and J. R. DeMember, *Tetrahedron Lett.* 3885 (1975).

201. P. K. Claus, F. W. Vierhapper, and R. L. Willer, *J. Chem. Soc.*, *Chem. Comm.* 1002 (1976).

202. K. Pihlaja and P. Pasanen, *Suomen Kemistilehti* **46**, 273 (1973); K. Pihlaja, personal communication.

203. W. A. Szarek, D. M. Vyas, A.-M. Sepulchre, S. D. Gero, and G. Lukacs, *Can. J. Chem.* **52**, 2041 (1974).

204. D. M. Frieze and S. A. Evans, *J. Org. Chem.* **40**, 2690 (1975).

205. K. Arai, M. Fukunaga, H. Iwamura, and M. Ōki, *Tetrahedron Lett.* 1685 (1976).

206. C. Giordano and A. Belli, *Synthesis* 193 (1977).

207. G. W. Buchanan, J. B. Stothers, and G. Wood, *Can. J. Chem.* **51**, 3746 (1973); *erratum* **53**, 2359 (1975).

208. G. W. Buchanan, N. K. Sharma, F. de Reinach-Hirtzbach, and T. Durst, *Can. J. Chem.* **55**, 44 (1977).

209. G. Wood, J. M. McIntosh, and M. H. Miskow, *Can. J. Chem.* **49**, 1202 (1971).

210. P. Albriktsen, *Acta Chem. Scand.* **27**, 3889 (1973).

211. S. I. Featherman and L. D. Quin, *J. Am. Chem. Soc.* **95**, 1699 (1973).

212. S. I. Featherman, S. O. Lee, and L. D. Quin, *J. Org. Chem.* **39**, 2899 (1974). Prof. Quin has kindly informed the authors that the CS_2-TMS shift difference used in this paper is 192.5 ppm.

213. Compare B. E. Mann, *J. Chem. Soc. Perkin 2*, 30 (1972).

214. S. I. Featherman and L. D. Quin, *Tetrahedron Lett.* 1955 (1973).

215. J. J. Breen, S. O. Lee, and L. D. Quin, *J. Org. Chem.* **40**, 2245 (1975).

216. A. T. McPhail, J. J. Breen, J. H. Somers, J. C. H. Steele, Jr., and L. D. Quin, *Chem. Comm.* 1020 (1971).

217. L. D. Quin, A. T. McPhail, S. O. Lee, and K. D. Onan, *Tetrahedron Lett.* 3473 (1974).

218. W. G. Bentrude, K. C. Yee, R. D. Bertrand, and D. M. Grant, *J. Am. Chem. Soc.* **93**, 797 (1971); W. G. Bentrude and H. W. Tan, *J. Am. Chem. Soc.* **95**, 4666 (1973).

219. E. E. Nifant'ev, A. A. Borisenko, and N. M. Sergeev, *Proc. Acad. Sci. USSR, Phys. Chem. Sec.* **208**, 100 (1973).

220. M. Haemers, R. Ottinger, D. Zimmermann, and J. Reisse, *Tetrahedron* **29**, 3539 (1973).

221. V. L. Foss, Yu. A. Veits, V. V. Kudinova, A. A. Borisenko, and I. F. Lutsenko, *J. Gen. Chem. USSR* **43,** 994 (1973).

222. A. A. Borisenko and N. M. Sergeev, *J. Gen. Chem. USSR* **44,** 2733 (1974).

223. A. Okruszek, W. J. Stec, and R. K. Harris, *Org. Mag. Res.* **9,** 497 (1977).

224. A. A. Borisenko, N. M. Sergeyev, E. Ye. Nifant'ev, and Yu. A. Ustynyuk, *J. Chem. Soc. Chem. Comm.* 406 (1972).

225. M. V. Sigalov, V. A. Pestunovich, V. M. Nikitin, A. S. Atavin, and M. Ya. Khil'ko, *Bull. Acad. Sci. USSR, Chem. Ser.* **26,** 1086 (1977).

226. (a) J. Martin and J. B. Robert, *Org. Mag. Res.* **7,** 76 (1975); (b) J. Martin, J. B. Robert, and C. Taieb, *J. Phys. Chem.* **80,** 2417 (1976).

227. E. E. Nifant'ev, A. A. Borisenko, A. I. Zavalisthina, and S. F. Sorokina, *Proc. Acad. Sci. USSR Chem. Sec.* **219,** 839 (1974).

228. V. S. Blagoveshchenskii, O. P. Yakovleva, A. A. Borisenko, N. V. Zyk, T. S. Kisileva, and S. I. Vol'fkovich, *Proc. Acad. Sci. USSR Chem. Sec.* **230,** 604 (1976).

229. J. Martin and J. B. Robert, *Org. Mag. Res.* **9,** 637 (1977).

230. W. Egan and G. Zon, *Tetrahedron Lett.* 813 (1976).

231. R. Kinas, K. Pankiewicz, W. J. Stec, P. B. Farmer, A. B. Foster, and M. Jarman, *J. Org. Chem.* **42,** 1650 (1977).

232. P. E. Rakita, L. S. Worsham, and J. P. Srebo, *Org. Mag. Res.* **8,** 310 (1976).

233. J. D. Kennedy, W. McFarlane, and G. S. Pyne, *Bull. Soc. Chim. Belg.* **84,** 289 (1975).

234. S. Samaan, *Z. Naturforsch* **32B,** 908 (1977).

235. D. J. Loomes and M. J. T. Robinson, *Tetrahedron* **33,** 1149 (1977).

236. K. C. Rice and R. E. Wasylishen, *Org. Mag. Res.* **8,** 449 (1976).

237. P. Y. Johnson and D. J. Kerkman, *J. Org. Chem.* **41,** 1768 (1976).

238. (a) M. H. Gianni, J. Saavedra, and J. Savoy, *J. Org. Chem.* **38,** 3971 (1973); (b) M. H. Gianni, J. Saavedra, J. Savoy, and H. G. Kuivila, *J. Org. Chem.* **39,** 804 (1974).

239. G. L. Kamalov, N. G. Luk'yanenko, Yu. Yu. Samitov, and A. V. Bogatskii, *J. Org. Chem. USSR* **13,** 1005 (1977).

240. M. H. Gianni, M. Adams, H. G. Kuivila, and K. Wursthorn, *J. Org. Chem.* **40,** 450 (1975).

241. H. Faucher, A. Guimaraes, and J. B. Robert, *Tetrahedron Lett.* 1743 (1977).

242. A. C. Guimaraes and J. B. Robert, *Tetrahedron Lett.* 473 (1976).

243. J. P. Dutasta, A. C. Guimaraes, and J. B. Robert, *Tetrahedron Lett.* 801 (1977).

244. R. E. DeSimone, M. J. Albright, W. J. Kennedy, and L. A. Ochrymowycz, *Org. Magn. Res.* **6,** 583 (1974).

245. D. Lloyd, R. K. Mackie, H. McNab, K. S. Tucker, and D. R. Marshall, *Tetrahedron* **32,** 2339 (1976).

246. W. Grahn, *Z. Naturforsch.* **31B,** 1641 (1976).

247. For an excellent review, see F. A. L. Anet, *Topics in Current Chemistry* **45,** 169 (1974).

248. F. A. L. Anet and P. J. Degen, *Tetrahedron Lett.* 3613 (1972).

249. F. A. L. Anet, P. J. Degen, and J. Krane, *J. Am. Chem. Soc.* **98,** 2059 (1976).

250. J. A. Ladd, *J. Mol. Struct.* **36,** 329 (1977).

251. T. T. Nakashima and G. E. Maciel, *Org. Mag. Res.* **4,** 321 (1972).

252. (a) D. S. Milbrath and J. G. Verkade, *J. Am. Chem. Soc.* **99,** 6607 (1977); (b) J. C. Clardy, D. S. Milbrath, and J. G. Verkade, *Inorg. Chem.* **16,** 2135 (1977).

253. (a) G. Vitt, E. Hadicke, and G. Quinkert, *Chem. Ber.,* **109,** 518 (1976); (b) J. B. Lambert and S. A. Khan, *J. Org. Chem.* **40,** 369 (1975).

254. (a) M. Ohnishi M.-C. Fedarko, J. D. Baldeschwieler, and L. F. Johnson, *Biochim. Biophys. Res. Comm.* **46,** 312 (1972); (b) M.-C. Fedarko, *J. Mag. Res.* **12,** 30 (1973).

255. (a) D. Live and S. I. Chan, *J. Am. Chem. Soc.* **98,** 3769 (1976); (b) L. R. Sousa and M. R. Johnson, *J. Am. Chem. Soc.* **100,** 344 (1978).

256. A. C. Coxon and J. F. Stoddart, *J. Chem. Soc. Perkin 1,* 767 (1977).

257. N. Herron, O. W. Howarth, and P. Moore, *Inorg. Chem. Acta* **20,** L43 (1976).

258. S. W. Shalaby, G. E. Babbitt, and R. L. Lapinski, *Spectros. Lett.* **6,** 231 (1973).

259. K. L. Williamson and J. D. Roberts, *J. Am. Chem. Soc.* **98,** 5082 (1976).

260. H. Fritz, P. Hug, H. Sauter, T. Winkler, and E. Logemann, *Org. Mag. Res.* **9,** 108 (1977).

261. K. Tsuboyama, S. Tsuboyama, J. Uzawa, K. Kobayashi, and T. Sakurai, *Tetrahedron Lett.* 4603 (1977).

262. H. Sliwa and J. P. Picavet, *Tetrahedron Lett.* 1583 (1977).

263. J.-P. Dutasta, J. Martin, and J.-B. Robert, *J. Org. Chem.* **42**, 1662 (1977).

264. P. K. Claus, F. W. Vierhapper, and R. L. Willer, *J. Org. Chem.* **42**, 4016 (1977).

265. K. L. Servis, E. A. Noe, N. R. Easton, and F. A. L. Anet, *J. Am. Chem. Soc.* **96**, 4185 (1974).

266. F. A. L. Anet and I. Yavari, *Tetrahedron Lett.* 1567 (1975).

267. D. Zimmermann, J. Reisse, J. Coste, E. Plénat, and H. Christol, *Org. Mag. Res.* **6**, 492 (1974).

268. E. Kleinpeter, H. Kühn, and M. Mühlstadt, *Org. Mag. Res.* 312 (1977).

269. N. R. Easton, F. A. L. Anet, P. A. Burns, and C. S. Foote, *J. Am. Chem. Soc.*, **96**, 3945 (1974).

270. F. Khuong-Huu, M. Sangare, V. M. Chari, A. Bekaert, M. Devys, M. Barbier, and G. Lukacs, *Tetrahedron Lett.* 1787 (1975).

271. M. Sangaré, B. Septe, G. Berenger, G. Lukacs, K. Tori, and T. Komeno, *Tetrahedron Lett.* 699 (1977).

272. P. Mison, R. Chaabouni, Y. Diab, A. Laurent, R. Martino, A. Lopez, F. W. Wehrli, and T. Wirthlin, *Org. Mag. Res.* **8**, 90 (1976).

273. P. Sohar, J. Kuszmann, G. Horvath, and V. Z. Mehefalvi, *Kem. Kozl.* **46**, 480 (1976); *Chem. Abstr.* **87**, 67470t (1977).

274. Y. Kashman and A. Rudi, *Tetrahedron Lett.* 2819 (1976).

275. C. Symmes, Jr. and L. D. Quin, *Tetrahedron Lett.* 1853 (1976).

276. M. Christl, *Chem. Ber.* **108**, 2781 (1975).

277. Y. Kobayashi, S. Fujino, H. Hamana, I. Kumadaki, and Y. Hanzawa, *J. Am. Chem. Soc.* **99**, 8511 (1977).

278. R. A. Archer, R. D. G. Cooper, P. V. Demarco, and L. F. Johnson, *Chem. Comm.* 1291 (1970).

279. C. M. Dobson, L. O. Ford, S. E. Summers, and R. J. P. Williams, *J. Chem. Soc. Faraday 2*, 1145 (1975).

280. J. C. Duggan, W. H. Urry, and J. Schaefer, *Tetrahedron Lett.* 4197 (1971).

281. I. M. Skvortsov and O. A. Subbotin, *J. Org. Chem. USSR* **13**, 426 (1977).

282. A. Pelter, R. S. Ward, E. V. Rao, and K. V. Sastry, *Tetrahedron* **32**,

2783 (1976); A. Pelter, R. S. Ward, and C. Nishino, *Tetrahedron Lett.* 4137 (1977).

283. W. K. Musker and P. B. Roush, *J. Am. Chem. Soc.* **98,** 6745 (1976).

284. K. Weinges, G.-U. Schwarz, M. Weber, and G. Schilling, *Chem. Ber.* **110,** 2961 (1977).

285. Y. Takeuchi, P. J. Chivers, and T. A. Crabb, *J. Chem. Soc. Chem. Comm.* 210 (1974).

286. V. A. Denisenko, V. L. Novikov, V. V. Isakov, A. V. Kamernitskii, and G. B. Elyakov, *Bull. Acad. Sci. USSR, Chem. Sci.* **25,** 1782 (1976).

287. C. Symmes, Jr. and L. D. Quin, *J. Org. Chem.* **41,** 238 (1976).

288. H. Singh and P. Singh, *Ind. J. Chem.* **14B,** 902 (1976).

289. F. Ramirez, J. F. Marecek, and H. Okazaki, *Tetrahedron Lett.* 4179 (1977).

290. W. Adcock, B. D. Gupta, and W. Kitching, *J. Org. Chem.* **41,** 1498 (1976).

291. H. Booth and D. V. Griffiths, *J. Chem. Soc. Perkin 2,* 111 (1975).

292. H. Booth, D. V. Griffiths, and M. L. Jozefowicz, *J. Chem. Soc. Perkin 2,* 751 (1976).

293. G. Van Binst and D. Tourwe, *Heterocycles* **1,** 257 (1973).

294. (a) D. W. Hughes, H. L. Holland, and D. B. MacLean, *Can. J. Chem.* **54,** 2252 (1976); (b) M. Christl, *Org. Mag. Res.* **7,** 349 (1975).

295. G. Van Binst, D. Tourwe, and E. De Cock, *Org. Mag. Res.* **8,** 618 (1976).

296. R. T. LaLonde and T. N. Donvito, *Can, J. chem.* **52,** 3778 (1974).

297. Y. Arata, T. Aoki, M. Hanaoka, and M. Kamei, *Chem. Pharm. Bull.* **23,** 333 (1975).

298. M. Sugiura and Y. Sasaki, *Chem. Pharm. Bull.* **24,** 2988 (1976).

299. M. Sugiura, N. Takao, and Y. Sasaki, *Chem. Pharm. Bull.* **25,** 960 (1977).

300. F. W. Vierhapper and R. L. Willer, *J. Org. Chem.* **42,** 4024 (1977).

301. D. K. Dalling, D. M. Grant, and E. G. Paul, *J. Am. Chem. Soc.* **95,** 3718 (1973).

302. D. M. Roush and C. H. Heathcock, *J. Am. Chem. Soc.* **99,** 2337 (1977).

303. H. Kessler, H. Möhrle, and G. Zimmermann, *J. Org. Chem.* **42,** 66 (1977).

304. S. Iwase, T. Maeda, S. Hamanaka, and M. Ogawa, *Nippon Kagaku Kaishi* 1934 (1975); *Chem. Abstr.* **84,** 29989z (1976).

305. D. Quarroz, J.-M. Sonney, A. Chollet, A. Florey, and P. Vogel, *Org. Mag. Res.* **9,** 611 (1977).

306. K. Yoshikawa, K. Bekki, M. Karatsu, K. Toyoda, T. Kamio, and I. Morishima, *J. Am. Chem. Soc.* **98,** 3272 (1976).

307. Y. Nomura, N. Masai, and Y. Takeuchi, *J. Chem. Soc. Chem. Comm.* 288 (1974).

308. R. B. Wetzel and G. L. Kenyon, *J. Am. Chem. Soc.* **96,** 5189 (1974).

309. S. F. Nelsen, and G. R. Weisman, *J. Am. Chem. Soc.* **98,** 1842 (1976).

310. H. J. Reich and J. E. Trend. *J. Org. Chem.* **38,** 2637 (1973).

311. E. Kleinpeter, G. Haufe, M. Mühlstadt, and J. Graefe, *Org. Mag. Res.* **9,** 105 (1977).

312. Y. Senda, J.-i. Ishiyama, and S. Imaizumi, *Bull. Chem. Soc. Japan* **49,** 1359 (1976).

313. G. T. Pearce, W. E. Gore, and R. M. Silverstein, *J. Mag. Res.* **27,** 497 (1977).

314. (a) L. Simeral and G. E. Maciel, *Org. Mag. Res.* **6,** 226 (1974). (b) S. J. Daum, C. M. Martini, R. K. Kullnig, and R. L. Clarke, *J. Med. Chem.* **18,** 496 (1975).

315. K.-H. Pook, W. Schulz, and R. Banholzer, *Justus Liebigs Ann. Chem.* 1499 (1975).

316. H.-J. Schneider and L. Sturm, *Angew. Chem. Int. Ed.* (Engl.) **15,** 545 (1976).

317. M. Lounasmaa, P. M. Wovkulich, and E. Wenkert, *J. Org. Chem.* **40,** 3694 (1975).

318. P. Hanish, A. J. Jones, A. F. Casey, and J. E. Coates, *J. Chem. Soc. Perkin 2,* 1202 (1977).

319. V. J. Baker, I. J. Ferguson, A. R. Katritzky, and R. Patel, *Tetrahedron Lett.* 4735 (1976).

320. H.-J. Schneider, M. Lonsdorfer, and E. F. Weigand, *Org. Mag. Res.* **8,** 363 (1976).

321. A. Heumann and H. Kolshorn, *Tetrahedron* **31,** 1571 (1975).

322. J. A. Peters, J. M. van der Toorn, and H. van Bekkum, *Tetrahedron* **33,** 349 (1977).

323. J. R. Wiseman and H. O. Krabbenhoft, *J. Org. Chem.* **40,** 3222 (1975).

324. J. R. Wiseman and H. O. Krabbenhoft, *J. Org. Chem.* **41,** 589 (1976).

325. J. R. Wiseman, H. O. Krabbenhoft, and B. R. Anderson, *J. Org. Chem.* **41,** 1518 (1976).

326. S. F. Nelsen, G. R. Weisman, E. L. Clennan, and V. E. Peacock, *J. Am. Chem. Soc.* **98,** 6893 (1976).

327. S. Morris-Natschke and E. L. Eliel, unpublished observations.

328. P. H. McCabe and C. R. Nelson, *J. Mag. Res.* **22,** 183 (1976).

329. J. A. J. M. Vincent, P. Schipper, A. de Groot, and H. M. Buck, *Tetrahedron Lett.* 1989 (1975).

330. M. Barrelle, M. Apparu, and C. Gey, *Tetrahedron Lett.* 4725 (1976).

331. A. G. Anastassiou and E. Reichmanis, *J. Am. Chem. Soc.* **98,** 8267 (1976); see also A. G. Anastassiou, *Acc. Chem. Res.* **9,** 453 (1976).

332. A. R. Farminer and G. A. Webb, *Org. Mag. Res.* **8,** 102 (1976).

333. R. E. Wasylishen and B. A. Pettit, *Can. J. Chem.* **55,** 2564 (1977).

334. Y. Shahab, *Org. Mag. Res.* **9,** 580 (1977).

335. A. W. J. D. Dekkers, J. W. Verhoeven, and W. N. Speckamp, *Tetrahedron* **29,** 1691 (1973).

336. F. Ramirez, I. Ugi, F. Lin, S. Pfohl, P. Hoffman, and D. Marquarding, *Tetrahedron* **30,** 371 (1974).

337. M. Sangare, F. Khuong-Huu, D. Herlem, A. Milliet, B. Septe, G. Berenger, and G. Lukacs, *Tetrahedron Lett.* 1791 (1975).

338. L. Zetta and G. Gatti, *Tetrahedron* **31,** 1403 (1975); also *Org. Mag. Res.* **9,** 218 (1977).

339. T. T. Nakashima, P. P. Singer, L. M. Browne, and W. A. Ayer, *Can. J. Chem.* **53,** 1936 (1975).

340. S. Takeuchi, J. Uzawa, H. Seto, and H. Yonehara, *Tetrahedron Lett.* 2943 (1977).

341. S. W. Pelletier and Z. Djarmati, *J. Am. Chem. Soc.* **98,** 2626 (1976).

342. K. Jankowski, *J. Org. Chem.* **41,** 3321 (1976).

343. H. Fritz and D. Weis, *Tetrahedron Lett.* 1659 (1974).

344. E. Wenkert, M. E. Alonso, H. E. Gottlieb, E. L. Sanchez, R. Pellicciari, and P. Cogolli, *J. Org. Chem.* **42,** 3945 (1977).

345. K. Kleinpeter, H. Kühn and M. Mühlstadt, *Org. Mag. Res.* **9,** 90 (1977).

346. B. Donzel, B. Kamber, K. Wüthrich, and R. Schwyzer, *Helv. Chim. Acta* **55,** 947 (1972).

347. A. H. Andrist and M. J. Kovelan, *Spectrosc. Lett.* **10,** 127 (1977).

348. S. Sørensen and H. J. Jakobsen, *Org. Mag. Res.* **9,** 101 (1977).

349. J. ten Broeke, A. W. Douglas, M. A. Kozlowski, and E. J. J. Grabowski, *Tetrahedron Lett.* 4303 (1977).

350. H. Beierbeck and J. K. Saunders, *Can. J. Chem.* **55,** 3161 (1977).

351. J. P. Snyder, L. Lee, and D. G. Farnum, *J. Am. Chem. Soc.* **93,** 3816 (1971).

352. J. W. ApSimon, K. Yamasaki, A. Fruchier, and A. S. Y. Chau, *Tetrahedron Lett.* 3677 (1977).

353. T. G. Dekker, K. G. R. Pachler, and P. L. Wessels, *Org. Mag. Res.* **8,** 530 (1976).

354. K. G. R. Pachler, P. L. Wessels, J. Dekker, J. J. Dekker, and T. G. Dekker, *Tetrahedron Lett.*, 3059 (1976).

355. J. Dekker, J. J. Dekker, L. Fourie, T. G. Dekker, K. G. R. Pachler, and P. L. Wessels, *Tetrahedron Lett.* 1613 (1976).

356. H. Gilboa, A. Rüttimann, and D. Ginsburg, *Tetrahedron* **33,** 1189 (1977).

357. K. Weinges, J. Pill, K. Klessing, and G. Schilling, *Chem. Ber.* **110,** 2969 (1977).

358. K. Weinges, H. Baake, H. Distler, K. Klessing, R. Kolb, and G. Schilling, *Chem. Ber.* **110,** 2978 (1977).

359. H. Hiemstra and H. Wynberg. *Tetrahedron Lett.* 2183 (1977).

360. E. Juaristi, Ph.D. Dissertation, University of North Carolina, Chapel Hill, NC, 1976; E. Juaristi, unpublished data.

361. H. Fritz, E. Logemann, G. Schill, and T. Winkler, *Chem. Ber.* **109,** 1258 (1976).

362. J. Altman, H. Gilboa, and D. Ben-Ishai, *Tetrahedron* **33,** 3173 (1977).

4 HIGH-RESOLUTION ^{13}C NMR OF SOLID POLYMERS

JACOB SCHAEFER
E. O. STEJSKAL

CONTENTS

I.   LINE-NARROWING TECHNIQUES

A.   Dipolar Decoupling

The single-resonance, natural abundance ^{13}C nmr spectrum of a solid usually consists of a single, broad, featureless line, with a width of the order of 20 kHz. The major source of line broadening in this situation is the static dipolar interaction between carbons and nearby protons. These protons include covalently bonded methine-, methylene-, or methyl-protons, together with more distant, indirectly bonded protons. As illustrated in Figure 4.1, dipolar interactions, $\mathcal{H}_{dipolar}$, depend upon the orientation of the ^{1}H—^{13}C internuclear vector with the applied static magnetic field, H_0. Each dipolar interaction produces a characteristic splitting. In a crystal powder or an amorphous material, all orientations occur, resulting in a distribution of dipolar splittings. Thus, the observed ^{13}C resonance is inhomogeneously broadened; that is, it results from a sum of a large number of sharper lines produced by the various dipolar splittings.

This dipolar broadening can be removed in a straightforward way.[1] If the ^{13}C resonance is observed in the presence of a strong radiofrequency (rf) field, H_{1I}, at the Larmor frequency of the protons (I), the protons undergo rapid transitions, or spin flips, which cause the dipolar field

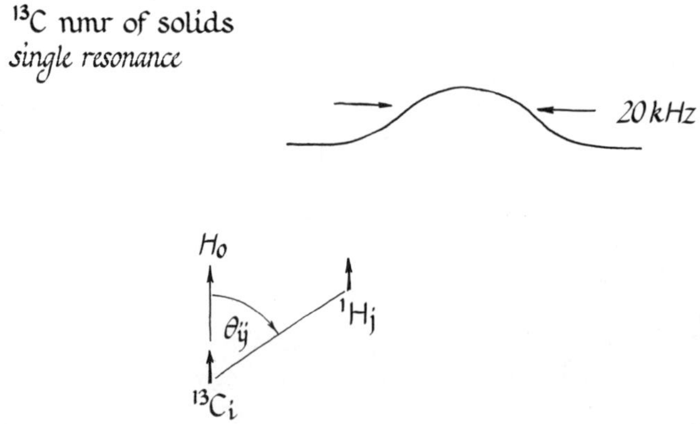

^{13}C nmr of solids
single resonance

20 kHz

H_0

θ_{ij}

^1H_j

$^{13}C_i$

$$\mathcal{H}_{dipolar} = \sum_{i<j}(\text{constants})\,r_{ij}^{-3}(3\cos^2\theta_{ij}-1)(\text{spin terms})$$

Figure 4.1
Schematic representation of a single-resonance ^{13}C lineshape broadened by dipolar interactions with protons in an amorphous solid.

generated by the protons at the carbon nucleus to disappear. In classical terms, the static dipolar interaction between carbons and protons becomes proportional to $\cos \theta$ where $\theta = \arctan(H_{1I}/H_{loc})$ and H_{loc} is the local dipolar field arising from the sum of various dipolar interactions. As H_{1I} becomes large, θ approaches $90°$, $\cos \theta$ approaches zero, and the static dipolar coupling between carbons and protons vanishes.[1,2] The entire process is known as dipolar decoupling, and is analogous to the more familiar scalar decoupling used to remove spin-spin J-splitting from the high-resolution ^{13}C nmr spectra of liquids.[3] The only difference is in the strength of the rf fields used in the two experiments. Dipolar decoupling requires rf fields greater than the local fields experienced by the protons arising from $^1H-^1H$ and $^1H-^{13}C$ interactions. For a typical solid this might require an rf field of 10 G, or about 10 times as large as is necessary to perform ordinary scalar decoupling.

B. Magic-Angle Spinning

With dipolar decoupling, the ^{13}C nmr spectrum of a solid begins to show signs of improved resolution. As illustrated in Figure 4.2, the width of a

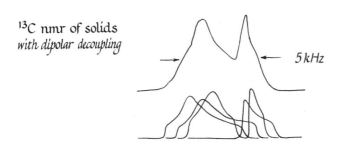

^{13}C nmr of solids
with dipolar decoupling

5 kHz

$$\mathcal{H}_{cs} = \sum_i (\text{constants}) \sigma_{izz} I_{iz} H_0$$

$$\sigma_{izz} = \lambda^2_{i1} \sigma_{i1} + \lambda^2_{i2} \sigma_{i2} + \lambda^2_{i3} \sigma_{i3}$$

$$\overline{\sigma}_{izz} = \sigma_i$$

Figure 4.2
Schematic representation of a double-resonance ^{13}C lineshape broadened by overlapping chemical shift anisotropies, the latter expressed in terms of the principal components of the chemical shift tensor σ_{ij} of carbon nucleus i and various direction cosines; $\sigma_{ij}(j = 1-3)$ refer to experimental values in order of increasing magnetic field, before assignment to molecular axes.

typical spectrum is now of the order of 5 kHz (at 20 MHz), with a few features clearly evident. The spectrum is not, however, of liquidlike high-resolution quality. The remaining broadening is due to chemical shift anisotropy (CSA). The magnetic field at a carbon nucleus depends upon the shielding or screening, $\mathcal{H}_{cs}$, afforded by the surrounding electron density. In general, the surrounding electron density is not symmetric. Thus, the chemical shift of, say, a carbonyl carbon in an ester group, depends upon whether the C—O carbonyl bond is lined along the magnetic field, or happens to be perpendicular to it, or is in some other orientation. In a liquid with rapid molecular motion only an average, or isotropic chemical shift, σ_i, is observed. In a single crystal, a single chemical shift may be observed, but its value depends upon the orientation of the crystal relative to H_0. In an amorphous solid or crystal powder, on the other hand, a complicated lineshape is observed that arises from the sum of all possible chemical shifts.[4]

This lineshape, or CSA, can be described[5] in terms of what are known as the principal values of the chemical shift tensor of carbon nucleus i, $\sigma_{ij}(j = 1-3)$, together with the direction cosines, $\lambda_{ij}^2(j = 1-3)$. The former contain information of electronic bonding and the latter describe the orientation of the chemical shift tensor relative to H_0. The most general line shape is tentlike, with the extremes of the tent defining σ_{i1} and σ_{i3}, and the apex of the tent, usually off center, defining σ_{i2}. Molecular symmetry can cause the apex of the tent to appear at one of the extremes of the CSA pattern, so that there are only two unique principal values of the chemical shift tensor.[6] This is the situation, for example, for methyl carbons having rapid, internal rotation creating a C_{3v} molecular axis of symmetry.

The line-broadening effect of the CSA is now apparent. For a typical solid with a variety of chemically different carbons, the ^{13}C nmr spectrum is a sum of different CSA patterns, having somewhat different shapes, and most importantly, having different isotropic centers. Overlapping CSA patterns for carbonyl carbons, aromatic carbons, and aliphatic carbons can, as illustrated in Figure 4.2, destroy the resolution one had hoped to gain from dipolar decoupling.

Fortunately, there is a method to regain this lost resolution. The broadening arises from restrictions placed on molecular motion in the solid. Experimentally, we can supply a kind of molecular motion ourselves. We can do this by mechanically rotating the sample.[7] Under a rigid-body rotation, the direction cosines, λ_{ij}^2, transfrom according to the spherical harmonics addition law. This means the CSA, σ_{izz}, can be expressed conveniently in terms of the mechanical rotation angle.[5] As shown in Figure 4.3, this expression has a simple form. Furthermore, σ_{izz}

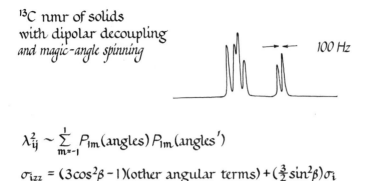

^{13}C nmr of solids
with dipolar decoupling
and magic-angle spinning

$\longrightarrow\!\!\longleftarrow$ $100\,Hz$

$$\lambda_{ij}^2 \sim \sum_{m=-1}^{1} P_{1m}(\text{angles})\, P_{1m}(\text{angles}')$$

$$\sigma_{izz} = (3\cos^2\beta - 1)(\text{other angular terms}) + (\tfrac{3}{2}\sin^2\beta)\sigma_i$$

β is angle between spinning axis and H_0

Figure 4.3
Schematic representation of a double-resonance ^{13}C lineshape under high-speed magic-angle spinning. At the magic angle $\beta = 54.7°$, $3\cos^2\beta - 1 = 0$, $\sigma_{izz} = \sigma_i$.

has the marvelous property of reducing to σ_i when the rotation angle, β, is chosen to be 54.7° and the rotation rate is fast compared with the CSA. This value of β makes $(3\cos^2\beta - 1)$ equal to zero and $\tfrac{3}{2}\sin^2\beta$ equal to one. In other words, overlapping CSA patterns reduce to their isotropic averages, resulting in genuine high-resolution ^{13}C nmr spectra of solids. Quite understandably, this rotation angle is called the "magic angle."

Experimentally, magic-angle spinning can be achieved in a variety of ways. One of the earliest techniques, illustrated in the upper right-hand corner of Figure 4.4, was used by Andrew[5,7] in the late 1950's. He employed a design used in an ultracentrifuge introduced by Beams about 25 years earlier.[8] The analytic sample is cylindrical and attached to a conical rotor. The rotor just fits into a stationary piece called a stator (not shown), and an air bearing is created by pumping air between the rotor and stator. Rotation is achieved by directing air against flutes cut in the side of the rotor. This design is useful also for slow spinning in ordinary high-resolution experiments on liquids. Andrew achieved the magic angle using a vertical spinning axis, tilting his magnet, and then supporting the magnet with wooden blocks. The Beams design can also be used in ordinary iron magnets, and in superconducting solenoids, by fitting the stator with a gear mechanism so that the rotation axis can be tilted relative to H_0 (Figure 4.4, lower left and lower right).

A second design was introduced by Lowe,[9] also in the late 1950's (Figure 4.4, upper left). Lowe used a cylindrical rotor, supported by two

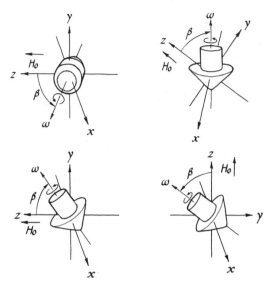

Figure 4.4
Magic-angle rotors of the Lowe geometry (upper left) and the Beams geometry. The Beams rotors are shown with a vertical spinning axis for use with a tilted iron magnet (upper right), and with a tilted spinning axis for use with a normal iron magnet (lower left) or a superconducting solenoid (lower right).

thin phosphor-bronze axles (not shown). The axle holes in the rotor body were a few thousandths of an inch greater than the axle itself, and this permitted a self-generating air bearing to form once the rotor was up to speed. Spinning was achieved by directing an air jet on flutes machined on the body of the rotor.

Lowe used an ordinary iron magnet and achieved the magic angle by rotation of the axle supports in the horizontal plane. Each of the two spinning designs has its unique advantages. The Lowe-design rotor is simpler to construct, is easy to thermostat, and makes efficient use of available rf power since the rf coil is orthogonal to H_0. The Beams-design rotor is generally easier to spin since the air bearing is not self-pumped and has a better filling factor for nmr sensitivity since the receiver coil can be wound with its axis parallel to the spinning axis.

The dipolar-decoupled ^{13}C nmr spectra of Lowe-design rotors made from polysulfone and poly(ether sulfone), two rigid engineering polymers, both with and without 3-kHz magic-angle spinning, are compared to the standard spectra of the two polymers in solution in Figure 4.5.[10] This

polysulfone

poly(ether sulfone)

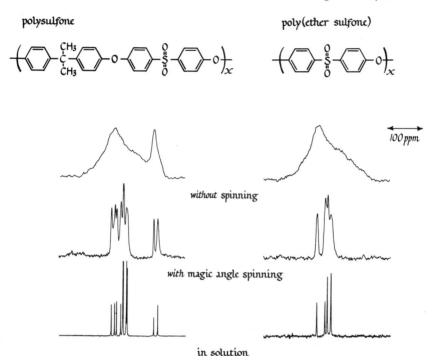

without spinning

with magic angle spinning

in solution

Figure 4.5
Dipolar-decoupled CP ^{13}C nmr spectra of polysulfone and poly (ether sulfone), with and without magic-angle spinning. The CP spectra are compared to FT spectra of the polymers in solution (with solvent lines omitted for clarity of presentation). There are ten lines in the FT spectrum of polysulfone and eight in the magic-angle CP spectrum. Spinning sidebands are visible in the CP spectrum of polysulfone.

figure dramatizes the importance of spinning in achieving high resolution in the solid-state ^{13}C nmr spectra of polymers. Except for the high-field line in the spectrum of polysulfone, the spectra of the two nonspinning polymers are virtually identical. With magic-angle spinning, however, differences in the aromatic-carbon region are clearly established. In fact, the magic-angle spinning spectra of the solid polymers are almost as detailed as the solution spectra. This makes it easy to see, for example, that there are no significant differences between the bulk and solution isotropic chemical shifts for either polymer. The line narrowing achieved by spinning is remarkably good.[11] The linewidths of the spectra of the spinning solids are about a factor of 50 less than those of the corresponding stationary solids. This remains true even when the magic-angle spinning rate is less than the ^{13}C CSA patterns. We will discuss at the end of Section II.F the reason why magic-angle spinning (in combination with dipolar decoupling) is so successful in ^{13}C line narrowing.

II. SENSITIVITY ENHANCEMENT BY CROSS POLARIZATION

A. Hartmann-Hahn Condition

Even with the resolution achieved by a combination of dipolar decoupling and magic-angle spinning, a Fourier transform (FT) experiment on a solid polymer still has a serious limitation. Namely, a delay time of several ^{13}C spin-lattice relaxation times (T_1) must be tolerated before data sampling can be repeated. These repetitions are necessary to provide a suitably strong signal by a time-averaging process. Since some ^{13}C T_1s for solid polymers are on the order of tens of seconds, the time-averaging process becomes tedious. These delays can be avoided, however, if the carbon polarization can be generated at the expense of some other spins (the protons) that recover more rapidly. This procedure is called cross polarization (CP).

The CP experiment was introduced by Hartmann and Hahn[12] in 1962, and first applied to ^{13}C nmr by Pines et al.[6] some 10 years later. As illustrated in Figure 4.6 (top), transfer of polarization from protons to carbons under normal conditions in H_0 is slow. It is slow because of the frequency mismatch of the Larmor resonance frequencies of the two nuclei. This mismatch means that the mutual ^{1}H—^{13}C spin flips required for a CP transfer are not energy-conserving. The energy difference must be made up from lattice phonons; the latter involve high-frequency T_1 processes which, as we noted above, are not always efficient in the solid.

Hartmann and Hahn realized that by performing the CP transfer as a double rotating-frame spin-lock (SL) experiment the energy mismatch could be avoided. Let us postpone for a moment the problem of how one achieves a spin-locked condition in a rotating field of, say, 20 G (while the sample physically remains in the 20 kG static field), and simply accept that both protons and carbons can be placed, maintained, and manipulated in rf fields just as though the static field did not exist. The advantage of the Hartmann-Hahn scheme is that the proton (I) and carbon (S) rotating field amplitudes, H_{1I} and H_{1S}, can be separately adjusted experimentally to satisfy the frequency condition $\gamma_I H_{1I} = \gamma_S H_{1S}$, where the γ_S are the respective gyromagnetic ratios of the nuclei. This condition permits exactly matched energy-conserving mutual spin

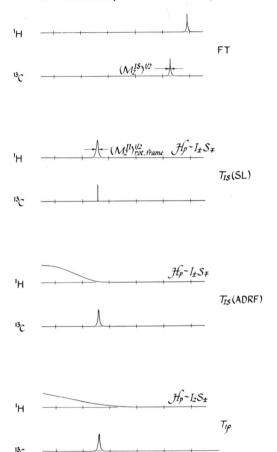

Figure 4.6
Spectral overlap between protons and carbons in four nmr experiments. Linewidths on the logarithmic frequency scale are schematic. The spin part of the coupling interaction, H_p, is identified for the CP experiments.

flips between carbons and protons (Figure 4.6, second from top). Thus, polarization of the hot carbons is achieved from the cold protons, spin-locked in their own rf field, via static dipolar interactions, in a time $T_{IS}(SL)$.

We will show a little later that this transfer is a T_2 process, and so for solid polymers generally requires no more than 100 μsec. Most importantly, the transfer can be repeated and more data accumulated after allowing the protons to repolarize in the static field. For most polymers near room temperature, this is more efficient than ^{13}C repolarization by spin-lattice processes and generally occurs in less than a half second. More elaborate transfer schemes are usually not required.

B. Spin Locking

Let us return now to the problem of achieving a spin lock.[13] Consider a proton spin system polarized by a large static field H_0 aligned along the z axis shown in Figure 4.7. We apply an rf field H_{1y} at the Larmor frequency of the protons. As viewed in a coordinate system rotating with H_{1y}, the proton magnetization, M_0, rotates to $M_\perp$, when the rf field is left on for a length of time equal to that of a 90° pulse, 90_y. A 90° pulse usually requires a few microseconds to complete. We now change the phase of the rf field by 90° so that the applied field is H_{1x}. This phase shift can be achieved either by a digital manipulation of the synthesizer generating the rf, or by an analog method such as re-routing of the rf along an extra quarter wavelength of cable. If H_{1x} is greater than any local dipolar fields, the proton magnetization is "spin locked" along its H_{1x}. The magnetization is under the influence of an effective field, $H_{eff} = [(H_0 - \omega/\gamma)^2 + H_{1x}^2]^{1/2}$, where ω is the frequency of the rotating coordinate system and γ the gyromagnetic ratio. At resonance ω is equal

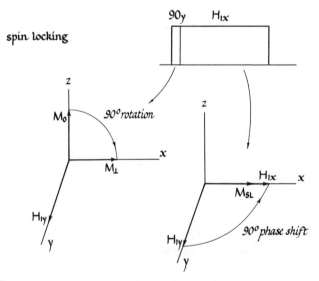

Figure 4.7
Spin locking a magnetization, M_0, by a resonant rf field, H_{1x}. The insert shows the pulse sequence. A left-handed coordinate system is used.

to ω_0 the Larmor frequency and the term in parentheses becomes zero. Thus the magnetization precesses about H_{1x}. The final result is that a large proton magnetization can be aligned along an applied rf field of only, say, 5–10 G. This permits the variable field, low-frequency experiments illustrated in Figure 4.6. The magnetization stays spin locked for a characteristic time, $T_{1\rho}$, about which we will have more to say in Section IV.E.

C. Matched Cross-Polarization Transfer

A classical picture of the spin lock CP transfer is shown in Figure 4.8. The static field, H_0, is aligned along the shared z axis of both carbon (S) and proton (I) spin systems. The carbon coordinate system is stationary as viewed by an observer rotating about the z axis with frequency, ω_{0S}; the same is true for the proton coordinate system but with frequency, ω_{0I}.

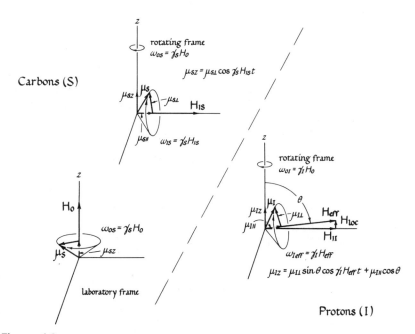

Figure 4.8
A classical representation of the mutual stationary components of proton and carbon rotating-frame magnetizations, μ_{Iz} and μ_{Sz}, respectively, under the Hartmann-Hahn condition. The insert at lower left shows the carbon magnetization in the laboratory frame. In the presence of dipolar decoupling, the static component of μ_{Iz} is removed eliminating a static interaction with the laboratory μ_{Sz}.

The protons are spin locked (as described in Figure 4.7) by H_{1I}. We assume H_{1I} is larger than H_{loc} so that H_{eff} is nearly along H_{1I} and θ is almost 90°. In the presence of H_{1I}, the carbons are dipolar decoupled and so they are exactly aligned along their applied rf field, H_{1S}. At this point, before any CP transfer has taken place, the net carbon magnetization may be nil. Nevertheless, each individual carbon spin behaves according to the picture shown in the upper left of Figure 4.8. Carbons and protons have components of magnetization, μ_{Sz} and μ_{Iz}, respectively, along the common z axis. These components are fluctuating with time according to the expressions shown in Figure 4.8. If the Hartmann-Hahn condition is satisfied ($\gamma_S H_{1S} = \gamma_I H_{1I}$), these z-axis components have the same time dependence. Thus, in a doubly rotating coordinate system, the carbons and protons have a common *stationary* component of magnetization.[14] It is this component with a time dependence common to both spin systems that allows CP transfers.

Of course, carbons and protons have a common stationary component of magnetization as described above only if they are undergoing no microscopic molecular motion. In the absence of molecular motion the simple classical pictures of Figure 4.8 are useful in understanding the spin dynamics of spin-lock CP transfers. More generally, however, the dipolar coupling interaction responsible for cross relaxation between I and S spins should be considered. In the laboratory reference frame this interaction over all spins N_I and N_S can be expressed[15] as

$$\mathcal{H}_{IS} = 2\gamma_I\gamma_S\hbar^2 \sum_{i=1}^{N_I} \sum_{m=1}^{N_S} r_{im}^{-3} P_2(\cos\theta_{im}) I_{iz}S_{mz}$$

where r_{im} is the distance between $i(I)$ and $m(S)$ spins with θ_{im} the angle between the vector r_{im} and H_0. The operators I_{1z} and S_{mz} are the usual spin operators and $P_2(\cos\theta) = (3\cos^2\theta - 1)/2$. The interaction H_{IS} transforms approximately in the doubly rotating reference frame to

$$\mathcal{H}_p = \sin\theta \sum_{i,m} b_{im} I_{ix}S_{mx}$$

where

$$b_{im} = 2\gamma_I\gamma_S\hbar^2 r_{im}^{-3} P_2(\cos\theta_{im})$$

and

$$\theta = \arctan(H_{1I}/H_{loc})$$

We see that the spin operator part of this interaction involves I_xS_x and so $I_{\pm}S_{\mp}$. This is reasonable since the spin-lock CP transfer results from a mutual spin flip between carbons and protons produced by the combination of spin-raising and spin-lowering operators (16). The requirement of a CP transfer for a strong static interaction is contained in b_{im}; this term averages to zero in the presence of rapid, isotropic molecular motion.

An approximate expression[6] for the matched spin-lock CP transfer rate is

$$[T_{IS}(SL)]^{-1} = \frac{C_{IS} M_2^{IS}}{(M_2^{II})^{1/2}}$$

where C_{IS} is a constant depending on geometrical factors and M_2^{IS} and M_2^{II} are the ^{13}C and ^1H second moments, respectively. In rigid solids M_2^{IS}

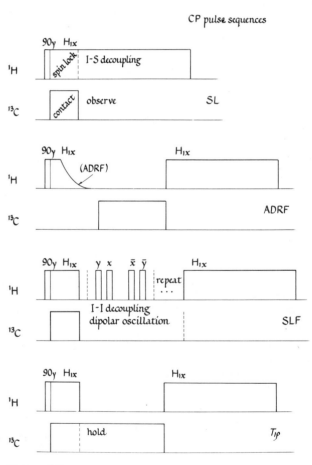

CP pulse sequences

Figure 4.9
Pulse sequences for four cross-polarization experiments.

is substantial, leading to a fast CP transfer rate. Note that a large M_2^{II} results in a reduced rate since (on a qualitative basis) protons compete with one another in performing mutual spin flips and so interfere with proton-carbon CP spin flips.

Because cross polarization involves no spin lattice processes directly, signal enhancements depending on nuclear Overhauser effects are not observed. Instead, by spin thermodynamic arguments[6] which we will not discuss, it is possible to show that the optimum carbon magnetization following a single spin-lock CP transfer is H_{1S}/H_{1I} (or γ_I/γ_S, a factor of 4) greater than the magnetization that would be observed in a fully relaxed FT experiment with gated dipolar decoupling. Actually, this ideal is not reached in practice. Losses in efficiency occur in the spin locking of the protons (approximately 25%) and in the fact that H_{1I} is seldom much greater than H_{loc} so that H_{1I} does not quite equal H_{eff}.

The pulse sequence for the single-contact spin-lock CP experiment (designated SL) is particularly simple (Figure 4.9. top). It involves a spin lock of the protons as described earlier, followed by a contact between carbons and protons, after which the carbon magnetization is observed with dipolar decoupling of the protons.

D.   Adiabatic Demagnetization in the Rotating Frame (Mismatched) Transfer

The procedure for CP transfer using adiabatic demagnetization in the rotating frame[12,14,17] (ADRF) is also shown in Figure 4.9 (second from top). Following a spin lock of the protons, H_{1I} is slowly (compared to the ^{1}H inverse linewidth or T_2) reduced to zero thereby transferring the dipolar order of the spin lock from the applied rf field to the local dipolar fields surrounding each proton. A CP transfer can now be made during the contact time with the dual advantage that first, no particular attention need be paid to satisfying the Hartmann-Hahn condition, and second, the greater one makes H_{1S}, the larger the observed carbon magnetization.[6,18] (The carbons achieve a greater polarization corresponding to the increased carbon rf field relative to the field locking the protons). Finally, the carbon magnetization is observed with dipolar decoupling as before. The price one must pay for the improved sensitivity in the ADRF scheme is in the length of time required to complete the CP transfer; that is, T_{IS} (ADRF) is longer than T_{IS}(SL), usually by orders of magnitude.

The reason for the much slower CP transfer rate in the ADRF experiment can be appreciated from examination of Figure 4.6. In the ADRF situation (third from top), despite the width of the carbon line in the absence of dipolar decoupling, the spectral overlap between I and S spin systems is limited to the tails of the two resonances.[14] This is in sharp contrast to the effective overlap under the Hartmann-Hahn condition (second from top). Clearly, with a limited frequency match, the stationary components of magnetization necessary to cross polarization are severely reduced. As a result, the CP transfer rate is slow. Demco et al.[15] have developed a quantitative theoretical analysis that describes the dependence of the CP transfer rate on frequency mismatch in both the SL and (ADRF) is actually longer than the proton and carbon spin-lattice $T_{1\rho S}$ making the ADRF experiments impractical.[10] The proton spin lock is lost before the CP transfer can be made. However, when the experiment is practical, it can have useful applications, as we will illustrate later in Section IV.D.

E. Experimental Determination of Cross-Polarization Transfer Rates

The buildup of carbon magnetization as a function of spin-lock CP contact with polarized protons is shown for poly(methyl methacrylate) in Figure 4.10.[10] These spectra were taken with magic-angle spinning and are directly comparable to spectra of the same polymer obtained under nonspinning conditions, presented earlier.[19] The two sets of results are similar. The low-field carbonyl-carbon builds up most slowly and the

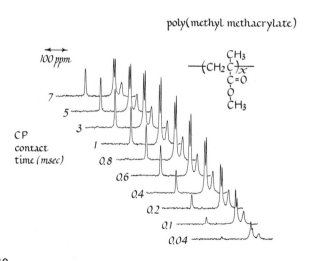

Figure 4.10
Dipolar-decoupled magic-angle CP ^{13}C nmr spectra of poly(methyl methacrylate) as a function of spin-lock contact time.

protonated-carbon lines much more rapidly. (The slowly polarizing line in the mid-field group is due to the main-chain quaternary carbon.) Magic-angle spinning increases the time constant describing the carbonyl-carbon polarization in these experiments by much less than a factor of 2: spinning has no detectable effect on the time constant associated with the methyl- and methylene-carbon cross polarization. After about 3-msec CP contact, the intensity of all the carbon lines decreases. The time constant describing this decay is the proton $T_{1\rho}$ which, for poly(methyl methacrylate) at room temperature, is approximately 7 msec.

The rate of polarization transfer during the first 50 μsec is taken as the average $T_{IS}(\text{SL})$. [We ignore the minor contribution of the carbon $T_{1\rho}$ to the initial slope of $T_{IS}(\text{SL})$ plots.] Determination of this value for amorphous polymers is not complicated by pronounced transient oscillations, as is sometimes the situation in crystalline materials.[14,20] The initial rate of polarization transfer is determined by plotting the contact time (over the first 50 μsec against the difference of the observed ^{13}C magnetization and the intercept of the proton $T_{1\rho}$ slope extrapolated to zero contact time. The final slope at long contact times yields the proton $T_{1\rho}$. For poly(methyl methacrylate), the $T_{IS}(\text{SL})$ for the methylene carbon is approximately 20 μsec, for the α-methyl carbon approximately 80 μsec, and for the carbonyl carbon about 300 μsec. Cross-polarization transfer in the ADRF experiment is slightly more complicated. The initial transfer rate depends on both the spin-spin T_{IS} (ADRF) and the spin-lattice part of the carbon $T_{1\rho}$. The steady-state polarization transferred depends on the competition between these two rates, with this value decreasing at long contact times as determined by the lifetime of the proton polarization in the local dipolar field, T_{1D}.

In general, we find that plots based on data such as shown in Figure 4.10, fitted by a sum of exponentials with parameters matched to the initial and final slopes, are only qualitatively represented in the intermediate region.[10] That is, a thermodynamic analysis,[6,17] with the flow or transfer between spin reservoirs characterized by a first-order rate constant, is not strictly correct, and a more involved, many-body dynamic description is required.[15]

F.   Cross-Polarization Transfer in the Presence of Magic-Angle Spinning

A possible conflict would seem to arise between the requirements of line narrowing by rotation and those of the spin-exchange process by which the rare spins are polarized in CP experiments. The latter depends on an effectively static dipole-dipole interaction H_{IS} between rare and abundant spins, while sample rotation can render this interaction oscillatory. High-speed spinning at the magic angle is known to modify significantly the rate of polarization transfer from abundant to rare spins in CP experiments on adamantane.[21] However, we have just seen that not only are strong spin-lock CP nmr signals observed for protonated and non-protonated carbons of poly(methyl methacrylate) under 3-kHz magic-angle spinning (Figure 4.10), but also the CP transfer rates for all carbons of that polymer are essentially independent of spinning.[10,11]

The crucial difference between the experiment on adamantane and that on poly(methyl methacrylate) is that in the former situation the spinning speed was comparable to the static dipolar interaction among the *abundant* protons, while in the latter situation the spinning speed was much less than this interaction.[21] We can gain some insight into the physics involved in this distinction from the illustrations of Figure 4.11.

The top illustration in Figure 4.11 we have seen before. It represents the frequency match in a spin-lock CP experiment under the Hartmann-Hahn condition. The ^{13}C resonance is very sharp because dipolar decoupling has removed static $I-S$ interactions and rare-spin $S-S$ dipolar interactions are weak. The ^1H resonance is not narrow because it is subject to the influence of strong $I-I$ dipolar interactions. The effect of these interactions is to produce "dipolar fluctuations," or mutual energy-conserving proton-proton spin flips.[14] Thus, the $I-I$ dipolar interaction is not truly static, but has dynamic character. The proton resonance has a width determined by the rate of these spin flips.

Under low-speed magic-angle spinning (Figure 4.11, middle), all $I-S$ dipolar interactions are, in addition to rf stirring, subject to an amplitude modulation as static C—H internuclear vectors are rotated at frequency ω_s relative to H_0. Amplitude modulation produces sidebands at $\pm\omega_s$ and $\pm2\omega_s$, with an intensity ratio of $\sqrt{2}$ for an isotropic powder.[21] The AM pattern is independent of the size of the $I-S$ interaction. The effect of spinning on the $I-I$ interaction is to modulate the rate of the proton spin flips. This corresponds to a frequency modulation, with a modulation index determined qualitatively by the ratio of the proton line width (in the rotating frame) to the spinning frequency.[21] When the ratio is large (low-speed spinning), a large number of FM sidebands are produced, but few of them are outside the original proton line width, which remains essentially unchanged.

When the ratio of the proton line width to the spinning frequency is low (0.5 or less, corresponding to high-speed spinning), there are fewer modulation sidebands, with Bessel amplitudes given by the Jacobi-Auger identities, but these sidebands fall outside the original proton line width. As suggested in Figure 4.11 (bottom), a spin-lock CP transfer is still

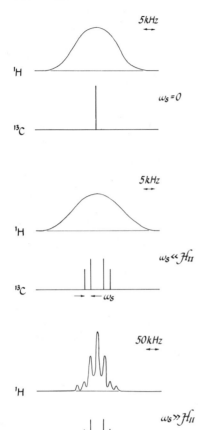

Figure 4.11
Spectral overlap between protons and carbons in a spin-lock cross-polarization experiment as a function of magic-angle spinning frequency.

possible under the Hartmann-Hahn condition, although a more effective frequency match between carbons and protons can be achieved if the H_1s are mismatched by the spinning frequency. This phenomenon has, in fact, been observed in spinning experiments on adamantane.[21] As a general rule for polymers, however, one can ignore the effect of spinning on spin-lock CP transfer rates; the spinning frequency is seldom even comparable to the proton line width, the FM modulation index is therefore high, and the frequency match of the Hartmann-Hahn condition is as effective with spinning as without.

As we have just seen, in the presence of truly high-speed coherent mechanical spinning, an efficient spin-lock CP transfer can always be made by mismatching the Hartmann-Hahn condition by the spinning frequency. In the presence of fast, *incoherent* random molecular motion no such correction can be made. This is true because the incoherent motion produces a broad distribution of sidebands no component of which makes a significant match with ^1H spectral density. In more familiar terms, rapid random isotropic motion eliminates the stationary components of magnetization needed for a CP transfer. The situation for random molecular motion of *intermediate* frequency is far more complicated. Qualitatively, based on notions of the AM and FM produced by motion, we might expect CP transfer rates to be reduced for exactly matched spin-lock experiments, and increased for mismatches, especially rather large mismatches. These increases would disappear as the frequency of the motion increased and static interactions were removed. A quantitative theory has not yet been developed.

Spinning at 3 kHz can have a devastating effect on ADRF CP experiments.[22] This occurs not because the instantaneous transfer rate is reduced, but rather because the *sign* of the polarization transferred fluctuates. As the sample rotates the ^1H–^1H internuclear vector changes orientation relative to H_0. Thus, for rotations at the magic angle, the proton spin temperature can actually change sign.[23] Unless the CP transfers are properly synchronized with the spinning (see Section III.C) the net carbon magnetization tends to cancel itself out. This problem does not arise with spin-lock transfers since, first, H_1s larger than local dipolar fields maintain a constant spin temperature, and, second, most spin-lock transfers are complete before a rotor has undergone even a quarter of a revolution.

Apart from the physics of cross-polarization processes, Figure 4.11 helps us to understand why magic-angle spinning (in combination with dipolar decoupling) is so effective at narrowing ^{13}C lines subject to broadening from chemical shift anisotropy. Recall that in Figure 4.5, we observed line narrowing of a factor of 50, even though the spinning speed was only comparable to the CSA pattern. The reason for the effectiveness of the spinning is that the CSA broadening is totally static in origin.[10] The CSA pattern is composed of a collection of sharp lines, not a single broad line. Thus, under spinning, each frequency component of the CSA pattern behaves as though amplitude modulated and so is completely displaced from its original resonance position (^{13}C pattern, Figure 4.11, middle). The result of all of these sharp-line displacements is a multiple-sideband spectrum, with the relative intensities of the sharp sidebands determined by the original CSA and the spinning frequency.[24,25] The sidebands are always sharp because there are no zero-frequency components to contribute to a line width.

This narrowing behavior is in sharp contrast to that observed in *single*-resonance FT ^{13}C nmr spinning experiments on polymers.[10] In the absence of dipolar decoupling, the *I–S* interaction is not totally static, but instead is subject to averaging fluctuations impressed by *I–I* spin flips. Consequently, the partially narrowed single-resonance ^{13}C linewidth is partly homogeneous in character. It behaves, under spinning, much like a proton line (^{1}H pattern, Figure 4.11, middle). For example, 3-kHz spinning narrows the carbonyl-carbon line of poly(methyl methacrylate) by only a factor of about 3, even though the spinning speed is comparable to both the CSA and the *I–S* interaction.[10,26] This is exactly the type of incomplete line narrowing observed in the past in attempts to narrow single-resonance 15–20 kHz proton lines with 10–15 kHz spinning. One quickly reaches the conclusion that complicated (and even dangerous) ultra-high-speed spinning for single-resonance nmr is not very handsomely repaid. The double-resonance ^{13}C nmr experiment is much more rewarding. With 40-kHz dipolar decoupling to remove the strong *I–S* interaction, technically simple means can be used to produce the 3-kHz spinning needed to eliminate completely the effects of CSA broadening.

III.   SPECIFICATIONS OF A CROSS-POLARIZATION SPECTROMETER

A.   Probe and Receiver

It may be of some value if, in a general way, we describe what we feel are the necessary ingredients for an effective cross-polarization spectrometer. These comments are based on our experience over the last 3 years, up to the time of this writing, August 1977. As with most such prescriptions, this one will, no doubt, soon become painfully outdated.

We have used both single- and double-coil probes. The double-coil probe has the advantage that ^{1}H and ^{13}C tuning circuits can be separately optimized. (One coil is used in both the transmitter and receiver functions for one nucleus.) A cylindrical coil is used for the ^{13}C channel and a Helmholtz saddle-shape coil for the ^{1}H channel. The two coils are approximately orthogonal. Typically, the ^{13}C coil may be 11 mm in diameter, 13 mm in height, and consist of 8 or 9 turns of #24 wire, pressed into ribbon. The single-turn ^{1}H coil is slightly larger in both dimensions. The two coils are usually separated by a thin glass support for insulation. The biggest problem we have had with crossed coils is their sensitivity to rf breakdown, especially, for reasons we don't fully understand, during spinning experiments. The breakdown can occur either between coils, or at the crossover point of the Helmholtz coil.

The doubly-tuned single coil eliminates this problem. Double tuning can be achieved by schemes similar to those suggested by Waugh[27] and by Vaughan,[28] and their respective coworkers. We have used a single-coil probe for the last year and prefer it to the double-coil designs. Our coil is 12 mm in diameter, 15 mm in height and consists of 9 turns. The loaded-coil ^{13}C Q is 150. With 200 W (CW) available for both ^1H (90 MHz) and ^{13}C (22.6 MHz) channels, this coil can produce 10 and 40 G, respectively.

Our receiver has a gain of about 100 dB. The preamplifier section, manufactured by Miteq Corp. (Hauppauge, NY) has a noise figure of 1.2 dB. The preamplifier is protected by the usual shorting diodes and is preceded by a 6-MHz bandpass filter. This helps ensure isolation from the ^1H transmitter. A low-pass filter would probably serve as well. We use a 2-MHz intermediate frequency and quadrature detection at that frequency. The use of quadrature detection is essential, not only for the gain in sensitivity, but more importantly, for the slower data acquisition rate possible.[29] For example, in order to obtain an 8-kHz-wide frequency-domain spectrum, the free induction decay is digitized with a sampling period of 62.5 μsec using single-phase detection, but 125 μsec using quadrature detection. The end of the first sampling period is crucial in determining the baseline of the transformed spectrum.[30] Since we are often interested in characterizing broad CSA patterns, we cannot tolerate curved baselines. On the other hand, in CP experiments we are usually observing weak, rare-spin signals at low frequency with high-Q receivers, a combination that results in relatively long recovery times following strong rf pulses. Thus, the longer sampling period that goes with quadrature detection is most welcome in the effort to achieve high-quality spectra of solids with flat baselines.

The demands on the main-frame digital data acquisition system are quite modest. Digitizer rates of 50 kHz are more than adequate together with a 2K main-frame storage capacity. Access to an interactive disk system for long-term data storage is desirable; the latter also permits communication with larger computers simplifying the problem of performing two-dimensional Fourier transforms (see Section IV.B).

B. Transmitter, Pulse Logic, and Gating

The rf transmitters we use are broadband, solid-state Class A amplifiers manufactured by Electron Navigation Industries (Rochester, NY). As mentioned above, with 200 W of CW power, we can produce a 10-G H_1 at 90 MHz in an 12-mm diameter coil. This is sufficient to perform all but a few CP experiments. A 15-G H_1 for protons is desirable for experiments[31] on proton-rich partially crystalline polymers such as polyethylene, as well as for experiments[32] involving multiple-pulse *I–I* decoupling (Section IV.C). Amplifiers with greater rf power generally employ a tube in the final stage. Unfortunately, this has the distinct disadvantage of causing the gain characteristics of the transmitter to change as the tube ages. Consequently, reproducing from one day to the next matched and mismatched Hartmann-Hahn conditions to say, 0.1 dB becomes a nuisance. We use high-power filters on the outputs of our rf transmitters. These are passband filters with 10% bandwidth and low insertion loss, manufactured by Cir-Q-Tel (Kensington, MD).

The pulse logic employed should be capable of generating pulse sequences (see Figure 4.9) with digital control of period, and accurate (but not necessarily digital) control of the length, amplitude, and phase of four independent pulses for both ^{1}H and ^{13}C channels. This flexibility not only makes possible involved multiple-pulse decoupling experiments, but also essential quadrature data routing[29] (to remove FT echos) and spin-temperature alternation[30] (to ensure flat baselines in CP experiments). The latter scheme, for example, alternates the sign of the 90° phase shift in the ^{1}H spin-locking part of the CP experiment (Figures 4.7 and 4.9). Since the sign of the ^{13}C CP signal follows the sign of this phase shift, but the sign of artifacts or transients generated by the intense ^{13}C pulse do not, the artifacts can be removed by alternately adding and subtracting new data entering the sum accumulated in the time-averaging computer. The pulse logic and rf gates required to perform these experiments are available from stand-alone computer systems designed explicitly for nmr experiments (Nicolet Instrument Corp., Madison, WI), from custom-built logic systems, from commercially available, digitally controlled pulse generators and rf components (for example, Ortec, Inc., Oak Ridge, TN; Merrimac, Inc., West Caldwell, NJ), or from combinations of all of these.

C. Spinning

Magic-angle spinning is an indispensable part of the ^{13}C nmr of solid, amorphous polymers, and of noncrystalline solids in general. We feel that both Lowe- and Beams-design rotors will play a part in future spinning experiments. The Beams-design rotor is particularly well suited to the spinning of powders, or other intractable materials such as coals and oil shales,[33] which cannot themselves be easily machined into dynamically balanced rotors.

All spinning experiments should be capable of synchronization. By means of a marking on the rotor illuminated by light transmitted through a flexible light pipe, the position and speed of the rotor can be monitored by a photosensor attached to a second light pipe. This permits two powerful types of experiments. First, data sampling can be synchronized with the spinning speed. Maricq and Waugh[25] have shown that sideband-free magic-angle spectra of solids can be obtained even when the spinning speed is much less than the CSA dispersion. In this situation one would normally expect the Fourier-transformed spectrum to be cluttered by spinning sidebands. However, synchronizing the acquisition of the free induction decay with the spinning speed eliminates the sidebands. If the spinning speed varies slightly from one experiment to another, lines not folded over in the centerband spectrum can be easily separated from those which are. The latter change resonance frequency with the spinning speed, and the former do not. Having two sets of data synchronized with different spinning rates permits a simple separation of the two. This technique may prove to be of great value for experiments performed at high magnetic fields in superconducting solenoids, where practical spinning speeds are unable to clear a spectrum of unwanted sidebands.

The second type of synchronized spinning experiment involves coordinating pulse sequences with the spinning speed. Lippmaa et al.[24] have described an approach to the problem of high resolution in solids without the loss of the chemical shift anisotropy. They suggest coordinating 180° ^{13}C rf pulses with the magic-angle spinning frequency. In the rotating frame, the ^{13}C magnetization is undergoing coherent oscillations due to the presence of spinning sidebands, even with fast magic-angle spinning. If the 180° pulses are applied after each half-revolution of the rotor, the effect of the oscillations can be refocussed into a unidirectional decrease of the carbon magnetization, by a process analogous to, but the reverse of, the familiar Carr-Purcell refocusing.[34] As a result, the information in the chemical shift anisotropy, transcribed to the sidebands, can be displayed in the intensity of the centerband as a function of the number of applied synchronized 180° pulses. Other synchronized pulsing schemes can be imagined that will averge local dipolar fields (see Section II.F) corresponding to specific orientations in an amorphous solid.[22] Such schemes rely upon ^{13}C rf pulses being coordinated with the physical orientation of the moving rotor with the applied field, H_0.

D. Magnet and Sensitivity

The selection of the magnet for a CP spectrometer can be a difficult one. Superconducting solenoids have advantages in high sensitivity and use probes having fast rf recovery, but have disadvantages in the difficulty of making large rf fields at high frequency and in restrictions on sample access. The latter is particularly important if, for example, one wants to manipulate mechanically a specimen such as a polymer while performing nmr experiments. Magic-angle spinning is also easier at a lower field since lower spinning speeds are required to remove completely the sidebands of a field-dependent CSA pattern from a transformed spectrum. As discussed above, however, this may not be a crucial limitation. Perhaps for unconvincing reasons, we prefer iron magnets.

Finally, we present a rough guide to the kind of sensitivity one can expect from a ^{13}C CP spectrometer. We observe a peak-to-peak signal-to-noise ratio of about 10:1 for the fully relaxed ^{13}C free induction decay (single-shot, 90° pulse, 3-kHz 4-pole Butterworth filter) from a neat sample of liquid poly(ethylene oxide), —$(CH_2CH_2O)_x$—, of molecular weight 600. This signal is measured at 22.6 MHz with decoupling and an Overhauser enhancement of nearly 3. We observe a comparable single-shot signal-to-noise ratio in a 1-msec matched spin-lock CP experiment (using a proton H_1 of 10 G) on a stationary solid plug of Delrin, poly(oxymethylene), —$(CH_2O)_x$—. The plug of Delrin has approximately the dimensions of the single coil described above. The probe and receiver systems in the two tests are identical.

IV.   APPLICATIONS: STRUCTURE AND MOTION OF SOLID POLYMERS

A.   Residual Linewidths in Magic-Angle Spinning Experiments

The magic-angle CP ^{13}C nmr spectra of poly(phenylene oxide) are actually more detailed than the corresponding FT spectra of the polymer in solution.[10] An unexpected feature of these spectra (Figure 4.12) is that the second line from the right (which can be unambiguously assigned to the protonated carbons of the aromatic rings), while a singlet in solution, is a doublet in the solid state. Insepction of models of the chain shows that the two protonated carbons of the ring need not be equivalent because the nonlinearity of the C—O—C bond leads to conformational isomerism within the chain. These two carbons become equivalent in the event of internal rotation of the phenyl group. This may occur in solution but does not occur in the solid state.

Fine structure, similar to the doublet, is also hinted at in the spectra of some of the other glassy polymers, poly(ether sulfone) in particular. In fact, unresolved fine structure, arising from asymmetric chain configurations, may also explain why the methyl-carbon resonances of polymers such as polcarbonate and polysulfone are consistently broader than the quaternary- and protonated aromatic-carbon resonances of those two polymers.[10] Resolved and partially resolved fine structure in the solid-state spectra of the poly(aryl ethers) show that, for these polymers, main-chain motions (whose presence are indicated by a variety of

poly(phenylene oxide)

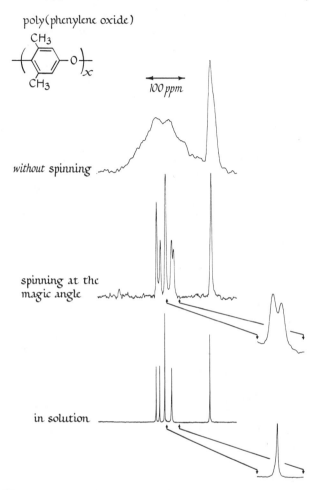

Figure 4.12
Cross-polarization ^{13}C nmr spectra of poly(phenylene oxide), with and without magic-angle spinning. The CP spectra are compared to a FT spectrum of the polymer in solution (with solvent lines omitted for clarity or presentation).

mechanical, dielectric, and nmr experiments) cannot involve large-amplitude conformational jumps, since such jumps would, of course, average out any fine structure.

Thus, from magic-angle solid-line shapes alone, we are led to two conclusions, which are probably applicable to most glassy polymeric solids.[10] First, dipolar-decoupled magic-angle ^{13}C line widths are determined by distributions of isotropic shifts arising from conformational isomerism unique to the solid state. Second, and more importantly, main-chain motion in solid glassy polymers is primarily torsional oscillation, perhaps requiring the cooperativity of several neighbors, but necessarily restricted to small-amplitude excursions. We will discuss other experiments to characterize this motion in Section IV.E.

B.   High-Resolution Spectra Without Spinning: The Separated Local Field Approach

Crystalline polyethylene can be highly oriented by drawing so that the extended, all-*trans* planar zig-zag chains are aligned in the same direction. Such samples have been examined by Opella and Waugh[35] using a high-resolution technique that does not involve magic-angle spinning. In this method, CP and dipolar decoupling are combined with *un*decoupled nmr spectroscopy in such a way as to provide *separate* dipolar split spectra (and hence resolution) for dilute spins having different chemical shifts.

The pulse sequence used in this approach, called separated local field (SLF) spectroscopy, is illustrated in Figure 4.9 (third from top). The first part of the experiment is a matched spin-lock CP transfer. The carbon H_1 is turned off and the carbon magnetization allowed to develop in time while all $I-I$ dipolar interactions are removed by multiple-pulse irradiation of the protons.[32] This development will depend on $I-S$ coupling, which is free from complications due to modulation by $I-I$ spin flips (Section II.F). The last part of the experiment is observation of the ^{13}C free induction decay with dipolar decoupling to remove $I-S$ interactions.

The results of CP experiments performed on oriented polyethylene with the draw axis perpendicular to H_0 are shown in Figure 4.13. The chemical shift CSA power pattern is shown at the top of the figure. Only two major lines are observed since oriented polyethylene is, of course, not a normal powder. Thus a signal from carbons in chains with the chain axis parallel to H_0 is not present. (This signal can be observed in a separate experiment by rotating the chain axis relative to H_0). The two major equal-intensity lines arise from carbons in chains with H_0 along the two directions shown in Figure 4.13 as σ_{aa} and σ_{bb}. Signals from carbons in chains with orientations intermediate between σ_{aa} and σ_{bb}, but still with the draw axis perpendicular to H_0, form the tails of the two major lines.

The assignment of the major lines can be made on the basis of the SLF spectrum (Figure 4.13, middle). This spectrum is the result of a two-dimensional Fourier transform.[32,35] The development of each frequency point in the normal spectrum is measured as a function of the time during

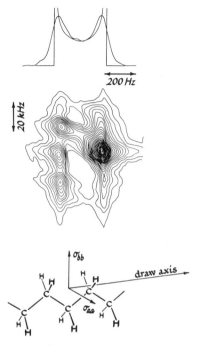

Figure 4.13
Dipolar-decoupled ^{13}C chemical shift anisotropy powder spectrum (top) and SLF spectrum (bottom) for oriented polyethylene with the draw axis perpendicular to H_0. The upper spectrum is consistent with the expected ideal powder pattern convolved with a symmetrical broadening function of full-width 1.1 ppm at half-maximum. A portion of the all-*trans* polyethylene chain is shown at the bottom of the figure. Two of the principal axes of the shielding tensor, σ_{aa} and σ_{bb}, can be identified when the draw axis is perpendicular to H_0, and the third, σ_{cc}, when the axis is parallel to H_0.

which the carbon magnetization is under an I–S dipolar interaction, but with no I–I interaction. These time responses are Fourier transformed separately and then added together to produce the SLF spectrum (displayed as contours of equal absolute intensity). Since one of the lines in the normal spectrum is unsplit in the SLF spectrum, it unambiguously can be assigned to those carbons whose C—H vector is at the magic angle relative to H_0 (σ_{bb}, Figure 4.13, bottom). The other C—H vector makes the angle $(\pi - \chi)/2$, where χ is the H—C—H bond angle. The signal from this carbon is a 1:1:1 triplet with a 38 kHz splitting.

SLF spectra can be used to examine phenomena such as the orientation of lamellae and the perfection of chain alignment in partially crystalline polymers.[35] In addition, for polyethylene, the observed absence of a temperature dependence for the anisotropic ^{13}C chemical shift and the SLF ^{13}C—^{1}H dipolar splittings is strong evidence that the intrachain motions present at temperatures as high as 90°C necessarily involve 180° flips of the —CH$_2$— group about the chain axis (with or without a translation along that axis). Although this conclusion is not new, the SLF results are far more direct and simple to interpret than earlier experiments involving ^{1}H nmr spin-lattice times and second moments,[36] or than experiments involving mechanical loss,[37] dielectric dispersion,[37] infrared spectroscopy,[38] etc.

Although it can be argued that crystalline polyethylene is not exactly a typical system, the SLF approach toward achieving high-resolution ^{13}C nmr spectra of polymers is nevertheless well illustrated by the examination of this particular polymer. The SLF experiment will work on less highly ordered systems. Besides, as with most experiments on polyethylene, the SLF results have some intriguing minor discrepancies. Theoretically, the —CH$_2$— triplet observed in Figure 4.13 should have intensity ratios of $1:2:1$ (two overlapping dipolar doublets) with a 45-kHz splitting. These predictions are based on results of experiments on the same sample, but with the orientation axis parallel to H_0, together with crystallographically determined bond lengths and angles. It is easy, but not very satisfying, to ascribe the intensity anomaly to imperfections in the nmr experiment and data processing. Opella and Waugh[35] have suggested (but under a strong disclaimer) that the reduced dipolar splitting could be accommodated by a 3° reduction in the H—C—H bond angle together with 15° main-chain torisional oscillation (at a frequency comparable to, or greater than 20 kHz). Further progress in understanding these spectra will have to await future experiments.

Naturally, for highly complex amorphous materials, the SLF experiment can be used in combination with magic-angle spinning. The information in the CSA pattern can, in principle, be maintained by means of synchronized pulsing-spinning techniques (see Section III). However, this would lead to a three-dimensional Fourier transform. In practice, we suspect the information in the CSA will be abandoned. Two-dimensional SLF spectra will then arise from isotropic chemical shifts under different I-S interactions. Although usually there will be no further improvement in resolution over normal dipolar-decoupled magic-angle spectra, the ability to characterize directly different static I–S interactions for different carbons is important. These values can then be compared with parameters such as T_{IS} and the ^{13}C $T_{1\rho}$, as well as with the ^{1}H line width and second moment, in order to lead to a better understanding of the details of slow motions of polymers in the solid state (cf, Section IV.E).

C. Practical Use of Low-Resolution Spectra

In the last section we described an approach towards high-resolution ^{13}C nmr applicable in situations where magic-angle spinning is either undesirable or impossible. The technique, in general, and the data manipulation in particular, were rather involved. It is reasonable to ask whether any CP ^{13}C experiments on solid polymers are worthwhile without some elaborate line-narrowing or line-splitting trick. In this section we will discuss

CP of some biological solids

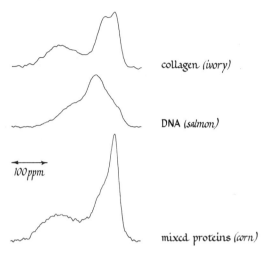

collagen (ivory)

DNA (salmon)

100 ppm

mixed proteins (corn)

Figure 4.14
Dipolar-decoupled CP ^{13}C nmr spectra of ivory, lyophilized salmon testes DNA, and dehydrated mixed corn proteins.

two examples where useful information could be obtained from ordinary dipolar-decoupled CP spectra of stationary samples.

Figure 4.14 shows the dipolar-decoupled CP ^{13}C nmr spectra of some complicated biological solids.[39] The observed widths of the lines of approximately 2 kHz arise from overlapping CSA patterns. Despite the absence of high-resolution in the conventional sense, the ^{13}C spectra of these materials are still indicative of chemical structure. Now it is true that the CP spectra of solid proteins always consist of three lines, a low-field line arising predominantly from carbonyl carbons, a mid-field line due to methine carbons of various types, and a combination high-field line arising from aliphatic methylene and methyl carbons. However, these three lines change in relative intensity as the amino-acid composition of the particular proteins change (Figure 4.14, top and bottom). Furthermore, the CP spectrum of solid DNA is substantively different (Figure 4.14, middle), consisting of a low-field part due to the aromatic carbons of the nucleotides, and a strong mid-field line due to sugar carbons. Solid starch produces a strong resonance in this position as well.[40]

The resonances of these solids are sufficiently different that they can be used semiquantitatively to determine the composition of materials containing an unknown mixture. For example, we have obtained the ^{13}C CP spectra of immature soybean ovules, cultured in a 10-mm diam nmr tube for a month on a medium containing ^{13}C-enriched glucose as the sole carbon source.[41] Naturally, the system was too delicate to attempt spinning, even if sterile conditions could be maintained. Nevertheless, we were able to observe the incorporation of label into the ovule and its synthesis into structural and storage protein and starch from CP spectra of the stationary system. Furthermore, scalar-decoupled FT spectra allowed us to characterize the formation of lipids and fats, as well as various intermediates in solution.[42] Thus, we were able to account for the majority of all carbon involved in primary synthesis in the growing soybean ovule.[41] These results were in good agreement with rather tedious classical wet chemical separations and analyses, performed in parallel experiments on other ovules. The advantage of the ^{13}C experiment, of course, is that it is performed in real time on a single ovule. Conclusions as to the effects of varying growth parameters (culture conditions, environmental stresses, etc.) need not, therefore, involve large numbers of replications, statistical analysis, and inevitable biological variability.

As a second example of the practical use of low-resolution spectra, Figure 4.15 shows the dipolar-decoupled FT and CP ^{13}C nmr spectra of intact bovine nasal cartilage.[39] Water comprises 75% of the intact cartilage with the remainder composed of equal amounts of collagen and proteoglycan. The FT spectrum consists primarily of sharp lines, which can be reasonably associated with what amounts to the bulk of the proteoglycan component essentially in solution.[43] The CP spectrum of intact cartilage shows several broad components not present in the FT spectrum, as well as some of the sharp lines observed in the FT spectrum, but at reduced intensities. The latter result is common for polymer chains with substantial segmental freedom. For example, the CP spectra of solid elastomers above their glass transition temperature are quite similar to the FT spectra, but of reduced intensity. As we have discussed, the reason for the reduction is that microscopic rotational motions of mobile chains tend to eliminate the near static dipolar interactions necessary for the transfer of polarization during the carbon-proton CP contact period (Section II.C). The net result is an inefficient transfer.

The appearance of the broad resonance lines in the CP spectrum of intact cartilage can be attributed to the lineshapes observed for the isolated components of cartilage, also shown in Figure 4.15. The cartilage CP spectrum displays a low-field line similar to that observed for the collagen component, and a broad central line (underlying the sharper lines) similar to that observed for the isolated lyophilized proteoglycan complex. We make no attempt at establishing a quantitative determination of the levels of the various components of cartilage present in

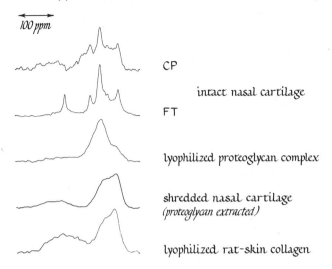

Figure 4.15
Dipolar-decoupled CP and FT ^{13}C nmr spectra of intact bovine nasal cartilage. The FT spectrum is shown at reduced vertical display. The CP ^{13}C nmr spectra of the two organic components of cartilage, proteoglycan complex and collagen, are also shown, together with the CP spectrum of a rat-skin collagen, the latter for purposes of comparison.

solid-like states compared to those present in solution. This would require extensive relaxation experiments. Nevertheless, it is reasonable to conclude that in a qualitative sense, comparison of the FT and CP spectra of intact nasal cartilage shows that a large part of the proteoglycan component is effectively in solution and presumably only weakly interacting with the collagen, that a much smaller part may be still solidlike, and that most, if not all, of the collagen component is also solidlike. These conclusions are consistent with the notion that a large fraction of the proteoglycan chains are segmentally mobile and are not involved in any type of interaction with collagen which substantially affects their molecular motion. It seems reasonable to suppose that the proteoglycan association with collagen is one of physical entrapment or entanglement within the rigid collagen network.

D. Motionally Modified Chemical Shift Anisotropies

Motions in solid polymers generally affect nmr spectra in two ways. First, the line shapes and relaxation parameters dependent on static interactions are modified, and second, spin-lattice parameters are altered. We will discuss the the first phenomenon in this section, and deal with the second in the next section.

We have already seen that mechanical sample rotation can have a profound effect on the ^{13}C CSA patterns of dipolar-decoupled spectra (Figure 4.5). In the situation that the rotation axis is at the magic angle relative to H_0 we observe dramatic line narrowing. If, however, the rotation axis is at some other angle, we observe instead of sharp lines, motionally modified, or incompletely averaged CSA patterns. This occurs because the $(3\cos^2\beta - 1)$ dependence of the expression for the CSA (Figure 4.3) does not vanish, and the explicit angular dependence of the direction cosines of the CSA is important. For example, as shown in Figure 4.16, the CSA of the carbonyl carbon of poly(methyl methacrylate) is a strong function of the choice of spinning axis.[26] The full CSA is developed for $\beta = 0$; a comparable but reduced pattern is observed for $\beta = 45°$; and apparently reversed patterns are observed for $\beta = 63°$ and 75°.

The same averaging of CSA patterns occurs if the motion is internal molecular motion rather than mechanical sample rotation. Thus, the CSA patterns of methyl and phenyl groups undergoing rapid internal rotation about a molecular symmetry axis display not the general tent-like lineshape (Figure 4.2, middle, left side), but rather averaged, narrowed and simplified patterns (Figure 4.2, middle, right side).

Since the lineshape is a rich source of information about motion in solid polymers, it is obviously desirable to have an experimental technique that can provide the high resolution of magic-angle spinning while preserving the CSA. In principle, such a technique could be used to determine the extent of angular torsional excursions of the phenyl side groups in polystyrene, or to confirm the absence of internally rotating phenyl groups in the main chain of poly(phenylene oxide) (Section IV.A).

Synchronized spinning and pulsing[24] has this capability (Section III.C), as does *slow*-spinning magic-angle ^{13}C nmr.[26] As illustrated in Figure 4.17, high resolution in the dipolar-decoupled ^{13}C spectrum of solid Delrin is achieved with quite low spinning frequencies, of the order of 100 Hz. The spinning frequency need only be large compared to the line width observed under strong dipolar decoupling and rapid magic-angle spinning, which is approximately 50 Hz in this case. This is true despite the fact that the chemical shift dispersion of the CSA is much larger than 100 Hz. As we discussed earlier, slow-speed spinning sidebands are sharp because the CSA is totally static in origin (cf. the end of Section II.F).

Asymmetry in the intensities of sidebands is often prominent for slow spinning experiments (Figure 4.17). We intend no detailed analysis here,

poly(methyl methacrylate)

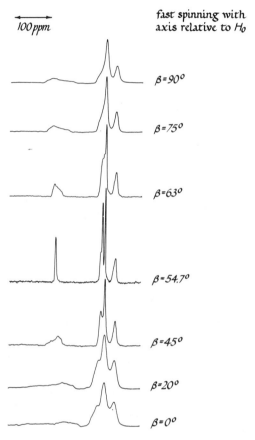

fast spinning with
axis relative to H_0

$\beta = 90°$

$\beta = 75°$

$\beta = 63°$

$\beta = 54.7°$

$\beta = 45°$

$\beta = 20°$

$\beta = 0°$

100 ppm

Figure 4.16
Dipolar-decoupled CP ^{13}C nmr spectra of a poly(methyl methacrylate) rotor of
the Lowe geometry as a function of the angle of the spinning axis relative to the
applied static field.

but it is clear that the intensity pattern of the spinning sidebands reflects
the chemical shift anisotropy observed in a nonspinning experiment.
Lippmaa et al.[24] have observed similar patterns of sidebands in spinning
experiments on hexamethyl benzene, and have presented an analysis for
the case of an axially symmetric chemical shift anisotropy. The complete
chemical shift dispersion can be characterized by slow-spinning magic-
angle spectra involving just a few spinning frequencies, perhaps using

$\{CH_2O\}_x$

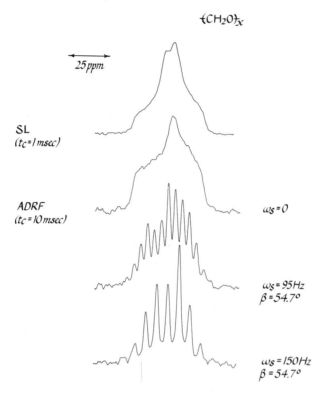

SL
$(t_C = 1\, msec)$

ADRF
$(t_C = 10\, msec)$

$\omega_s = 0$

$\omega_s = 95\, Hz$
$\beta = 54.7^o$

$\omega_s = 150\, Hz$
$\beta = 54.7^o$

Figure 4.17
Dipolar-decoupled CP ^{13}C nmr spectra of a Delrin rotor of the Lowe geometry.
Both spin-lock (top) and adiabatic demagnetization in the rotating frame (second
from top) cross polarization schemes were used. Spinning sidebands with intensities following the CSA pattern are developed in slow-speed magic-angle spinning experiments (bottom two spectra).

synchronized data acquisition (Section III.C). Thus, in the same experiment, it is possible to have both the high resolution inherent in magic-
angle spinning and the information contained in the chemical shift anisotropy.

In some situations in solids, local molecular internal motion is so
extensive it is close to isotropic in nature. The CSA then collapses to its
isotropic average just as it does for mobile liquids. This occurs for
adamantane,[44] for example, and for rubbery polymers[45] well above their
glass transition temperature. Delrin, in fact, has about a 30% rubber
content. In spin-lock CP experiments, the polarization transfer efficiency

between protons and carbons in the rubbery phase is still good enough, despite extensive averaging of static dipolar interactions by chain motions, that a strong signal from the rubber component at the isotropic center of the CSA pattern is observed (Figure 4.17, top). For the less-efficient CP transfer of ADRF experiments, no rubber-component signal is observed (Figure 4.17, second from top). In general, therefore, comparison of the two types of CP spectra provides a means of characterizing motionally modified CSA patterns and line shapes, assuming, of course, the ADRF experiment is indeed practical (Sections II.D and II.F).

E. Molecular Motion and the ^{13}C $T_{1\rho}$ Experiment

The lifetime of the ^{13}C magnetization spin locked along a carbon H_1 (in the absence of a proton rf field) is designated as the ^{13}C $T_{1\rho}$.[10] The pulse sequence used to measure $T_{1\rho}$ is shown at the bottom of Figure 4.9. The experiment begins with a CP transfer to produce conveniently a spin-locked carbon magnetization. This magnetization is held for a variable time with the proton H_1 turned off. In the final part of the experiment the proton H_1 is turned back on, and the remaining carbon magnetization observed with dipolar decoupling. A semilog plot of the observed carbon magnetization as a function of the hold time yields the carbon $T_{1\rho}$ as a slope.

Unlike the corresponding ^1H $T_{1\rho}$ experiment, the ^{13}C $T_{1\rho}$ is not confused by spin diffusion between carbons because their low natural abundance ensures a physical separation within the solid and hence a slow spin diffusion or transfer rate. Thus, individual carbons have their own $T_{1\rho}$s. The possibility exists, however, that these $T_{1\rho}$s are determined by spin-spin processes. This can occur if the carbon H_1 is less than the local dipolar field.[14] Then, as illustrated in Figure 4.6 (third from top), a sufficiently good frequency match may exist to enable a cross-polarization transfer from carbons to protons. This is the reverse of the transfer we have discussed earlier, but it is still characterized by the same time constant, T_{IS} (ADRF).

As the carbon H_1 is increased, T_{IS} (ADRF) becomes long. A spin-lattice mechanism can now dominate $T_{1\rho}$. That is, when the carbon H_1s are comparable to or greater than the local dipolar fields, the *nonsecular* part of the $I-S$ interaction ($I_z S_\pm$) which produces ^{13}C spin flips induced by molecular motion becomes important.[10] Of course, spin-lattice relaxation only occurs when the solid on which the experiment is being performed actually has some microscopic molecular motions with components near the rotating-frame Larmor frequency defined by H_1. This is the situation for most organic glassy polymers near room temperature, a fact that can be established by a wide variety of mechanical, dielectric, and ^{1}H $T_{1\rho}$ nmr experiments.[37] In the presence of such motions, the spin-lattice mechanism is expected to be a more-efficient relaxation process than the spin-spin CP transfer mechanism because the spectral density distribution describing mid-kilohertz motion generally has a Lorentzian shape with a long tail, rather than the Gaussian shape with a short tail one associates with low-frequency dipolar fluctuations[15] (Figure 4.6, bottom illustrations). Thus, for large H_1s, spin-lattice processes should provide a greater frequency match between carbon and proton spin systems than spin-spin processes.

This expectation has been confirmed by both calculations and measurements. A theoretical estimate of a methylene-carbon T_{IS} (ADRF) for a crystalline rigid polymer (with a proton line width of 10 G and a 36-G carbon H_1) is approximately 20 msec[10,15]; experiments on polyoxymethylene produce values of about 10 msec.[22] On the other hand, main-chain ^{13}C $T_{1\rho}$s of approximately 2 msec at 36 G are observed for glassy, mechanically lossy polymers such as polystyrene and poly(vinyl chloride), as measured from the initial slopes of $T_{1\rho}$ plots for these polymers under nonspinning conditions.[22] The average $T_{1\rho}$s are much shorter than the observed T_{IS} (ADFR)s.[22] Hence, these $T_{1\rho}$ parameters must be determined by spin-lattice rather than spin-spin processes.

Displays of $T_{1\rho}$ relaxation rates (at 36 G) for two specialty polycarbonates, molded at 100,000 psi, machined into Lowe-geometry rotors, and spinning at 3 kHz, are shown in Figure 4.18.[22] Structural formulas for the two polymers are also shown in the figure. (Polycarbonate itself has the same structure as the chlorinated polymer illustrated at the bottom of the figure, except protons replace the chlorines.) We will concern ourselves only with the signals arising from the protonated carbons of the aromatic rings. For norborenylbisphenol-polycarbonate (NBP-PC), these are the second and third lines from the left, while for the tetrachlorobisphenol-polycarbonate (Cl$_4$BPA-PC), it is the third line from the left. The high resolution achieved by magic-angle spinning makes this identification easy and we will not bother with the details here.

The relaxation rates of the protonated-aromatic and methyl carbons in NBP-PC are about the same as in ordinary polycarbonate, while in Cl$_4$BPA-PC they are about five times slower. The latter major difference cannot be explained in terms of any difference in ADRF spin-spin transfers between carbons and protons because the temperature-insensitive ^{1}H line width of polycarbonate (slighly less than 5 G) is comparable to that of the chlorinated polycarbonate.[46] Thus, since *static* dipolar C—H and H—H interactions are the same, the faster relaxation

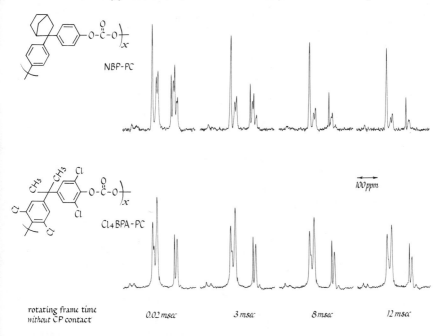

Figure 4.18
Dipolar-decoupled magic-angle CP ^{13}C nmr spectra of two specialty polycarbonates as a function of the time the carbon magnetization was held in the rotating frame without CP contact; $(H_1)_{carbon} = 36$ kHz.

rates of polycarbonate relative to chlorinated polycarbonate must be due to molecular motion and spin-lattice processes. We are, therefore, led to the conclusion that the mid-kilohertz motion responsible for the effective $T_{1\rho}$ relaxation of polycarbonate, while not influenced by the structural substitution of the bulky norborenyl group, is dramatically affected by the substitution of the polar chlorines.

This conclusion is consistent with our previous notion[10] that the main-chain motion in polycarbonate involves an operating unit consisting of two aromatic rings connected by a relatively rigid quaternary carbon. Several such units, linked by flexible carbonyl carbons, engage in cooperative torsional oscillations having angular excursions less than necessary to promote conformational changes in the chain. It seems reasonable to us that adding to the bulk of the rigid connection should make little difference to this motion, while inhibiting the freedom of the flexible link by severe polar repulsions should make a big difference.

Although the above application may have seemed straightforward enough, there are, in general many pitfalls in the use of the ^{13}C $T_{1\rho}$ relaxation parameter. We will mention just a few of them.

1. There is not one $T_{1\rho}$ for a given carbon in a polymer chain, but many $T_{1\rho}$s. This is true because of the dynamic heterogeneity of glassy polymers.[10,47] Molecular motion can be different in one part of a nominally homogeneous glass relative to another part, because of differences in local inter- and intrachain steric interactions. Of course, there is no averaging of these relaxation times by spin-spin diffusion. Thus, the observed $T_{1\rho}$ is a sum of relaxation rates, producing a nonlinear appearance to a standard semilog plot. Initial and final slopes of such plots can differ by two orders of magnitude. It is conceivable therefore that while the initial slope reflects a $T_{1\rho}$ dominated by spin-lattice processes, the final slope yields a $T_{1\rho}$ dominated by spin-spin relaxation processes. It is crucial in discussions of $T_{1\rho}$ experiments to compare similar parts of the dispersions of relaxation times.

2. The ^{13}C $T_{1\rho}$ depends upon magic-angle spinning. There is a second contribution to the dispersion of $T_{1\rho}$s that we did not mention in the previous paragraph. This contribution is completely static in nature and arises from the fact that the efficiency of the ^{1}H—^{13}C dipolar interaction depends upon the angle of the H—C internuclear vector with the applied static magnetic field. With the exception of those few vectors that happen to be aligned at the magic angle, differences in static orientations give rise to differences in $T_{1\rho}$s.[48] Under a 3-kHz sample rotation, significantly different orientations are realized after about 50 μsec. Consequently, those carbons, which under nonspinning conditions would have been relatively slow relaxers, have an opportunity with spinning to undergo faster relaxation when their C—H internuclear vector is at a more favorable angle relative to H_0. Naturally, some fast relaxers are made slower, so that, the average relaxation time (which determines the initial slope of a $T_{1\rho}$ plot) is not substantially affected even though the dispersion is reduced. In addition, a reduction in $T_{1\rho}$s can arise when H_1 is comparable to local dipolar fields. Since sample rotation causes the local dipolar fields to change signs (Section II.F), magic-angle spinning introduces a time dependence to H_{eff} (Figure 4.8). This can be shown to improve the efficiency of both $I_zS_{\pm}$ and $I_{\pm}S_{\mp}$ relaxation mechanisms.[49,50] Finally, despite our earlier position,[10] we now believe that dilational stress variations within a solid spinning rotor can give rise to differences

in local packing and consequently variations in molecular motion and relaxation rates.[51] Obviously, comparison of $T_{1\rho}$s of stationary and spinning samples requires caution.

3. Useful ^{13}C $T_{1\rho}$s often require dangerously small H_1s. To be sensitive to the low-frequency motions of the solid polymer, the ^{13}C $T_{1\rho}$ demands the use of a carbon H_1 having a frequency close to that of the molecular motion.[10] For many polymers this means H_1s of 20–30 kHz. However, the use of small H_1s raises the specter of spin-spin contributions to $T_{1\rho}$ and hence confusions in molecular interpretations. For a given (small) H_1, confirmation of the absence of such contributions should be made. This can be done either by direct measurement of T_{IS} (ADRF) (if practical), or by the observation that protonated and nonprotonated carbons of the same motional unit, near the same types of protons, have $T_{1\rho}$s that scale with (that is, are proportional to) their C—H static dipolar interactions. In the latter test, failure to scale means the width of the rotating-frame carbon resonance is crucial, which therefore suggests a frequency overlap with a restricted proton frequency distribution (spin-spin mechanism, Figure 4.6, third from top) rather than with an extended one (spin-lattice mechanism, Figure 4.6, bottom).

4. There is no simple nmr theory to interpret ^{13}C $T_{1\rho}$s. Because H_1s are generally not much greater than local dipolar fields, ordinary weak-collision theory is not strictly applicable.[52] That is, the rotating-frame carbon lineshape is not a delta-function but has a width (Figure 4.6, bottom). This complicates extracting quantitatively a molecular correlation time from an nmr relaxation time.

5. There is no simple molecular dynamics theory to interpret ^{13}C $T_{1\rho}$s. Ignoring the complication mentioned in the previous paragraph, we are still faced with the as yet unsolved problem of constructing a motional model of the polymeric amorphous solid state that will allow us to even attempt to connect nmr relaxation times with molecular correlation times. For example, at present, we are usually ignorant of whether a certain $T_{1\rho}$ relaxation rate implies a $10°$ torsional angular excursion at 20 kHz or a $20°$ excursion at 10 kHz.

Despite all these pitfalls and limitations, ^{13}C $T_{1\rho}$ experiments can still be useful. As one example, we have established[10] a predictive correlation between main-chain ^{13}C $T_{1\rho}$s and the mechanical toughness of glassy polymers. This correlation is rationalized in terms of energy dissipation by low-frequency cooperative motions (and ultimately flow) that are determined by the same inter- and intrachain steric interactions that influence the $T_{1\rho}$ relaxation parameters. Of course, the empirical correlation remains valid despite our ignorance as to what all the parameters really mean.

REFERENCES

1. F. Bloch, *Phys. Rev.* **11,** 841 (1958).
2. A. Abragam, *The Principles of Nuclear Magnetism*, Oxford University Press, London, 1961, p. 571.
3. See, for example, Ref. 2, p. 527.
4. N. Bloembergen and T. J. Rowland, *Acta Metal.* **1,** 731 (1953).
5. E. R. Andrew, *Prog. Nucl. Mag. Res. Spectrosc.* **8,** 1 (1971).
6. A. Pines, M. G. Gibby, and J. S. Waugh, *J. Chem. Phys.* **59,** 569 (1973).
7. E. R. Andrew, A. Bradbury, and R. G. Eades, *Nature* **182,** 1659 (1958).
8. J. W. Beams, *Rev. Sci. Instr.* **1,** 667 (1930).
9. I. J. Lowe, *Phys. Rev. Lett.* **2,** 285 (1959).
10. J. Schaefer, E. O. Stejskal, and R. Buchdahl, *Macromolecules* **10,** 384 (1977).
11. J. Schaefer and E. O. Stejskal, *J. Am. Chem. Soc.* **98,** 1031 (1976).
12. S. R. Hartmann and E. L. Hahn, *Phys. Rev.* **128,** 2042 (1962).
13. See, for example, M. Goldman, *Spin Temperature and Nuclear Magnetic Resonance in Solids*, Oxford University Press, London, 1970, p. 34.
14. D. A. McArthur, E. L. Hahn, and R. E. Walstadt, *Phys. Rev.* **188,** 609 (1969).
15. D. E. Demco, J. Tagenfeldt, and J. S. Waugh, *Phys. Rev. B.* **11,** 4133 (1975).
16. See Ref. 2, p. 104.
17. F. M. Lurie and C. P. Slichter, *Phys. Rev.* **133,** A1108 (1964).
18. A. Pines and T. W. Shattuck, *J. Chem. Phys.* **61,** 1255 (1974).
19. J. Schaefer, E. O. Stejskal, and R. Buchdahl, *Macromolecules* **8,** 291 (1975).

20. R. K. Hester, J. L. Ackerman, V. R. Cross, and J. S. Waugh, *Phys. Rev. Lett.* **34**, 993 (1975).

21. E. O. Stejskal, J. Schaefer, and J. S. Waugh, *J. Mag. Res.* **28**, 105 (1977).

22. E. O. Stejskal, J. Schaefer, and T. R. Steger, *Faraday Symp. Chem. Soc.*, 1978, in press.

23. See Ref. 13, p. 46; also M. Goldman, *C. R. Hebd. Séanc. Acad. Sci. Paris* **246**, 1058 (1963).

24. E. Lippmaa, M. Alla, and T. Tuherm, *Proc. XIX Congress Ampère*, Heidelberg (1976).

25. M. Maricq and J. S. Waugh, *Chem. Phys. Lett.* **47**, 327 (1977).

26. E. O. Stejskal, J. Schaefer, and R. A. McKay, *J. Mag. Res.* **25**, 569 (1977).

27. V. R. Cross, R. K. Hester, and J. S. Waugh, *Rev. Sci. Instrum.* **47**, 1486 (1976).

28. M. E. Stoll, A. J. Vega, and R. W. Vaughan, *Rev. Sci. Instrum.* **48**, 800 (1977).

29. E. O. Stejskal and J. Schaefer, *J. Mag. Res.* **14**, 160 (1974).

30. E. O. Stejskal and J. Schaefer, *J. Mag. Res.* **18**, 560 (1975).

31. D. L. Vanderhart, *J. Mag. Res.* **24**, 467 (1976).

32. R. K. Hester, J. L. Ackerman, B. L. Neff, and J. S. Waugh, *Phys. Rev. Lett.* **36**, 1081 (1976).

33. V. J. Bartuska, G. E. Maciel, J. Schaefer, and E. O. Stejskal, *Fuel* **56**, 354 (1977).

34. H. Y. Carr and E. M. Purcell, *Phys. Rev.* **94**, 630 (1954).

35. S. Opella and J. S. Waugh, *J. Chem. Phys.* **66**, 4919 (1977).

36. H. G. Olf and A. Peterlin, *J. Polym. Sci.* **8**, 771 (1970).

37. See, for example, N. G. McCrum, B. E. Read, and G. Williams, *Anelastic and Dielectric Effects in Polymeric Solids*, Wiley, New York, 1967.

38. B. Ewen, E. W. Fischer, W. Piesczek, and G. Strobe, *J. Chem. Phys.* **61**, 5265 (1974).

39. J. Schaefer, E. O. Stejskal, C. F. Brewer, H. D. Keiser, and H. Sternlicht, *Arch. Biochem. Biophys.* in press.

40. J. Schaefer and E. O. Stejskal, *J. Am. Oil Chem. Soc.* **52**, 366 (1975).

41. L. D. Kier, J. Schaefer, and E. O. Stejskal, in preparation.

42. J. Schaefer, E. O. Stejskal, and C. F. Beard, *Plant Physiol.* **55**, 1048 (1975).

43. C. F. Brewer and H. Keiser, *Proc. Nat. Acad. Sci.* **72,** 3421 (1975).

44. A. Pines, M. G. Gibby, and J. S. Waugh, *J. Chem. Phys.* **56,** 1776 (1972).

45. M. W. Duch and D. M. Grant, *Macromolecules* **3,** 165 (1970).

46. L. J. Garfield, *J. Polym. Sci. C.* No. 30, 551 (1970).

47. J. Schaefer, E. O. Stejskal, and R. Buchdahl, *J. Macromol Sci.—Phys.* **B13,** 665 (1977).

48. M. G. Gibby, A. Pines, and J. S. Waugh, *Chem. Phys. Lett.* **16,** 296 (1972).

49. J. F. J. M. Pourquié and R. A. Wind, *Phys. Lett.* **55A,** 347 (1976).

50. G. P. Jones, *Phys. Rev.* **148,** 332 (1966).

51. J. Schaefer, T. R. Steger, E. O. Stejskal, and R. A. McKay, *Macromolecules,* in review.

52. C. P. Slichter and D. Ailion, *Phys. Rev. A* **135,** 1099 (1964); D. Ailion, *Adv. Mag. Res.* **5,** 172 (1972).

5 HIGH-POWER DOUBLE-RESONANCE STUDIES OF FIBROUS PROTEINS, PROTEOGLYCANS, AND MODEL MEMBRANES

D. A. TORCHIA
D. L. VANDERHART

CONTENTS

I. INTRODUCTION

During the past decade, ^{13}C pulsed nmr spectroscopy has developed so rapidly that ^{13}C spectra of complex biological molecules in solution can now be routinely obtained on a variety of commercial instruments. The literature on this subject is extensive and has been ably reviewed recently.[1-6] In spite of the advances in instrumentation, there have been few studies of ordered biological macromolecules (molecular weight $> 5 \times 10^4$) either in solution or in the solid state. The reason for this is the well-known fact that when C—H bonds do not reorient isotropically with

325

correlation times τ less than 10^{-6} sec, the ^{13}C—^{1}H dipolar interaction severely broadens lines, resulting in poor resolution and sensitivity. In a conventional pulsed Fourier transform (FT) experiment, protons are irradiated with a field of ~1 G, to remove ^{13}C—^{1}H scalar couplings. Since ^{13}C—^{1}H and ^{1}H—^{1}H dipolar couplings in slowly tumbling or fixed molecules can exceed 20 kHz, application of the 1-G field ($\gamma H_2/2\pi \sim$ 4 kHz) normally has no effect on the dipolar broadened lines. Application of a strong resonant field, $H_2 \gtrsim 12$ G, at the proton resonance frequency, normally removes the dipolar coupling between protons and ^{13}C nuclei.[7-10]

Although high-power decoupling eliminates the dipolar contribution to the ^{13}C line widths, poor sensitivity, arising from the large ^{13}C T_1 values in rigid solids, often precludes measurement of dipolar decoupled spectra. A clever cross-polarization method of sensitivity enhancement has been developed[11] that takes advantage of the smaller T_1 values of the normally abundant protons in solids by using a matched Hartman-Hahn contact[12] to transfer polarization from protons to ^{13}C nuclei. This technique provides up to a fourfold enhancement of ^{13}C polarization per contact. After ^{13}C polarization is established, contact between the spin systems is broken and the ^{13}C free induction decay is detected with high-power proton decoupling.[11] Since the rate at which ^{13}C polarization accumulates during contact is proportional to the square of the secular ^{13}C—^{1}H dipolar interaction,[13] the contact time T_{cp}, can often be chosen to enhance signals of nuclei in ordered portions of a sample relative to those in mobile regions.*

Since ^{13}C signals in a dipolar decoupled spectrum do not experience proton dipolar broadening and the low natural abundance of ^{13}C insures that homonuclear dipolar interactions are ≤ 50 Hz, the lineshape in a dipolar decoupled spectrum often provides direct information about the ^{13}C chemical shift anisotropy.[10,11,14-19] The chemical shift anisotropy and residual ^{13}C homonuclear dipolar interactions can be removed by spinning the sample at the magic angle,[11,20,21] 54.7°, usually at a rate of a few kilohertz in the case of ^{13}C dipolar decoupled signals, producing spectra having resolution approaching that found in liquids.[22,23]

To date, high-power double resonance techniques have been used primarily to study molecular structure and dynamics in organic

---

* Following the precedent of Schafer et al.,[26] we have normally used the term dipolar decoupled spectrum to refer to a spectrum in which the ^{13}C magnetization is generated using a 90°-t pulse sequence, whereas the term cross-polarization spectrum is used to refer to a spectrum obtained using a matched Hartmann-Hahn contact to establish ^{13}C magnetization. We realize that this may cause confusion since dipolar decoupling is employed in both experiments. For this reason the pulse sequence used is indicated in each figure legend.

solids[10,11,14,-19,24,25] and in bulk synthetic polymers.[22,26-28] It is our aim in this article to show that these techniques are well suited for studying structural macromolecules of biological relevance.

II. ELASTIN

Elastin is a major protein constituent of connective tissues such as skin, lung, vessel, and ligament, conferring upon these tissues macroscopic rubberlike properties of high extensibility and small elastic modulus.[29-31] At the molecular level, elastin and rubber (cis-polyisoprene) have similar structural features. First, proelastin is a long single polypeptide chain containing about 850 residues, with a molecular weight of about 70,000.[29] Second, the proelastin chains are cross-linked (by desmosine and isodesmosine) and form an insoluble three-dimensional elastic network.[29]

A major difference between elastin fibers and rubber is that the former must be swollen to exhibit elasticity. Unswollen elastin is brittle and, in further contrast to rubber,[32,33] has dipolar broadened line widths that are so large that a scalar decoupled ^{13}C spectrum of dried elastin has not been reported. The apparent rigidity of the unswollen elastin network is thought to arise from interactions beween the polar NH and C'O moieties in the peptide backbone. In rubber the weaker nonpolar dispersion forces permit segmental chain motion at room temperature.

Elastin chains differ from those of rubber in another important respect, namely, available sequence data[29] indicates that elastin contains two chemically distinct regions. The major portion of the chain consists largely of valyl, glycyl, and prolyl residues, which often occur in repeating short sequences, whereas the sequences near the lysine derived crosslinks contain alanyl residues, almost exclusively.[29] It may be that the reported[34,35] nonuniform swelling of elastin is due to chemical heterogeneity.

The possibility that elastin swelling is not uniform at the molecular level, together with the sensitive volume dependence of water swollen elastin on temperature and stress, have been the source of much of the controversy surrounding the interpretation of thermoelastic and mechanical data.[30,36-45] Different interpretations of the data have led to two general models[43] of elastin structure: the network model and the two-phase model. In the network model, the chains are assumed to be more or less uniformly swollen and to be segmentally mobile with a high degree of configurational entropy. In the two-phase model, the chains are supposed to pack into more ordered conformations that minimize contacts between the nonpolar side chains (~90% of elastin sidechains are nonpolar) and water. These models have been evaluated in terms of the variety of experimental procedures used to study elastin.[30,31,46] Our present discussion is restricted to ^{13}C magnetic resonance data of elastin.

Scalar decoupled spectra of elastin are obtained at 37° (Figure 5.1) when intact calf ligamentum nuchae fibers are swollen in suitable solvents[47,48] As is clear from Figure 5.1, elastin line widths are sensitive functions of the swelling solvent, with formamide and dimethylsufoxide producing sharper resonances than 0.15 M NaCl. Mixed solvent systems (0.15 M NaCl—formamide, 5:4 by volume and 0.15 M NaCl—ethanol,

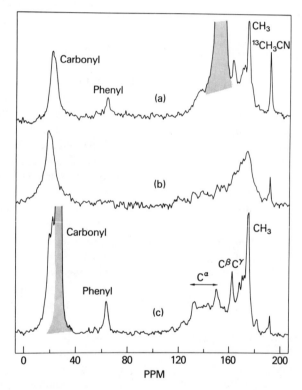

Figure 5.1
Comparison of intact calf ligamentum nuchae scalar decoupled spectra obtained in three solvents at 37°C, using $90° - \tau$ pulse sequences: (a) dimethyl sulfoxide, $\tau = 3.2$ sec; (b) 0.15 M NaCl, $\tau = 3.5$ sec; (c) formamide, $\tau = 1.0$ sec. 8192 transients were accumulated in each case. Acetonitrile, 90% enriched at the methyl carbon was used as an internal reference. Chemical shift scale is in ppm from external CS_2. The solvent resonances are shaded.

1:1 by volume) also yield spectra that are better resolved than the spectrum obtained in 0.15 M NaCl. It should be kept in mind that the intensity observed in scalar decoupled ^{13}C spectra arises from highly mobile carbons whose resonances are motionally narrowed by fast ($\tau \leq 10^{-6}$ sec) nearly isotropic motions. Intensity measurements of the carbonyl resonances have shown that approximately $80\% \pm 15\%$ of the elastin backbone carbons exhibit such motional narrowing. Average T_1, Δ, and NOE values have been determined for ligamentum nuchae elastin for C^α and C' backbone resonances in several solvent systems,[48] and these data have been analyzed using a $\log \chi^2$ distribution of correlation times defined by

$$G_p(s) = \frac{p(ps)^{p-1} \exp(-ps)}{\Gamma(p)} \tag{1}$$

with

$$s = \log_b [1 + (b-1)\tau/\bar{\tau}] \tag{2}$$

$G_p(s)$ is the probability of finding correlation time τ corresponding to $s(\tau)$ and the width of the distribution is governed by the parameters p and b. Schaefer[49] found that a broad distribution, defined by $p = 14$, $b = 1000$, provided a self-consistent analysis of T_1, Δ, and NOE values of solid *cis*-polyisoprene and this same distribution has provided an approximately self-consistent analysis of the elastin data.[48] The nmr parameters calculated for this distribution (dashed curves) are plotted as a function of $\tau = \bar{\tau}$ (the correlation time that defines the approximate maximum of the distribution) in Figure 5.2 where they can be compared with parameters calculated for a single correlation time model[50] (solid curves).

When the dashed curves (Figure 5.2) were used to calculate $\bar{\tau}$ in a given solvent, it was found[48] that each measured nmr parameter yielded a different value of $\bar{\tau}$ with the individual $\bar{\tau}$ values deviating from their mean value by approximately a factor of two. These inconsistencies in the derived values of $\bar{\tau}$ were ascribed to the large uncertainties in the data combined with the sensitive dependence of calculated values of $\bar{\tau}$ on measured T_1 and NOE values. These discrepancies also arose since it is probably not possible to account for the complex segmental motion in polymers in terms of a few parameters.

In spite of these shortcomings the $\log \chi^2$ distribution does account for the measured nmr parameters in a semiquantitative fashion and the consistency of the calculated correlation times is much better than that obtained using the single correlation time model. A comparison of the average value of $\bar{\tau}$, obtained in each solvent with that obtained for

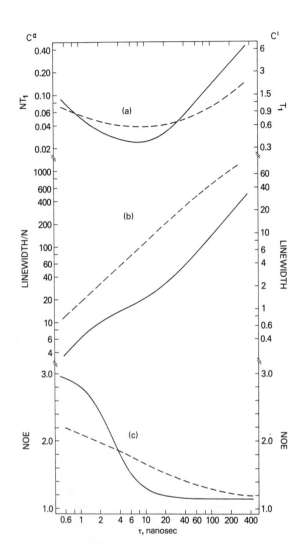

Figure 5.2

Nmr parameters at $H_0 = 14$ kG: (a) T_1; (b) $\Delta = 1/(\pi T_2)$; (c) NOE; plotted (1) as a function of $\tau = \tau_R$ in the single correlation time model (solid curves) and (2) as a function of $\tau = \bar{\tau}$ in the $\log \chi^2$ distribution of correlation times model, have $b = 1000$, $p = 14$ (dashed curves). The $\log \chi^2$ distribution and the parameters p, b, and $\bar{\tau}$ are defined in Eqs. 1 and 2. The scales on the left apply to a C^α carbon, dipolar relaxed by the N-hydrogens to which it is bonded, with $r_{CH} = 1.09$ Å. The scale on the right applies to a C' carbon, dipolar relaxed by proximal protons (see ref. 48 for further explanation of calculations).

330

Table 5.1

Comparison of $\bar{\tau}$ Values (nsec) Obtained for *cis*-Polyisoprene[49] and Elastin[48] Using a Log χ^2 Distribution[a] of Correlation Times having $p = 14$, $b = 1000$

Sample	Solvent	$\bar{\tau}$
cis-Polyisoprene	None	0.4
Elastin	0.15 M NaCl—formamide[b]	1.8
Elastin	0.15 M NaCl—ethanol[c]	2.5
Elastin	Dimethyl sulfoxide	2.7
Elastin	0.15 M NaCl	80

[a] Defined in Eqs. 1 and 2.
[b] 5:4, v/v.
[c] 1:1, v/v.

cis-polyisoprene (Table 5.1) leads to several interesting conclusions about elastin chain motion. Swelling solvents that contain formamide, the most polar solvent used, produce the smallest value of $\bar{\tau}$, suggesting that polar interactions between peptide moities play a major role in determining elastin chain mobility. However, it seems clear from the line widths observed as a function of solvent composition in 0.15 M NaCl—ethanol mixtures that polar interactions alone cannot account for solvent influences on chain mobility. The sharpest elastin resonances in this solvent system were observed at 1 : 1 by volume 0.15 M NaCl—ethanol composition, where $\bar{\tau}$ was found to be approximately 30 times less than in 0.15 M NaCl (Table 5.1). This result has been explained by suggesting that ethanol disrupts hydrophobic interactions between the (almost exclusively) nonpolar side chains in elastin. Comparable $\bar{\tau}$ values were obtained (Table 5.1) for *cis*-polyisoprene[49] (0.4 nsec) and elastin swollen by solvents containing a polar organic component (1.8–2.7 nsec) in agreement with a variety of studies that have shown that elastin behavior in these solvents approximates that of an ideal rubber. A considerably larger $\bar{\tau}$ (80 nsec) was found for elastin swollen by 0.15 M NaCl. The restricted elastin chain motion suggested by this $\bar{\tau}$ value is in accord with macroscopic dynamic measurements that show that water-swollen elastin at 37°C has dynamic behavior like low-temperature rubber in the viscoelastic transition zone.[46] Thus the nmr data indicate that a viscoelastic rubberlike network is an appropriate model for the mobile elastin component observed in the scalar decoupled spectra. A model in which elastin chains are more highly ordered has been proposed by Urry and collaborators[51] on the basis of ^{13}C studies of model polypeptides.

Although the signal intensities obtained from scalar decoupled spectra enable one to infer that a fraction of elastin carbons experience strong dipolar couplings, the dipolar-broadened resonances are so broad that they have not been observed in scalar-decoupled spectra. Signals from the dipolar-broadened component in elastin were first observed in proton free induction decays by Ellis and Packer.[34] Subsequently it was shown[52] that the integrated aliphatic carbon signal intensity in the dipolar decoupled spectrum of chick aortic elastin was 35–50% larger than the aliphatic signal intensity in the scalar decoupled spectrum. The signal increase was difficult to measure accurately because of curvature in the baseline. A similar enhancement of the aliphatic signal intensity in the dipolar de-coupled spectrum of calf ligamentum nuchae elastin can be seen by comparing Figures 5.3a and 5.3b. The relative areas of the aliphatic resonances in these spectra are equal to the relative numbers of elastin carbons that contribute to each signal, since nonelastin material has been removed using the procedure of Lansing et al.[53] and the NOEs have been suppressed using gated decoupling. The excess aliphatic signal intensity

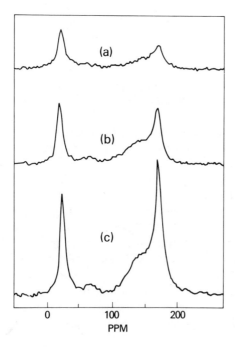

Figure 5.3
Comparison of spectra of calf ligamentum nuchae elastin obtained using two decoupling fields: (a) $H_2 = 0.8\,G$; (b) $H_2 = 14\,G$; (c) $H_2 = 14\,G$, with protons saturated $(H_2 \sim 1\,G)$ during the 5 sec delay between 90° pulses. The sample was swollen in 0.15 M NaCl at a temperature of 37°C. In each case a $90° - \tau$ pulse sequence with $\tau = 5$ sec was employed and 8192 transients were accumu-lated. Chemical shift scale is in ppm from external CS_2. Collagen and other contaminants were removed from the tissue using the procedure of Lansing et al.[53]

observed in the dipolar spectrum is due to the fraction of the elastin carbons that experience restricted mobility, resulting in severe dipolar broadening in the scalar-decoupled spectrum. An increase in the integrated signal intensity in the downfield region of the dipolar-decoupled spectrum is not observed since (as will be shown) the motionally restricted carbonyl and aromatic carbons have large chemical shift anisotropies and possibly large T_1 values.

In addition to increasing the observed signal intensity, dipolar decoupling enhances the resolution of the elastin spectrum so that one can measure T_1 and NOE values for the four major groups of signals (carbonyl, 20 ppm; aromatic, 50–70 ppm; C^α, 120–150 ppm; CH_3, 170 ppm). NOEs of 1.6 and 2.2 are found for backbone (C' and C^α) and methyl carbons, respectively, whereas T_1 values for C', C^α, and CH_3 carbons are 1.2, 0.08, and 0.25 sec, respectively.[54] Although these nmr parameters can readily be derived from dipolar decoupled elastin spectra, their interpretation in the case of aliphatic carbons is complicated by the fact that signals from the mobile and motionally restricted fractions are not separated.

One can effect a partial resolution of these fractions by taking advantage of the fact that carbons in the less mobile fraction will cross-polarize* more rapidly than the mobile carbons. This is illustrated in Figure 5.4, which shows spectra[54] of ligamentum nuchae obtained as a function of cross-polarization contact time T_{cp}. For $T_{cp} \leq 100$ μsec the broad signals, particularly in the −50 to 100 ppm region of the spectra (Figure 5.4a,b) indicate that there is little motional narrowing of chemical shift anisotropy of carbons in domains of restricted elastin mobility. In contrast, when $T_{cp} \geq 500$ μsec, carbonyl and methyl resonances are seen that clearly exhibit motional narrowing of their chemical shift anisotropies. The observation that the carbons in the mobile chains do cross-polarize indicates that even they experience a small residual dipolar interaction, and implies that chain motions on the time scale of $\leq 10^{-5}$ sec do not produce an exact isotropic average of the dipolar interactions.

The decay of intensity of the broad and narrow signals (seen in Figure 5.4 when T_{cp} is large) cannot be simply described because T_{cp}, and the measured[54] ^{13}C and ^1H $T_{1\rho}$ values are comparable. $T_{1\rho}$ values differ for the broad and narrow components as do the cross-polarization signal enhancements. Thus, the relative areas of the broad and narrow resonances do not provide a direct measure of the relative number of carbons in each component.

---

* All cross-polarization experiments were performed with alternation of spin temperature.[54a]

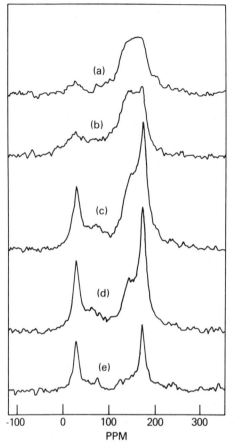

Figure 5.4
Cross-polarization spectra of calf ligamentum nuchae elastin obtained as a function of T_{cp}, the matched Hartmann-Hahn contact time: (a) T_{cp} 0.05 msec, (b) $T_{cp} = 0.1$ msec; (c) $T_{cp} = 0.5$ msec; (d) $T_{cp} = 1$ msec; (e) $T_{cp} = 3$ msec. The sample was swollen in 0.15 M NaCl at a temperature of 37°C and 8192 transients were accumulated. Chemical shift scale in ppm from external CS_2. Collagen and other contaminates were removed using the procedure of Lansing et al.[53]

Although the broad resonances of the elastin carbons that experience restricted mobility can be directly detected using cross polarization, it does not appear possible to assign these signals to specific regions of the protein sequence using natural abundance spectra. Various authors[31] have suggested that the repeating sequence regions and/or the crosslinked regions in elastin have ordered structure. One method for assigning the broad resonances would be to incorporate ^{13}C labeled amino acids into elastin. A preliminary study using chick aorta in tissue culture suggests the ^{13}C labeled glycine incorporates into the tissue,[52] but it still remains to be shown that highly purified intact labeled elastin can be isolated from the tissue. Purified reconstituted collagen fibrils containing ^{13}C labeled glycine have been prepared from chick calvaria cultures.[55] As we now show, these samples illustrate that dynamical information about a specific region of a protein (the collagen backbone in this instance) can be obtained when labeled samples are available.

III. COLLAGEN

Like elastin, collagen is a structural protein that is widely distributed in extracellar tissues and is in fact the major protein constituent of soft connective tissue and bone.[56] In contrast with elastin fibers, collagen fibers have high tensile strength and little elastic extensibility. It is generally accepted that the mechanical properties of the collagen fiber result from a cross-linked assembly of ordered, helical molecules. The collagen helix consists of three polypeptide chains (α chains) each composed of triplets Gly—X—Y. This triple helix is the structural feature common to all collagens and is stabilized by interchain hydrogen bonds between NH and C′O moieties on the backbone of each chain. The interactions that stabilize the higher levels of collagen structure (microfibrils, fibrils, etc.) are less well understood and are being actively investigated.[57–61]

Recently, high-power double resonance techniques have been used to study interactions in fibrous collagens.[55,62] The dipolar decoupled ^{13}C spectrum of an unoriented sample of calf achilles tendon (Figure 5.5a) exhibits two broad signals having maxima at about 20 ppm (carbonyl and aromatic carbons) and 170 ppm (methine, methylene, and methyl carbons). The integrated intensity measured from the tendon spectrum showed that at least 75% of the ^{13}C nuclei in each sample had T_1 values of the order of 2 sec or less. This result suggested that the majority of collagen C—H bond vectors reorient anisotropically with $\tau \leq 10^{-6}$ sec. The motion in collagen is anisotropic since scalar-decoupled collagen line widths are so broad that they have not been detected using high-resolution techniques.[47] This implies that reorientation is slow ($\tau \geq 10^{-4}$ sec) over certain directions. To study the motion of backbone carbons in detail, samples of reconstituted chick calvaria collagen[55,63] were prepared that were enriched with ^{13}C at either the C' or C^α position in glycyl residues. Enrichment levels of 35% at the C^α position and 45% at the C' position were determined by measuring the intensity of the ^{13}C satelites of the glycine α protons in the 220 MHz spectra of hydrolyzed collagen samples. The spectra of the C^α and C' labeled native samples (Figure 5.5b,c) show major signals having maxima at ~150 ppm and ~20 ppm, the respective isotropic chemical shifts of glycine C^α and C' resonances. Integrated intensities (obtained with $t > 5T_1$) showed that

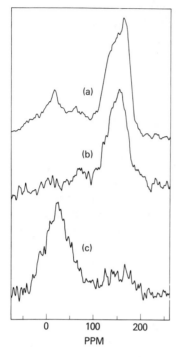

Figure 5.5
Comparison of dipolar-decoupled ^{13}C spectra of various collagens using $90°\text{-}t$ pulse sequences. (a) Calf achilles tendon $t = 3.0$ sec, 19,500 scans; (b) reconstituted chick calvaria collagen, ^{13}C enriched at the glycine C^{α}, $t = 1.0$s, 18,350 scans. (c) reconstituted chick calvaria collagen, ^{13}C enriched at the glycine C', $t = 1.0$s, 18,350 scans. Chemical shift scale in ppm from external CS_2.

within experimental error ($\pm 20\%$) all carbons in each labeled sample contributed unsaturated signal intensity to the spectrum. The large difference in T_1 values measured for the C^{α} ($T_1 \sim 0.1$ sec) and C' ($T_1 \sim 1.5$ sec) carbons is consistent with a ^1H—^{13}C dipolar relaxation mechanism.

The small T_1 value measured at 20° for each type of carbon indicates that anisotropic backbone motion takes place in the reconstituted fibrils. The anisotropic molecular motion is frozen out when the samples are dried and the temperature is reduced to −95°. Spectra obtained under these conditions (Figure 5.6b,c) have line widths that are two to three times larger than observed at 20°, as a consequence of the full chemical shift anisotropies and ^{14}N—^{13}C dipolar couplings exhibited by the rigid molecules at the low temperature. This interpretation is supported by the observation that the line shape of a spectrum of dry polyglycine, at −95° (Figure 5.6a) is well approximated by the sum of the two low-temperature labeled collagen spectra.

It is difficult to extract quantitative information about the anisotropic motion from the line width observations since the chemical shift tensor assignments have not been made for glycyl residues. This fact also

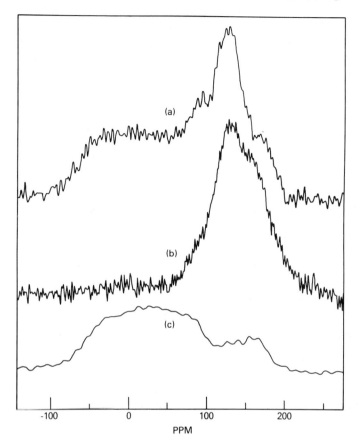

Figure 5.6
Comparison of low temperature, $-95°$, cross-polarization spectra of polyglycine
with dried chick calvaria collagen samples labeled with ^{13}C glycine: (a) polyglycine, $T_{cp} = 3$ msec; (b) calvaria collagen, labeled with ^{13}C$^{\alpha}$ glycine, $T_{cp} =$
0.7 msec; (c) calvaria collagen labeled with ^{13}C$'$ glycine, $T_{cp} = 6$ msec. Chemical
shift scale in ppm from external CS_2.

prevents inclusion of the ^{13}C—^{14}N dipolar interaction in a lineshape
calculation. Nevertheless, the low-temperature broad C' resonances observed in Figure 5.6a,c correspond reasonably well to what one would
expect from a chemical shift anisotropy of 140 ppm, measured for the
single crystal glycine C' resonance,[64] and a superimposed ^{13}C—^{14}N
broadening having a calculated[65] rms value of 46 ppm. Thus, the narrowing at 20° (Figure 5.5c,) indicates that anisotropic backbone motion is
taking place in these fibrils at frequencies greater than a few kilohertz.

Since the ^{13}C spin lattice relaxation in the reconstituted fibrils is due to an intramolecular dipolar mechanism, the measured T_1 and NOE values can be used to estimate the rate and spatial extent of the anisotropic motion, once a model for backbone motion is assumed. The dimensions of the collagen helix (15×3000 Å) suggest that rapid diffusive motion of the helix backbone is limited to rotation about the long axis.

Expressions for the dipolar T_1, in the case of axial diffusion, have been derived using available theory.[66-69] In the case of axial jump diffusion among three ($\Delta\phi = 120°$) or six ($\Delta\phi = 60°$) equivalent sites, the correlation time τ, is related to the dipolar T_1 according to

$$\frac{1}{T_1} = \frac{\hbar^2 \gamma_C^2 \gamma_H^2}{r^6} \frac{9}{64} \tau \left[\frac{f_1(\psi, \theta')}{1 + (\omega_H - \omega_C)^2 \tau^2} + \frac{f_2(\psi, \theta')}{1 + \omega_C^2 \tau^2} + \frac{f_3(\psi, \theta')}{1 + (\omega_H + \omega_C)^2 \tau^2} \right] \quad (3)$$

In Eq. 3, γ_C and γ_H are the respective carbon and proton magnetogyric ratios, ω_C and ω_H are the respective carbon and proton Larmor precession frequencies, r is the ^{13}C—^{1}H internuclear distance, ψ is the angle between the helix axis (rotation axis) and the external field H_0, and θ' is the angle between the helix axis and the C—H internuclear vector. The functions $f_i(\psi, \theta')$ have the following explicit forms:

$$f_1(\psi, \theta') = \sin^2 2\psi \sin^2 2\theta' + \sin^4 \psi \sin^4 \theta' \quad (4)$$

$$f_2(\psi, \theta') = (\cos^2 2\psi + \cos^2 \psi) \sin^2 2\theta' + (\sin^2 \psi + 0.25 \sin^2 2\psi) \sin^4 \theta' \quad (5)$$

$$f_3(\psi, \theta') = (4 \sin^2 \psi + \sin^2 2\psi) \sin^2 2\theta' + [(1 + \cos^2 \psi)^2 + 4 \cos^2 \psi] \sin^4 \theta' \quad (6)$$

Expressions 3–6 reduce to the results of Gibby et al.[24] in the special case, $\theta' = 90°$. An analytic expression was also derived for the NOE.

The predicted behavior of these parameters, as a function of ψ and τ, is presented graphically in Figure 5.7 for the special case $\theta' = 60°$, which corresponds to the angle between the C^α—H bonds and the helix axis for both glycyl α-protons in collagen. The general dependence of T_1 and the NOE on τ is similar to that predicted for isotropic motion (see solid curves in Figure 5.2). It is noted in passing that, even though motion is restricted to axial reorientation, the minimum T_1 values for $\theta' = 60°$ lie in the range of $20 - 30$ msec (Figure 5.7), in close agreement with the minimum T_1 value of 25 msec obtained when motion is isotropic (Figure 5.2a). In spite of similarities in T_1 minima, the detailed behavior of the T_1 and NOE parameters in the case of axial diffusion differs markedly from that predicted when reorientation is isotropic. As seen in Figure 5.7(b), NOE values are not limited to the range 1.15–3 as predicted for isotropic motion and observed in solution. Furthermore, for axial reorientation, T_1 and NOE explicitly depend not only on τ but also on ψ and θ'. The anisotropy in the nmr parameters is particularly large when $\theta' = 90°$ and

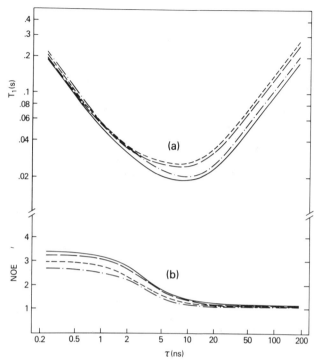

Figure 5.7

Nmr parameters: (a) T_1, (b) NOE, calculated for axial jump reorientation at $H_0 = 14$ kG, $\theta' = 60°$, $\psi = 0°$ (——), $\psi = 30°$ (— —), $\psi = 60°$ (— — — —), $\psi = 90°$ (—·—·). θ' is the angle between the rotation axis and the CH bond vector, and ψ is the angle between the rotation axis and the external field. In the calculation it is assumed that the carbon is dipolar relaxed by a single proton ($r_{CH} = 1.09$ Å) and that axial reorientation occurs by random jumps among three or six sites having respective angular separations of 120 or 60°. τ is the correlation time for the jump diffusion.

$\omega_C \tau \gg 1$.[24] Under these conditions, T_1 and NOE parameters are predicted and observed to change sevenfold as ψ is varied from 0 to 90°.[24] In contrast, when $\theta' = 60°$, it is seen (Figure 5.7a) that the T_1 and NOE values change by less than 50% as ψ increases from 0 to 90°. The lack of observed T_1 and NOE anisotropy in the collagen spectra is thus consistent with the assumed model of motion, from which $\theta' = 60°$ is derived. Due to the small T_1 and NOE anisotropy, one can use the measured T_1 and NOE values and Figure 5.7 to estimate a jump correlation time that is approximately correct for all ψ. Since the glycyl C^α is relaxed by two protons and the two C^α—H bonds make the same angle ($\theta' = 60°$) with respect to the helix axis,[55] the $C^\alpha NT_1$ value (0.2 sec) can be used to calculate τ directly from Figure 5.7a. As is true with isotropic motion, Figure 5.7a shows that two τ values, one on each side of the T_1 minimum, are compatible with the 0.2 sec NT_1 value. However, the measured NOE[55] (1.35 ± 0.15) is only compatible with the larger jump correlation time, $\tau \approx 150$ nsec. In the case of jumps among three equivalent sites,[66,68] the jump rate between sites is $(1/3\tau) = 2.2 \times 10^6$ sec^{-1}.

The continuous diffusion model is an alternative to the jump diffusion model, which is particularly useful when one wishes to compare the correlation time in the fibril with that predicted for the helix in solution. When diffusion is slow compared with ω_C and anisotropy is small, conditions which we have seen apply to reorientation in the collagen fibrils, the general expression for the continuous diffusion case becomes

$$R_1 = 2.5 \times 10^5 \Big/ \left(T_1 \sum_i^N r_i^{-6} \sin^2 \theta_i' \right) \tag{7}$$

In Eq. 7 R_1 is the axial diffusion coefficient, r_i has units of angstroms, T_1 has units of seconds and the summation extends over the N protons which relax the carbon in question. It may be noted that ψ does not appear in Eq. 7 because the angular functions have been averaged over ψ, since anisotropy is assumed to be small. A value of $R_1 = 2.3 \times 10^6$ sec^{-1} is obtained using Eq. 7 with $T_1 = 0.1$ sec, $\theta' = 60°$, $r_i = 1.09$ Å and $N = 2$. This value of R_1 is in close agreement with the value 3×10^6 sec^{-1} calculated[55,70,71] for an ellipsoid of revolution (15×3000 Å) in aqueous solution at 20°, and is equal to the rate for jump diffusion among three equivalent sites calculated earlier.

Thus far it has been assumed that reorientation, be it continuous or in discrete jumps, covers the full range of azimuthal angle. However, reorientation over a limited range of azimuthal angle is also consistent with the data. We consider a model in which the C^α—H bond axes jump between two azimuthal orientations ϕ_1, ϕ_2. Averaging over ψ and ϕ, a T_1 is calculated which agrees with the measured value provided that (a) $R_j \omega_C = 1$, where R_j is twice the jump rate and $\omega_C/2\pi = 15.1$ MHz, (b) the sites have equal populations, and (c) $\phi_1 - \phi_2 = \Delta\phi = 30°$. $\Delta\phi$ must be greater than 30° if the sites have different populations or if $R_j \omega_C \neq 1$.

The foregoing calculations show that rates for continuous or jump diffusion exceed 10^6 sec^{-1} and in the latter model $\Delta\phi \geq 30°$. Since the calculated hydrodynamic diffusion constant for azimuthal reorientation of the entire helix in solution is 3×10^6 sec^{-1} it appears that barriers to azimuthal reorientation, resulting from intermolecular interactions in fibrils are similar to those in solution. This surprising conclusion suggests that the intermolecular interactions that stabilize the structure of the reconstituted fibrils do not depend strongly on the azimuthal angle and, hence, do not arise from a unique set of interactions between the side chains. It should be possible to test this conclusion by studying collagen samples which contain amino acid residues labeled at specific side chain carbons.

IV. CARTILAGE

The extracellular matrix of hyaline cartilages[72] consists primarily of proteoglycans (60% of dry weight) and type II collagen[73] (40% of dry weight). The collagen helix in these cartilages consists of three identical α_1 chains[74] whereas the type I collagens in tendon, bone, and skin contain two α_1 chains and an α_2 chain. Intact type II collagen unlike type I, must be treated with enzymes or other agents that break covalent bonds to be solubilized.[74] In spite of these differences, types I and II collagen have similar functions and fibrous assemblies of triple helical type II molecules are the source of the tensile strength of cartilage.

The proteoglycans in cartilage are macromolecules having molecular weights of several millions in which about 100 chondroitin sulfate chains (MW $\sim 2 \times 10^4$) and 50 keratan sulfate chains (MW $\sim 6 \times 10^3$) are covalently attached to a core protein (MW $\sim 2 \times 10^5$).[75] These sulfated polysaccharides are designated glycosaminoglycans[72] and, together, account for about 90 wt.% of the bovine nasal cartilage proteoglycan. The 4 and 6 isomers of chondroitin sulfate, shown in Scheme 5.1, make up about 90 wt.% of the glycosaminoglycans in bovine nasal cartilage, with the 4 isomer accounting for about 90% of the chondroitin sulfate content.[76] In the cartilage matrix the proteoglycans form aggregates of 20 to 50 molecules in which the core protein of each molecule is noncovalently

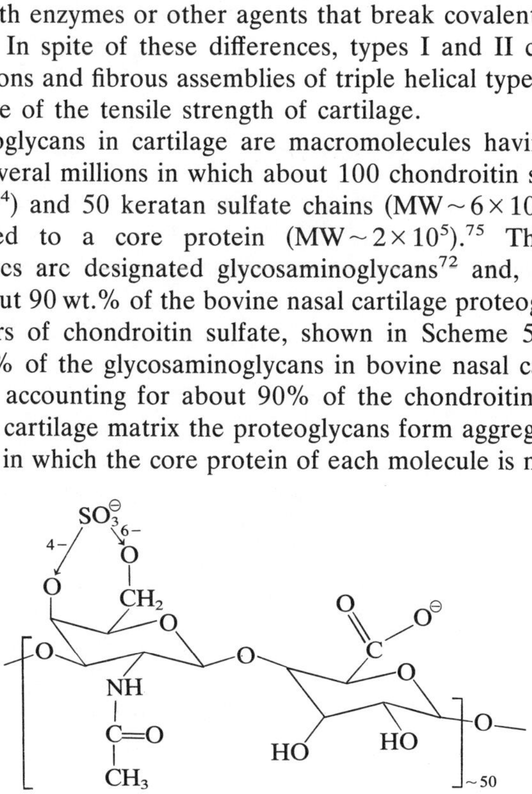

Scheme 5.1

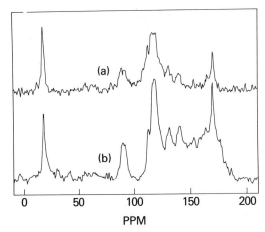

Figure 5.8
Comparison of bovine nasal cartilage spectra obtained at 37°C using (a) scalar decoupling with $H_2 = 0.8$ G; (b) dipolar decoupling with $H_2 = 14$ G. Each spectrum is the Fourier transform of 8192 transients accumulated using a $90° - t$ pulse sequence with $t = 2$ sec. Chemical shift scale in ppm from external CS_2.

bound to hyaluronic acid.[72] The branched negatively charged proteoglycan molecules have large hydrodynamic volumes[75,77] and are thought to provide the tissue with resilliency by their control of solvent flow under compressive loads.

The signals in the scalar decoupled spectrum of intact bovine nasal cartilage (Figure 5.8a) were first observed by Brewer and Keiser[78] and have been assigned to carbons in the glycosaminoglycan disaccharide repeat sequence.[76,78] In the dipolar decoupled spectrum (Figure 5.8b), a substantially larger signal was observed in the 120–190 ppm region. This broad signal in the upfield region has been assigned[76] to the type II collagen carbons, whose lines, like those of tendon collagen, were too broad to detect using scalar decoupling. From a measurement of collagen signal intensity in the dipolar decoupled spectrum (Figure 5.8b) it is found that most, and possibly all, aliphatic collagen carbons contributed full signal intensity to the dipolar decoupled spectrum. This result shows that most type II collagen carbons have small T_1 values and very broad lines, indicating that type II collagen in intact cartilage exhibits the highly anisotropic motion found in type I collagens.

The highly anisotropic motion of the cartilage collagen contrasts with the motion of the glycosaminoglycan chains. Reorientation of the polysaccharide C—H bond axes must be very rapid ($\tau < 10^{-6}$ sec) over almost all solid angle (i.e., virtually isotropic fast motion as found for the mobile component in elastin) to produce the motionally narrowed line widths of protonated carbons ($\Delta < 200$ Hz) observed in the scalar decoupled spectrum[76,78] (Figure 5.8a). Furthermore, intensity measurements, under a variety of conditions, have shown that at least 80% of the glycosaminoglycan carbons have motionally narrowed lines.[76] The protonated carbon T_1 (~ 60 msec) and NOE values[76] (1.35) also indicate that the correlation times for reorientation are in the neighborhood of 10^{-8} sec (Figure 5.2). In spite of this evidence that reorientation is rapid, several observations suggest that glycosaminoglycan motions do not completely average the static dipolar interaction. In the first place at least two line widths had to be assumed[76] to reproduce the experimentally observed glycosaminoglycan spectra (Figure 5.8). Respective line widths of 120 and 40 Hz were assigned to 80 and 20% of the glycosaminoglycan signal intensity to simulate the scalar decoupled spectrum. The dipolar-decoupled spectrum was simulated using 60 and 40 Hz line widths for the two signal components. The reduction in line width observed to accompany high-power decoupling, shows that motions on the time scale of $\leq 10^{-6}$ sec do not produce full isotropic averaging of the ^{13}C—^1H dipolar interaction. Cross-polarization spectra[79] of cartilage provide further evidence that glycosaminoglycan ^{13}C nuclei experience a residual static dipolar interaction, since glycosaminoglycan resonances are seen after 1 msec of contact and more clearly so after 3 msec of contact (Figure 5.9a,b). As expected, the ^{13}C nuclei in the highly ordered collagen cross polarized much more rapidly than the ^{13}C nuclei in the mobile glycosaminoglycan. After 50 μsec of contact the protonated carbon collagen intensity had attained 0.7 of its maximum value.

The $\log \chi^2$ distribution (eq. 1, with $p = 14$, $b = 1000$) has provided a self-consistent analysis of glycosaminoglycan T_1 and NOE values measured in intact cartilage and in solution. However, the distribution failed to predict the line widths obtained from the simulated spectra. Although it is possible to modify the tail of the distribution and fit all the data (simple truncation works), there seems to be little point in introducing new parameters in the distribution since it does not appear possible to extract information about details of the motion from the precise form of the distribution function. The main value of the distribution is in the overall view of chain motion that it suggests; namely, that (a) there is a broad distribution of motions having correlation times in the neighborhood of $\omega_C^{-1} \sim 10^{-8}$ sec which account for the measured T_1 and NOE values, and (b) there is a small fraction of very slow motions (or a small amount of anisotropy) that qualitatively accounts for the scalar decoupled line widths and the results of high-power decoupling and cross-polarization experiments.

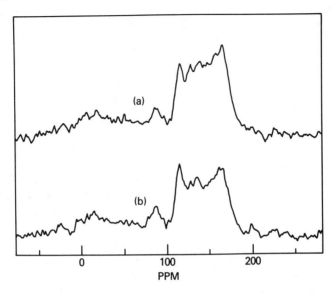

Figure 5.9
Comparison of bovine cartilage cross-polarization spectra obtained at 37°C with (a) $T_{cp} = 1$ msec and (b) $T_{cp} = 3$ msec. In each case 8192 transients were accumulated with a 2 sec interval between contacts. Chemical shift scale in ppm from external CS_2.

It is difficult to discuss, even qualitatively what interactions might be the source of the slow cooperative glycosaminoglycan motions or motional anisotropy in cartilage, because of the large number of different macromolecules in the tissue. A considerably simpler model system can be realized in solution by utilizing a trypsin-digested proteoglycan monomer isolated from Swarm rat chondrosarcoma. This material consists of oligopeptides to which an average of six chondroitin-4-sulfate chains (MW $\sim 2 \times 10^4$) are covalently attached.[80] Since over 95 wt% of the sample is chondroitin-4-sulfate[76] and the glycosaminoglycan is the only macromolecule in the sample, slow cooperative motions can be presumed to arise from interactions involving the polysaccharide chains.

Spectra[79] of the chondroitin-4-sulfate (Figure 5.10) carbons in the trypsin-treated proteoglycan (CS-T), at 0°C, show a fourfold reduction in protonated carbon line widths as the decoupling field is increased from 0.2 to 14 G ($\gamma H_2 = 60$ KHz). This result shows that slow motions ($\tau > \gamma H_2$) are present, a result that is supported by the cross-polarization spectra[79] (Figure 5.11). After 1 ms of contact the protonated carbon

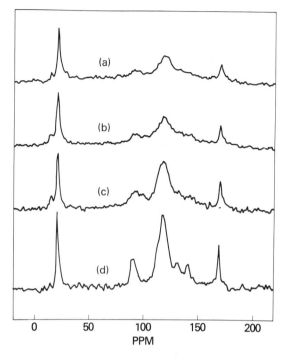

Figure 5.10
Effect of decoupling field strength on the linewidths of chondroitin-4-sulfate at 0°C. (a) $H_2 = 0.22$ G, (b) $H_2 = 0.7$ G, (c)$H_2 = 3.5$ G, (d) $H_2 = 14$ G. The sample concentration was 25 wt% in 0.1 M AcNa (pH = 7.4) and 8192 transients were accumulated in each case using a $90° - t$ pulse sequence with $t = 2$ sec. Chemical shift scale in ppm from external CS_2.

signal intensity equals that obtained in the dipolar decoupled Fourier transform spectrum (Figure 5.10d). Increasing decoupling power decreases line widths of the 25 wt% CS-T solution at 37°C, although less dramatically than at 0°C, and the cross-polarization time is about twice as long at the higher temperature. These results show that although rapid motions more effectively average the dipolar interactions at the higher temperature, a static component remains.

The line widths in the scalar-decoupled spectrum (Figure 5.10b) can be reduced fourfold by lowering the concentration to 5 wt%. Digesting the oligopeptides with papain reduces line widths by 25%. Increasing the temperature also decreases the line widths and, at 60°C, the line widths of the 25 wt% CS-T solution are about one-fifth of those found at 0°C. In

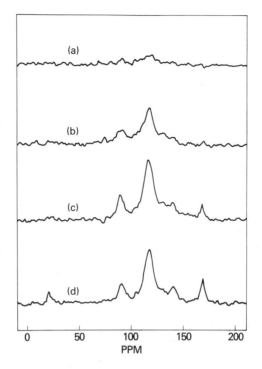

Figure 5.11
Cross-polarization spectra of chondroitin-4-sulfate at 0°C (a) $T_{cp} = 0.05$ msec; (b) $T_{cp} = 0.25$ msec; (c) $T_{cp} = 1$ msec; (d) $T_{cp} = 3$ msec. The sample concentration was 25 wt% in 0.1 M AcNa (pH = 7.4) and 4096 transients were accumulated in each case with 2 sec interval between contacts. Chemical shift scale in ppm from external Cs$_2$.

contrast with the line widths, the T_1 (~ 60 msec) and NOE (1.4) values obtained at 0°C change by less than 15% as temperature increases from 0 to 60°C. This behavior is in accord with predictions of the log χ^2 distribution (Figure 5.2), which predicts that these parameters are insensitive to changes in $\bar{\tau}$ when $\bar{\tau} \sim 10^{-8}$ sec.

Taken together the nmr data indicate that the chondroitin-4-sulfate chain motions in the concentrated CS-T solution (25 wt%), like the glycosaminoglycan chain motions in cartilage (approximately 80% of the cartilage glycosaminoglycan is chondroitin-4-sulfate) are characterized by a broad distribution of motions. The similar values of the chondroitin-4-sulfate line widths in the two environments at 0 and at 37°C shows that the interactions that determine the slow motions are similar in concentrated solution (25 wt%) and in cartilage. The motions in the soluble CS-T sample must arise primarily from interactions among the polysaccharide chains themselves, suggesting that such interactions determine the molecular dynamics of the glycosaminoglycan chains in cartilage.

Although the chondroitin-4-sulfate chain motions in the concentrated CS-T solution are a good model for chondroitin-4-sulfate chain motions in cartilage, this is not the case for free chondroitin-4-sulfate chains in more dilute solution (5 wt%). It has been shown that, in the latter circumstances, the mean correlation time for backbone reorientation, $\bar{\tau}$ is less than that obtained in intact cartilage by one to two orders of magnitude.[76]

The molecular structures responsible for the broad distribution of chain motions in intact cartilage and in concentrated solution remain to be determined. For instance, the slow cooperative motions could result from entanglement of unordered but concentrated chains or may be the consequence of short segments of ordered structure. X-ray diffraction studies have shown that various sulfated polysaccharides as well as a proteoglycan aggregate have substantial helical contents when prepared as oriented films.[72,81,82,83]

V. HEMOGLOBIN S

It has been noted that the structural proteins in connective tissues form fibrous three-dimensional networks that are essential to their function. In contrast the normal adult hemoglobin A (HbA) molecules do not form fibrous aggregates in either the oxygenated or deoxygenated states even though their concentration within the red cell is 38 wt%. The HbA molecule contains four similar polypeptide chains (α_2, β_2) each bearing a heme group. Its function is to bind O_2 in the lungs and deliver it to the tissues in the capillaries. The only difference in the sequence of HbA and hemoglobin S (HbS) is that position six in each β chain is a glutamic acid residue in HbA, whereas it is a valine residue in HbS.[84] This single substitution causes a marked reduction in the solubility of the deoxygenated molecule, and deoxygenated HbS reportedly[85] forms fibers within the cell. It is hypothesized that fiber formation causes a marked reduction in the flexibility of the cell, which is thought to cause sickle cell syndrome.

A reversible transition of deoxygenated HbS from an isotropic solution to a gel has been demonstrated *in vitro*. Careful kinetic studies in solution[86–89] have shown that above a critical temperature rapid gel formation follows a latent period. The critical temperature, the length of the latent period and the fraction of hemoglobin that is in the gel state are sensitive functions of concentration. Many other aspects of HbS behavior, in addition to gelation kinetics, have been investigated using a variety of techniques.[84] Our purpose here is to show that high-power double-resonance spectroscopy is a useful new method for studying the molecular structure and dynamics in the HbS gel not only *in vitro* but also within the intact cell.[90]

The cross-polarization spectrum of a 28 wt% *in vitro* sample of deoxygenated HbS[90] (Figure 5.12a) shows the broad resonances of a protein in which little motional averaging of chemical shift anisotropy occurs. In contrast the dipolar decoupled and scalar decoupled spectra[90] (Figure 5.12b,c) show only the much sharper peaks found in the spectrum of HbA at 37°C (Figure 5.12d). The observation that only the gelled hemoglobin contributes intensity to the cross-polarization spectrum shows that the isotropic motion of the HbS monomer is fast enough to average out the static dipolar interactions, i.e., $\tau \leq 10^{-5}$ sec. Of greater interest is the observation that the line widths of the HbS monomer (Figure 5.12c), in the presence of the gel, are not measurably greater than those of HbA (Figure 5.12d), in spite of the fact that the sample viscosity is virtually infinite in the former instance. This result shows that formation of the gel does not significantly alter the microscopic viscosity in the neighborhood of the HbS monomer.

The integrated signal intensities of the cross-polarization and scalar decoupled spectra (Figure 5.12a,c) are about equal. (Note that the signal amplitude is expanded by a factor of 2 in Figure 5.12a so that the broad signals can be more easily visualized.) Since we typically observe signal enhancements of 2.0–3.5-fold in fully cross-polarized spectra of rigid proteins, these signal intensities imply that approximately $27 \pm 6\%$ of the HbS is gelled and the remainder is monomer. Since the polymerized HbS accounts for only about one-quarter of the total sample, signal intensity in the dipolar decoupled spectrum is seen to be only slightly greater than the intensity measured in the scalar decoupled spectrum. The same result was obtained when the interval between the 90° pulses was 2 or 10 sec. However the intensity difference between scalar- and dipolar-decoupled spectra cannot be used to determine quantitatively the fraction of hemoglobin in the gel state until: (a) T_1s of the gelled HbS are measured and (b) spectra with better signal to noise are obtained.[90]

A comparison of Figure 5.13 with Figure 5.12 shows that a much larger fraction of the Hbs molecules in deoxygenated erythrocytes is in the gel state than is found *in vitro*. [Hemoglobin contains approximately 90% of the carbon in the erythrocyte.[91]] The relative intensities of the signals in Figure 5.13 indicate that at least 50 and possibly 75% of the HbS is

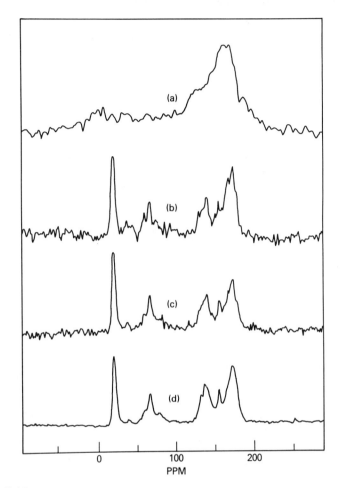

Figure 5.12
Comparison of spectra of hemoglobins in solution at 37°C: (a) deoxy HbS cross-correlation spectrum, $T_{cp} = 1$ msec, 2 sec interval between contacts; (b) deoxy HbS, dipolar decoupled spectrum, $90° - t$ pulse sequence, $t = 2$ sec; (c) deoxy HbS, scalar decoupled spectrum, $90° - t$ pulse sequence, $t = 2$ sec; (d) oxy HbA, scalar decoupled spectrum, $90° - t$ pulse sequence, $t = 2$ sec. 8192 transients were accumulated in (a–b) and 16384 transients in (c–d). Chemical shift scale in ppm from external CS_2.

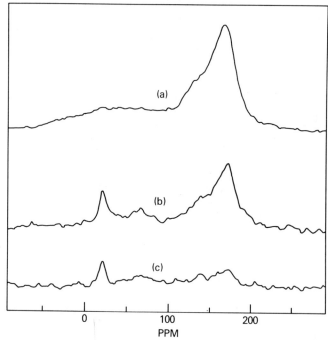

Figure 5.13
Comparison of spectra of deoxygenated HbS cells at 37°C: (a) cross-polarization spectrum, $T_{cp} = 1$ msec, 2 sec interval between contacts; (b) dipolar decoupled spectrum, $90° - t$ pulse sequence, $t = 10$ sec; (c) scalar decoupled spectrum, $90° - t$ pulse sequence, $t = 10$ sec. 8192 transients accumulated in each case. Chemical shift scale in ppm from external CS_2.

gelled. The result that a larger fraction of gelled HbS is found in the cells is expected since the HbS concentration in the cells (38 wt%) is greater than *in vitro* (28 wt%). Although the higher concentration of HbS in the cells produces a larger fraction of gelled HbS, the line profile of the gelled HbS observed *in vitro* (Figure 5.13a) is the same (within the uncertainty due to the noise) as in the erythrocyte, indicating that the gel structure formed *in vitro* is similar to that formed in the cell.

In spite of their preliminary nature, the above results indicate that high-power double resonance spectroscopy will provide useful information about molecular aspects of sickle cell disease. The technique should provide a way to measure the fraction of gelled Hbs in the intact cell and *in vitro*. Furthermore, one can utilize relaxation measurements and line shapes to compare molecular motions of HbS in the cell with those *in vitro* under a variety of conditions.

VI. MODEL MEMBRANES

Hydrocarbon chain motions in phospholipid bilayers have been investigated using electron spin resonance (esr) and nmr.[92] Although one can obtain information about anisotropic motion directly from the esr spectrum, there is some question that the spin label may perturb the hydrocarbon chain motions. While there is no such problem in the nmr experiments, ^{1}H and scalar decoupled ^{13}C nmr studies have usually been limited to sonicated samples composed of curved bilayers[93] (vesicles) since severe dipolar broadening occurs in unsonicated samples composed of planar bilayers. Dipolar broadening does not significantly effect quadrupole splittings observed in deuterium spectra of model membranes. Various investigators have used the relationship between the observed quadrupole splitting and the order parameters to obtain information about molecular structure and motion at specific sites in labeled bilayers in the liquid crystalline state.[94–98] Like ^{2}H nmr, high-power double resonance ^{13}C spectroscopy is not affected by dipolar broadening and has a sensitivity advantage over the former technique because the broadening due to chemical shift anisotropy is about a hundred times less than that due to the quadropole interaction. Probably the major obstacle to ^{13}C studies is the limited availability of ^{13}C labeled lipids. Nevertheless, as we now discuss, double resonance ^{13}C studies[99,100] have provided information about mobility on the kilohertz frequency scale in unsonicated bilayers over a wide range of temperatures.

High-power double resonance ^{13}C spectra of unsonicated aqueous dispersions of L-α-(β,γ-dipalmitoyl)-lecithin (DPL) at 22.5°C have shown[99] that the methylene carbons exhibit their full 35 ppm anisotropy. In contrast, the choline methyl signal exhibited little anisotropic broadening, indicating that nearly isotropic motion occurs for the methyl group in the gel phase. This motion was frozen out when the DPL was complexed with UO_2+, a cation that increases the transition temperature by 5°C. At 40°C, just above the transition temperature, the methylene spectrum sharpens, indicating the onset of anisotropic molecular motion for the methylene moieties, which partially removes the anisotropy of the chemical shift. At this temperature relative intensities of choline and terminal methyl groups were substantially decreased, indicating that substantial molecular motion of these groups had increased the time needed to transfer polarization to these carbon nuclei. Similar results were obtained for L-α-(β,γ-dimyristoyl)-lecithin (DML) dispersed in water at 30°C. Here, the polarization times for methylene, terminal methylene, and methyl carbons were 2, 10, and 50 msec respectively, indicating that, in the liquid crystal phase, molecular motion continuously increases as one progresses from interior methylene carbons to the terminal methyl, in qualitative agreement with results obtained in other magnetic resonances studies (see Ref. 96 and references cited therein.)

The molecular motion of cholesterols in aqueous DPL dispersions has also been studied using double resonance spectroscopy.[100] Two cholesterol samples, 90% ^{13}C labeled at the 4 and 26 positions respectively, were separately dispersed in water with DPL. A variety of DPL: cholesterol molar mixtures were studied over the temperature range 5–90°C. The proton decoupled spectrum of cholesterol labeled in the 26 position showed complete averaging of the chemical shift tensor over the whole temperature range. This result suggests that the terminus of the aliphatic tail of the cholesterol molecule undergoes isotropic motion on the time scale of 1 kHz. However, a decoupling field of 4 G ($\gamma H_2/2\pi = 15$ kHz) was needed to remove the ^{13}C—^{1}H dipolar broadening, indicating that motion is anisotropic on the time scale of 15 kHz.

The C^4-labeled sample exhibited considerably more complex behavior. To begin with, motion of the C^4 carbon is more restricted than motion of the C^{26} carbon as was indicated by the much greater line width of the C^4 resonance. The decrease in sensitivity which resulted from the broad C^4 resonance required the removal of the background natural abundance signal using difference spectra. The difference spectra of dispersions containing more than 20% cholesterol consisted of two signals. For a 2:1 cholesterol:DPL molar dispersion the two signals were observed up to 80°C. The two C^4 signals were assigned to cholesterol molecules in two distinct phases in the dispersion with phase composition dependent upon temperature and cholesterol concentration. The phase diagram is not completely consistent with that obtained from esr and electron microscopy studies. This may be due in part to the fact that esr and nmr experiments are sensitive to motions on vastly different time scales.

These studies of lipid bilayers show that because of chemical shift anisotropy, hydrocarbon line widths are often so broad that labeled samples must be studied if information about motion at specific sites is to be obtained. The chemical shift anisotropy can be eliminated by spinning the entire sample about an axes which makes an angle of 54.7° (the magic angle) with $\bar{H}_0$. Schafer et al.[26] were the first to combine high-power double resonance with magic-angle sample spinning (at a rate, ν_R, somewhat greater than the chemical shift anisotropy $\Delta\sigma$, of a few kilohertz). Using this technique several investigators[27,101] have largely removed chemical shift anisotropy from solid polymer spectra.

It has been recognized[23,102-106] that when $\nu_R < \Delta\sigma$ a narrow line is obtained at the isotropic value of the chemical shift with narrow sidebands at integral multiples of ν_R. When $\nu_R \sim \Delta\sigma$ the sideband intensities are small,[105] a result of significance since even at high fields, sample spinning at a few kilohertz is sufficient to yield highly resolved ^{13}C spectra without interfering sidebands. this has been shown by Haberkorn et al. who have recently obtained 75 MHz ^{13}C cross-polarization spectra of unsonicated DML and DPL dispersions in water (Figure 5.14b,c) with $\nu_R = 2.6$ kHz.[23] As is clear from Figure 5.14 the line widths with magic-angle spinning and full proton decoupling are much sharper than those in the spectrum of the decoupled, but unspun DML sample (Figure 5.14a). All lines that are observed when the lipids are dissolved in $CDCl_3$ are seen in Figure 5.14b,c. Thus, without recourse to isotopic enrichment, one achieves a substantial improvement in spectral resolution so that measurement of relaxation times for individual carbons is possible.

One disadvantage of spinning is that information contained in the chemical shift tensor is lost when the spinning rate is so large that only the center band is observed. The possibility of recovering chemical shift tensor information from data obtained at lower spinning speeds, has been discussed by Lippmaa et al.[103] and Waugh et al.[107]

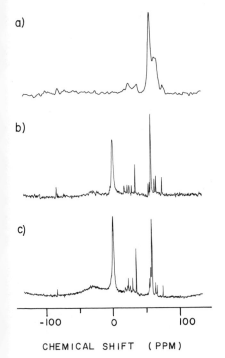

a)

b)

c)

-100   0   100

CHEMICAL SHIFT (PPM)

Figure 5.14
Comparison of dipolar decoupled model membrane spectra obtained, at 21°C, for stationary and rotating samples: (a) DML, $\nu_{rot} = 0$ (b) DML, $\nu_{rot} = 2.6$ kHz; (c) DPL, $\nu_{rot} = 2.6$ kHz. The samples were prepared by dispersing the phospholipids in excess water (≥ 50 wt% H_2O) and were not sonicated. Chemical shift scale in ppm from enternal delrin. The large peak at 0 ppm and the broad peak at -30 ppm arise from the delrin rotor. (This figure was courteously provided by Dr. R. G. Griffin.)

VII. CONCLUSIONS

The work that has been described illustrates that high-power double-resonance ^{13}C spectra can provide information about molecular dynamics and structure in biological solids that would be difficult to obtain by other means. Even when motional narrowing is almost complete, as is the case with the chondroitin sulfate chains in cartilage and in solution, high-power decoupled and cross-polarization spectra are sensitive to the residual ^{13}C—^{1}H dipolar interaction and thus provide information about slow motions and motional anisotropy. The cross-polarization spectra of HbS, elastin and the lipid bilayers show that the contact time can be adjusted to enhance the signal of the ordered component in a sample relative to the mobile component. In favorable circumstances, such as occur with HbS, only the ordered component is observed and thus, can be studied without interference from the mobile material. The uses of labeled samples to study motion at defined sites are illustrated by studies of collagen fibrils and cholesterol in DPL bilayers. However, it appears that, when sensitivity and the nature of the sample permit, spinning at the magic angle combined with double resonance will reduce the need to label. As has been discussed, T_1 data obtained from double-resonance experiments can be used to probe motions near $\omega_C^{-1} \sim 10$ nsec, whereas the line widths provide information about motion on the kilohertz time scale. Although we have not discussed $T_{1\rho}$ measurements, Schaefer et al.[22] have shown that ^{13}C $T_{1\rho}$ values can be readily measured using double-resonance techniques. When the measured ^{13}C $T_{1\rho}$ is determined by a spin-lattice relaxation mechanism, valuable information is obtained about motion in the 10^4–10^5 Hz regime.[22] Thus, using double-resonance techniques one can obtain information about motion in solids covering the range of frequencies from 10^3 to 10^9 Hz. Furthermore, as the theory of high-power decoupling and cross-polarization is extended to systems with motion and chemical shift tensor assignments become available it should be possible to derive more quantitative information from the spectra.

REFERENCES

1. T. L. James, *Nuclear Magnetic Resonance in Biochemistry*, Academic, New York, 1975.
2. K. Wüthrich, *NMR in Biological Research: Peptides and Proteins*, North Holland/American Elsevier, New York, 1976, Chapter V.

3. R. Deslauriers and I. C. P. Smith, in *Topics in Carbon-13 NMR Spectroscopy*, Vol. 2, G. C. Levy Ed., Wiley, New York, 1976, p. 1.

4. E. Wenkert, B. L. Buchwalter, I. R. Burfitt, M. J. Gasic, H. E. Gottlieb, E. W. Hagaman, F. M. Schell, P. M. Wowkulich and A. Zheleva, in *Topics in Carbon-13 NMR Spectroscopy*, Vol. 2, G. C. Levy Ed., Wiley, New York, 1976, p. 81.

5. R. A. Komoroski, I. R. Peat and G. C. Levy, in *Topics in Carbon-13 NMR Spectroscopy*, Vol. 2, G. C. Levy Ed., Wiley, New York, 1976, p. 180.

6. W. Egan, H. Shindo and J. S. Cohen, in Ann. Rev. Biophys, Bioeng., L. J. Mullins, W. A. Hagins, L. Stryer and C. Newton, Eds. *Annual Reviews*, Vol. 3, Palo Alto, 1977, p. 383.

7. F. Bloch, *Phys. Rev.* **111,** 841 (1958).

8. L. R. Sarles and R. M. Cotts, *Phys. Rev.* **111,** 853 (1958).

9. M. Mehring, A. Pines, W-K Rhim and J. S. Waugh, *J. Chem. Phys.* **54,** 3239 (1975).

10. M. Mehring, *High Resolution NMR in Solids*, Academic, New York, 1976.

11. A. Pines, M. G. Gibby and J. S. Waugh, *J. Chem. Phys.* **59,** 569 (1973).

12. S. R. Hartmann and E. L. Hahn, *Phys. Rev.* **128,** 2042 (1962).

13. D. E. Demco, J. Tegenfeldt and J. S. Waugh, *Phys. Rev. B* **11,** 4133 (1975).

14. A. Pines, M. G. Gibby and J. S. Waugh, *Chem. Phys. Lett.* **15,** 373 (1972).

15. S. Pausak, A. Pines and J. S. Waugh, *J. Chem. Phys.* **59,** 591 (1973).

16. A. Pines, J. J. Chang and R. G. Griffin, *J. Chem. Phys.* **61,** 1021 (1974).

17. S. Pausak, J. Tegenfeldt, and J. S. Waugh, *J. Chem. Phys.* **61,** 1338 (1974).

18. R. G. Griffin, A. Pines, S. Pausak, and J. S. Waugh, *J. Chem. Phys.* **63,** 1267 (1975).

19. D. L. VanderHart, *J. Chem. Phys.* **64,** 830 (1976).

20. E. R. Andrew, A. Bradbury and R. G. Eades, *Nature* **182,** 1659 (1958).

21. I. J. Lowe, *Phys. Rev. Lett.* **2,** 285 (1959).

22. J. Schaefer, E. O. Stejskal and R. Buchdahl, *Macromolecules* **10,** 384 (1977).

23. R. A. Haberkorn, J. Herzfeld and R. G. Griffin, submitted to *Nature*.

24. M. G. Gibby, A. Pines and J. S. Waugh, *Chem. Phys. Lett.* **16**, 296 (1972); S. Kaplan, H. A. Resing, and J. S. Waugh, *J. Chem. Phys.* **59**, 5681 (1973); J. J. Chang, A. Pines, J. J. Fripat and H. R. Resing, *Surface Sci.* **47**, 661 (1975).

25. D. L. VanderHart and L. L. Retcofsky, *Fuel* **55** 202 (1976).

26. J. Schaefer, E. O. Stejskal and R. Buchdahl, *Macromolecules* **8**, 291 (1975).

27. J. Schaefer and E. O. Stejskal, *J. Am. Chem. Soc.* **98**, 1031 (1976).

28. H. Resing and W. B. Moniz, *Macromolecules* **8**, 560 (1975). D. L. VanderHart, *J. Mag. Res.* **24**, 467 (1976).

29. L. B. Sandberg, in *International review of Connective Tissue Research*, Vol. 7, D. A. Hall and D. S. Jackson, Eds., Academic, New York, 1976, p. 160.

30. J. M. Gosline, in *International Review of Connective Tissue Research*, Vol. 7, D. A. Hall and D. S. Jackson, Eds., Academic, New York, 1976, p. 211.

31. For recent developments in elastin research see, *Advances in Experimental Medicine and Biology*, Vol. 79, L. B. Sandberg, W. R. Gray, and Carl Franzblau, Eds., Plenum, New York 1977.

32. J. Schaefer, *Macromolecules* **5**, 427 (1972).

33. M. W. Duch and D. M. Grant, *Macromolecules* **3**, 165 (1970).

34. G. E. Ellis and K. J. Packer, *Biopolymers* **15**, 813 (1976).

35. G. E. Ellis and K. J. Packer, in *Advances in Experimental Medicine and Biology*, Vol. 79, L. B. Sandberg, W. R. Gray, and C. Franzblau, Eds Plenum, New York, 1977, p. 663.

36. K. H. Meyer and C. Ferri, *Pflüegers Arch. Gesamte Physiol.* **238**, 78 (1937).

37. E. Wöhlish, H. Weitnauer, W. Gowning, and R. Rohrbuch, *Kolloid Z.* **104**, 14 (1943).

38. C. A. J. Hoeve and P. J. Flory, *J. Am. Chem. Soc.* **80**, 6523 (1958).

39. A. Oplatka, I. Michaeli, A. Katchalsky, *J. Polym. Sci.* **46**, 365 (1960).

40. T. Weis-Fogh and S. O. Andersen, *Nature* **227**, 718 (1970).

41. T. Weis-Fogh and S. O. Andersen, in *Chemistry and Molecular Biology of the Intracellular Matrix*, Vol. 1, E. A. Balazas, Ed., Academic, New York, 1970, 671.

42. F. Mistrali, D. Volpin, G. B. Garibaldo and H. Ciferri, *J. Phys. Chem.* **75,** 142 (1971).

43. C. A. J. Hoeve and P. J. Flory, *Biopolymers* **13,** 677 (1974).

44. W. Grut and N. G. McCrum, *Nature* **251,** 165 (1975).

45. D. P. Mukherjee, A. S. Hoffman and C. Franzblau, *Biopolymers* **13,** 2447 (1974).

46. K. L. Dorrington and N. G. McCrum, *Biopolymers* **16,** 1201 (1977).

47. D. A. Torchia and K. A. Piez, *J. Mol. Biol.* **76** 419, (1973).

48. J. R. Lyerla, Jr. and D. A. Torchia, *Biochemistry* **14,** 5175 (1975).

49. J. Schaefer, *Macromolecules* **6,** 882 (1973).

50. I. Solomon, *Phys. Rev.* **99,** 559 (1955).

51. D. W. Urry and M. M. Long, in *Advances in Experimental Medicine and Biology,* Vol. 79, L. B. Sandberg, W. R. Gray, and C. Franzblau, Eds., Plenum, New York, 1977, p. 685, and references therein.

52. D. A. Torchia and C. E. Sullivan, in *Advances in Experimental Medicine and Biology,* Vol. 79, L. B. Sandberg, W. R. Gray, and C. Franzblau, Eds., Plenum, New York, 1977, p. 655.

53. A. I. Lansing, T. Rosenthal, M. Alex, and E. Dempsey, *Anat. Rec.* **114,** 555 (1952).

54. D. A. Torchia, C. E. Sullivan, and W. W. Fleming, unpublished results.

54a. F. O. Stejskal and J. Schaefer, *J. Mag. Res.* **18,** 560 (1975).

55. D. A. Torchia and D. L. VanderHart, *J. Mol. Biol.* **104,** 315 (1976).

56. For recent developments in collagen research see, *Biochemistry of Collagen,* G. N. Ramachandran and A. H. Reddi, Eds., Plenum, New York, 1976.

57. A. Miller, in *Biochemistry of Collagen,* G. N. Ramachandran and A. H. Reddi Eds., Plenum, New York, 1976, Chapter 3.

58. L. W. Cunningham, H. A. Davies, and R. G. Hammonds, Jr., *Biopolymers* **15,** 483 (1976).

59. B. L. Trus and K. A. Piez, *J. Mol. Biol.* **108,** 705 (1976).

60. K. A. Piez and B. L. Trus, *J. Mol. Biol.* **110,** 701 (1977).

61. A. D. McLachlan, *Biopolymers* **16,** 1271 (1977).

62. J. Schaefer, E. O. Stejskal, C. F. Brewer, H. Keiser, and H. Sternlich, Chapter 3, this volume.

63. R. C. Siegal, *Proc. Nat. Acad. Asci.,* **71,** 4826 (1974).

64. R. G. Griffin, A. Pines, and J. S. Waugh, *J. Chem. Phys.* **63,** 3676 (1975).

65. C. P. Slichter, *Principles of Magnetic Resonance,* Harper and Row, New York, 1963, p. 58.

66. N. Bloembergen, *Phys. Rev.* **104,** 1542 (1956).

67. A. Abragam, *The Principles of Nuclear Magnetism,* Oxford, London, 1971, Chapt. 7.

68. D. Wallach, *J. Chem. Phys.* **47,** 5258 (1967).

69. D. E. Woessner, B. S. Snowden, Jr., and G. H. Meyer, *J. Chem. Phys.* **50,** 719 (1969).

70. F. Perrin, *J. Phys. Rad.* Ser. 7, **5,** 497 (1934).

71. D. E. Woessner, *J. Chem. Phys.,* **37,** 647 (1962).

72. For a recent review of proteoglycan structure see, H. Muir and T. E. Hardingham, in *Biochemistry of Carbohydrates,* Vol. 5, Series 1, W. H. Whalan, Ed. Butterworth, London, 1975, Chapter 4.

73. E. J. Miller and V. J. Matukas, *Proc. Natl. Acad. sci.* **64,** 1264 (1969).

74. K. A. Piez, in *biochemistry of Collagen,* G. N. Ramachandran and A. H. Reddi, Eds., Plenum, New York, 1976, Chapt. 1.

75. V. C. Hascall and S. W. Sajedra, *J. Biol. Chem.* **245,** 4920 (1970).

76. D. A. Torchia, M. A. Hasson, and V. C. Hascall, *J. Biol. Chem.* **252,** 3617 (1977).

77. S. G. Pasternack, A. Veis, and M. Breen, *J. Biol. Chem.* **249,** 2206 (1974).

78. C. F. brewer and H. Keiser, *Proc. Natl. Acad. Sci.,* **72,** 3421 (1978).

79. D. A. Torchia, M. A. Hasson, and V. C. Hascall, unpublished results. Similar FT and cross-polarization spectra of bovine nasal cartilage have been obtained by J. Schaefer, E. O. Stejskal, C. F. brewer, H. Keiser and H. Sternlich, private communication from J. Schaefer.

80. D. Heinegard and V. C. Hascall, *Arch. Biochem. Biophys.* **165,** 427 (1974).

81. S. Arnott, J. M. Guss, D. W. L. Hukins, and M. B. Mathews, *Science* **180,** 743 (1973).

82. E. D. T. Atkins, T. E. Hardingham, D. H. Isaac, and H. Muir, *Biochem. J.* **141,** 919 (1974).

83. D. H. Isaac and E. D. T. Atkins, *Nature New Biol.* **244,** 252 (1973).

84. For recent developments in Hemoglobin-S research see, *Proceedings*

of the Symposium on Molecular and Cellular Aspects of Sickle Cell Disease, J. I. Hercules, G. L. Cottam, M. R. Waterman and A. N. Schechter, Eds., DHEW, Bethesda, 1976.

85. S. J. Edelstein, R. Josephs, H. S. Jarosch, R. H. Crepeau, J. N. Telford and G. Dykes, in *Proceedings of the Symposium on Molecular and Cellular Aspects of Sickle Cell Disease,* J. I. Hercules, G. L. Cottam, M. R. Waterman, and A. N. Schechter, Eds., DHEW, Bethesda, 1976, p. 33.

86. K. Malfa and J. Steinhardt, *Biochem. Biophys. Res. Commun.* **59,** 887 (1974).

87. J. Hofrichter, P. D. Ross, and W. A. Eton, *Proc. Nat. Acad. Sci.* **71,** 4864 (1974).

88. W. A. Eaton, J. Hofrichter, P. D. Ross, R. G. Tschudin, and E. D. Becker, *Biochem. Biophys. Res. Commun.* **69,** 538 (1976).

89. J. Hofrichter, P. D. Ross, and W. A. Eton, *Proc. Nat. Acad. Sci.* **73,** 3035 (1976).

90. J. W. H. Sutherland, W. Egan, A. N. Schechter, and D. A. Torchia, in preparation.

91. J. W. H. Sutherland and A. N. Schechter, private communication.

92. For recent reviews see, S. Ohnishi, *Adv. Biophys.* **8,** 35 (1975). A. G. Lee, N. J. M. Birdsall, and J. C. Metcalfe, in *Methods of Membrane Biology,* Vol. II, E. D. Korn, Ed., 1974, p. 1.

93. For discussion of effects of vesicle formation on hydrocarbon chain mobility: G. W. Stockton, C. F. Polnaszek, A. P. Tulloch, F. Horsan, and I. C. P. Smith, *Biochemistry* **15,** 954 (1976); N. O. Petersen and S. I. Chan, *Biochemistry,* **16,** 2657 (1977).

94. J. Seelig and W. Niederberger, *J. Am. Chem. Soc.* **96,** 2069 (1974).

95. J. Seelig and W. Niederberger, *Biochemistry* **13,** 1585 (1974).

96. A. Seelig and J. Seelig, *Biochemistry* **13,** 4839 (1974).

97. G. W. Stockton and I. C. P. Smith, *Chemistry and Physics of Lipids* **17,** 251 (1976).

98. G. W. Stockton, K. G. Johnson, K. W. Butler, A. P. Tulloch, Y. Boulanger, I. C. P. Smith, J. H. Davis, and M. Bloom, *Nature* **269,** 267 (1977).

99. J. Urbina and J. S. Waugh, *Proc. Nat. Acad. Sci.* **71,** 5062 (1974).

100. S. J. Opella, J. P. Yesinowski, and J. S. Waugh, *Proc. Nat. Acad. Sci.* **73,** 3812 (1976).

101. A. N. Garroway, W. B. Moniz, and H. A. Resing, *Coatings and Plastics* **36,** 33 (1976).

102. D. Doskocilova, Dang Duc Tao, and B. Schneider, *Czech. J. Phys.* **B25,** 202 (1975).

103. E. Lippmaa, M. Alla and T. Tuhern, *Proc. XIX Congress Ampere,* Heidelburg, Germany, Sept. 1976, p. 113.

104. M. Maricq and J. S. Waugh, Chem. *Phys. Lett.* **47,** 327 (1977).

105. E. O. Stejskal, J. Schaefer, and R. A. Makay, *J. Mag. Res.* **25,** 569 (1977).

106. A. N. Garroway, *J. Mag. Res.* **28,** 365 (1977).

107. J. S. Waugh, M. Maricq, R. Cantor, and W. Rothwell, *Proc. XX Congress Ampere,* Tallinn, U.S.S.R., in press (1979).

6 ¹³C CIDNP AS A MECHANISTIC AND KINETIC PROBE

W. B. MONIZ

C. F. PORANSKI, JR.

S. A. SOJKA

CONTENTS

I. INTRODUCTION

A. General Background

A new dimension was added to nmr spectroscopy with the discovery of Chemically Induced Dynamic Nuclear Polarization, CIDNP, in 1967. The observation of enhanced or inverted nmr signals during certain chemical reactions carried out in magnetic fields heralded the birth of an important new tool for the study of radical reaction mechanisms and of magnetic properties of free radicals.[1] CIDNP was soon theoretically described in terms of the radical-pair model.[2-4] Kaptein developed a set of simple rules[5] that facilitated using CIDNP for solving mechanistic problems. With these rules, the sense of the signal polarization (enhanced absorption, emission, or multiplet effects) from diamagnetic reaction products gives information about the multiplicity of the radical-pair precursors, the type of reactions leading to formation and destruction of radical pairs, the signs of electron-nuclear hyperfine interaction constants, and the relative magnitudes of g-factors. Although most of the CIDNP work to date has involved protons, the phenomenon has been observed for a number of other nuclei, notably ^{19}F, ^{31}P, and ^{13}C.

Almost from the beginning workers realized the potential of CIDNP as a kinetic tool. As the mechanism of the CIDNP effect became more clearly understood, efforts were made to quantify experimental observations. Although a variety of approaches were followed, the common theme was to relate the time evolution of the CIDNP signal to the concentrations of the reactants and products.

Most of the early kinetic studies were done with proton nmr in the continuous wave (CW) mode. It was relatively easy to repeatedly sweep through the CIDNP signal and obtain an accurate intensity versus time profile of the reaction. No standard method of extracting the kinetic parameters from such data has become established. While some workers prefer the polarized intermediate model due to Walling and Lepley,[6] and others proceed via matrix formulations, both experimental and modelling aspects of CIDNP continue to evolve.

With the use of pulse Fourier transform (FT) techniques, one can obtain CIDNP spectra in a matter of seconds. This enables many spectra to be recorded on the reacting chemical system as a function of time. In this fashion the time evolution of CIDNP signals can be measured and kinetic information subsequently extracted. Ideally, by examining CIDNP signals from spectra obtained during a single experiment, one should be able to obtain product identification and knowledge of the reaction mechanism and kinetic parameters. Thus, the wealth of pertinent information obtained may obviate time-consuming kinetic studies or exhaustive separation schemes for reaction mixtures.

CIDNP reactions have been initiated chemically, thermally, and photochemically (including laser[7] and radiolytic[8] irradiation). Chemical or thermal initiation requires no instrumental modification, but a sample mixing or warmup time is needed, making it difficult to define the zero of reaction time. This is especially crucial in the determination of accurate kinetics. In addition, the reaction cannot be easily controlled once initiated. Photochemical excitation does not have these two difficulties; $t = 0$ is precisely defined and the reaction can be controlled by varying light intensity or irradiation time. However, some means must be devised to bring the radiation into the spectrometer probe.

Once the CIDNP spectrum has been obtained, the CIDNP signals must be assigned. This may not be a simple task. Large polarizations sometimes occur from trace reaction products that are difficult to detect in spent reaction samples.

The sign of the net polarization of a particular product helps unravel its origin. The net polarization can be analyzed in terms of Kaptein's equation,[5] $\Gamma_{net} = \mu \varepsilon \Delta g A$, where μ is the multiplicity of the radical pair (+ for the triplet pairs and pairs formed from free radical encounters, − for singlet pairs); ε denotes the type of product-forming reaction (+ for cage products, − for escape products); Δg is the sign of the spectroscopic splitting factor difference (+ for the radical with larger g value, − for the radical with smaller g value); and A is the sign of the electron-nuclear hyperfine interaction constant. Since Γ_{net} is observable any three terms on the right side of the equation will determine the sign of the fourth. The Kaptein equation can be used as a test of the proposed reaction mechanism. A reasonable scheme can be written and the sign of Γ_{net} predicted. If the prediction agrees with the experimental Γ_{net}, reasonable confidence can be placed in the proposed mechanism.

B. ^{13}C CIDNP

^{13}C CIDNP is an extremely powerful tool for studing free radical reactions in solution. It has a number of distinct advantages compared to proton CIDNP. First, the 25-fold larger chemical shift range of ^{13}C reveals subtle structural, conformational, and stereochemical effects. Second, while in proton CIDNP overlapping complex splitting patterns from a number of different species often confuse interpretation, for ^{13}C, proton broadband irradiation simplifies spectra by eliminating all proton—carbon coupling. Another advantage of ^{13}C CIDNP is the ability to study carbon species that do not have protons.

A potential problem in ^{13}C CIDNP is the low natural abundance of the ^{13}C isotope (1.1%). However, at least two factors partially compensate for this difficulty. In many cases ^{13}C spin-lattice relaxation times (T_1) are longer than proton spin-lattice relaxation times. Since the polarization decays by T_1, the ^{13}C CIDNP signal should be relatively longer lived. This will be especially true for the quaternary carbons. Second, the free electron resides directly on carbon and this proximity will maximize electron—carbon interaction (relatively larger electron—carbon hyperfine interaction constants) and, hence, contribute to larger enhancement factors. For these reasons, sizable carbon polarizations are expected (and observed) in ^{13}C CIDNP experiments.[9]

In this chapter we review the progress made in ^{13}C CIDNP over the past few years.[10] We have divided the discussion into two areas: mechanisms and kinetics. Clearly, the majority of ^{13}C CIDNP studies have been in the mechanism area, aimed at defining the paths leading to the CIDNP and judging their overall importance in the reaction scheme. The kinetic work has chiefly been limited to fairly well characterized systems with the aim of developing useful approaches to the extraction of kinetic data from CIDNP experiments.

II. MECHANISTIC APPLICATIONS OF ^{13}C CIDNP

A. Azo Compounds

Early studies done in the CW mode quickly revealed the power of ^{13}C CIDNP, especially compared to proton CIDNP. For example, Lippmaa and coworkers studied the thermal decomposition of triazo compound 1 at 170° by proton and ^{13}C CIDNP.[11] Proton CIDNP was observed only for polarized benzene, whereas ^{13}C CIDNP was seen also for anilines 2 and 3 and biphenyl.

$$\phi\text{-NH}-\text{N}{=}\text{N}-\phi \xrightarrow[\text{cycloheptanone}]{170°}$$

1   2   3

Likewise, thermal decomposition of diazo compound 4 in hexachloracetone (HCA) resulted in one polarized signal in proton CIDNP.[11a] In the ^{13}C CIDNP spectrum, however, all carbons but the trichloromethyl carbon of insertion product 5 were polarized.

The thermolysis of triazenes 6 has been studied, more recently, by proton and ^{13}C CIDNP.[12] Solvents used were 1,1,2,2-tetrachloroethane,

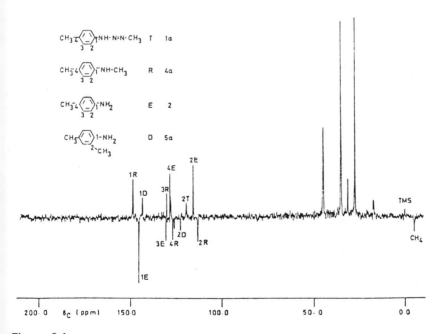

4

5

trans-decalin, and *o*-dichlorobenzene. The reaction was performed at 125–135°C. A typical ^{13}C spectrum they obtained is shown in Figure 6.1. It was found that triazenes **6** tautomerize to **7** before homolysis. This was suggested by the observation that anilines are major products. Thus, there was a preference for arylaminyl radical formation rather than alkylaminyl radical formation. Radical pair **9** is formed by loss of N_2 from radical pair **8**. The radicals of pair **9** may escape to form abstraction products such as the indicated hydrocarbons and anilines. All the carbons in toluene and in

Figure 6.1
^{13}C CIDNP FT spectrum obtained during the thermolysis of 3-methyl-1-*p*-tolyltriazine (*1a*) in *trans*-decalin (300 mg/2 ml). From Ref. 12.

the hydrocarbon products were polarized, as well as all the carbons of 4-methylaniline.

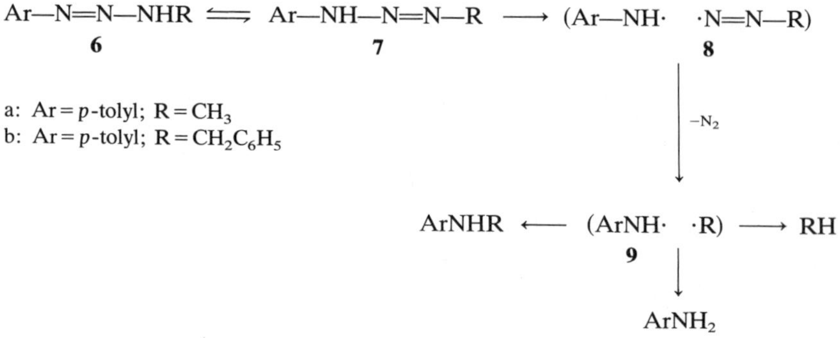

$$Ar—N{=}N—NHR \rightleftharpoons Ar—NH—N{=}N—R \longrightarrow (Ar—NH\cdot \quad \cdot N{=}N—R)$$
$$\mathbf{6} \qquad\qquad\qquad \mathbf{7} \qquad\qquad\qquad\qquad \mathbf{8}$$

a: Ar = p-tolyl; R = CH$_3$
b: Ar = p-tolyl; R = CH$_2$C$_6$H$_5$

$$\Big\downarrow {-}N_2$$

$$ArNHR \longleftarrow (ArNH\cdot \quad \cdot R) \longrightarrow RH$$
$$\mathbf{9} \quad\Big\downarrow$$

$$ArNH_2$$

Interestingly, the radicals of pair **9** can recombine in three ways. The first route produces N-alkylanilines with all carbons polarized. The second recombination pathway produces o-substituted anilines. Polarization was observed only for C-1, C-2 and 2-CH$_3$ carbons. Finally, recombination at C-4 produces imines **10**, which have not been produced by independent synthesis. They were only detected by CW-mode ^{1}H CIDNP; they were present at levels too low to be detected by ^{13}C FT CIDNP. The signs of the proton and carbon hyperfine interaction constants were determined for the p-tolyl-aminyl radical.[12] The esr spectrum of this radical has not been determined previously.

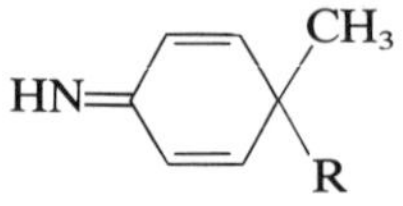

$$\mathbf{10}$$

The electrophilic attack of methoxycarbonylcarbene on ether oxygen has been studied by proton and ^{13}C CIDNP[13] (Figure 6.2). The reaction of this carbene with dibenzyl ether produced polarized insertion product **12** and toluene.

The CIDNP evidence established that a pathway exists that leads to production of **12** via radical pair **13**. Insertion product **12** must be produced by formation of a singlet radical pair followed by cage recombination. The authors favored initial formation of ylide **11** rather than direct radical pair formation (step b) because when benzyl ethyl was used as the substrate considerably more benzyloxyacetate than ethoxyacetate was produced. Since C—O bond dissociation will be determined essentially by carbon radical stabilities, the larger amount of ethyl elimination would not be consistent with direct radical-pair formation (step b).

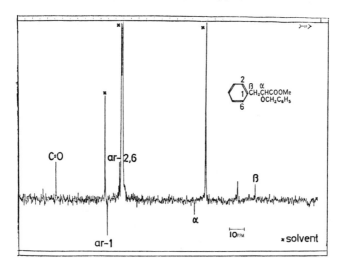

Figure 6.2
^{13}C CIDNP FT spectrum obtained during the thermolysis of a 15% solution of methyl diazoacetate in dibenzyl ether. From Ref. 13b.

Iwamura et al.[14] found evidence to refute a radical chain mechanism for the production of **15** during the photolysis of methyl diazoacetate in chloromethanes (Figure 6.3). Interpretation of the polarization signs for the indicated carbon atoms in insertion product **15** required the intermediacy of precursor radical-pair **14**. A chain mechanism would require that the polarization be remembered from the initial pair through

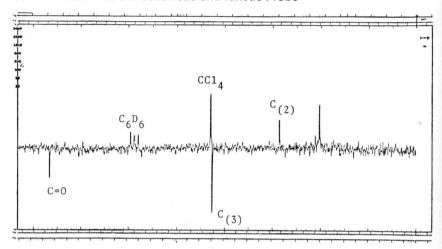

Figure 6.3
¹³C CIDNP FT spectrum obtained during the photolysis of an 8.5% solution of methyldiazoacetate in CCl₄. Polarization is observed for the carbons of **15**: Cl₃C(=C₍₃₎); CHCl(=C₍₂₎); and carboxyl. From Ref. 14.

many chain transfer steps. The authors contend that the polarization would be lost due to relaxation processes during these chain transfers.

$$N_2CHCO_2CH_3 \xrightarrow[h\nu]{CCl_4} (Cl_3C\cdot \ \cdot CHClCO_2CH_3) \longrightarrow Cl_3\overset{*}{C}\overset{*}{C}HCl\overset{*}{C}O_2CH_3$$

$$\hspace{6cm} \textbf{14} \hspace{4cm} \textbf{15}$$

A similar mechanism was used to explain the observed polarization from insertion products formed by photolysis of methyl diazoacetate in CBrCl₃ or CBr₂Cl₂.¹⁵ The attack of methoxycarbonylcarbene is mainly at the bromine atom producing radical pair **16**. Recombination produces polarized **17**. Since the polarizations are weak, the authors contend that both singlet and triplet (formed by free encounters) radical pairs contribute to the polarized signals. Polarization from the singlet pair slightly

$$(XCl_2C\cdot \ \cdot CHBrCO_2CH_3) \xrightarrow{S,T} \overset{*}{C}XCl_2\overset{*}{C}HBr\overset{*}{C}O_2CH_3$$

$$\hspace{3cm} \textbf{16} \hspace{3.5cm} \textbf{17}$$

$$N_2CHCO_2CH_3 \xrightarrow{CXBrCl_2}$$

a: X = Cl
b: X = Br

$$(XBrClC\cdot \ \cdot CHClCO_2CH_3) \xrightarrow{S,T} \overset{*}{C}XBrCl\overset{*}{C}HCl\overset{*}{C}O_2CH$$

$$\hspace{3cm} \textbf{18} \hspace{3.5cm} \textbf{19}$$

predominates. To a lesser extent, the carbene abstracts chlorine to give radical pair **18**. Collapse of **18** produces polarized **19**.

In the case of CBrCl$_3$, radical pair **20** formed by free radical encounter collapses to yield polarized **21**.

$$(Cl_3C\cdot \ \cdot CHClCO_2CH_3) \rightarrow Cl_3\overset{*}{C}CHCl\overset{*}{C}O_2CH_3$$

$$\qquad\quad \textbf{20} \qquad\qquad\quad \textbf{21}$$

B. Peroxides

The photochemical decomposition of t-butylhydroperoxide (t-BuOOH) in various alcohols has been studied by Moniz and coworkers.[16,17] In

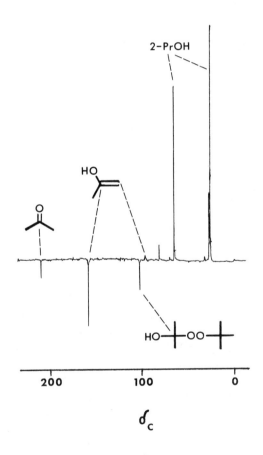

Figure 6.4
^{13}C CIDNP FT spectrum obtained during photolysis of a 10% solution of t-butylhydroperoxide in 2-propanol. From Ref. 16.

general, the reaction produced polarized ketones or aldehydes **22** corresponding to oxidized alcohol. A typical ^{13}C CIDNP spectrum obtained on these systems is shown in Figure 6.4.

$$\text{ROH} + t\text{-BuOOH} \xrightarrow{h\nu} \underset{\textbf{22}}{R_1-\overset{O}{\underset{*}{\overset{\|}{C}}}-R_2} + \underset{\textbf{23}}{t\text{-BuOO}\overset{R_3}{\underset{R_4}{\overset{|}{-}\overset{*}{\underset{|}{C}}-OH}}}$$

$$+ \underset{\textbf{24}}{R_5-\overset{O}{\overset{\|}{\underset{*}{C}}}\text{-OO}t\text{-Bu}} + \underset{\textbf{25}}{R_7R_8C=\overset{*}{C}(OH)R_6}$$

In some cases, t-butylperoxy compounds **23** and t-butyl peresters **24** were detected. Most interestingly, simple enols **25** were detected as transient intermediates by virtue of their polarized hydroxyl carbons. The ^{13}C chemical shifts of these reactive species were measured and substituent effects analyzed.[18] The enols are direct precursors of the oxidized products, as acid catalysis caused disappearance of the enol CIDNP signal with increase of the CIDNP signal from keto form **22**. This establishes an important aspect of the mechanism; the enol must be formed by disproportionation in the polarizing cage **26**. The t-butylperoxy radical preferentially removes an α-hydrogen rather than the hydroxyl hydrogen.

$$\left(t\text{-BuOO}\cdot \quad \underset{R}{\overset{R}{\diagdown}}\text{-OH} \right) \longrightarrow \mathbf{25}$$

$$\mathbf{26}$$

The thermal decomposition of benzoyl peroxide (BPO) has been extensively studied by proton and ^{13}C CIDNP and has been previously reviewed.[10] A photochemical study in chloroform gave similar results.[19] The time dependence of the CIDNP signals was analyzed, and it was shown that in the early part of the reaction, production of phenyl benzoate was very high.[19] This indicated additional cage collapse due to possible multiple-bond cleavage of BPO.

Similar products were reported for the thermal decomposition of acetylbenzoyl peroxide (ABP) at 80°C in cyclohexanone[20]; polarized CO_2, methyl benzoate **27** and toluene **28** were detected.

$$\phi-\overset{O}{\overset{\|}{C}}\text{-OOAc} \xrightarrow{80°} \overset{*}{C}O_2 + \underset{\textbf{27}}{\text{Ph}-\overset{O}{\underset{OCH_3}{\overset{\|}{\underset{*}{C}}}}} + \underset{\textbf{28}}{\text{Ph}-\overset{*}{C}H_3}$$

^{13}C CIDNP has demonstrated its mechanistic usefulness in reactions where the products are the same as the starting materials. Thus, evidence was obtained for the reversible formation of σ-complexes **29** during the photolysis of perfluorobenzoyl peroxide in the presence of chlorobenzene. This interpretation was suggested by the observation that the intensities of the ortho and para carbons of chlorobenzene were augmented, while those of meta and substituted carbons were diminished during photolysis. These polarizations also provided evidence to rule out reversible attack at C-1.

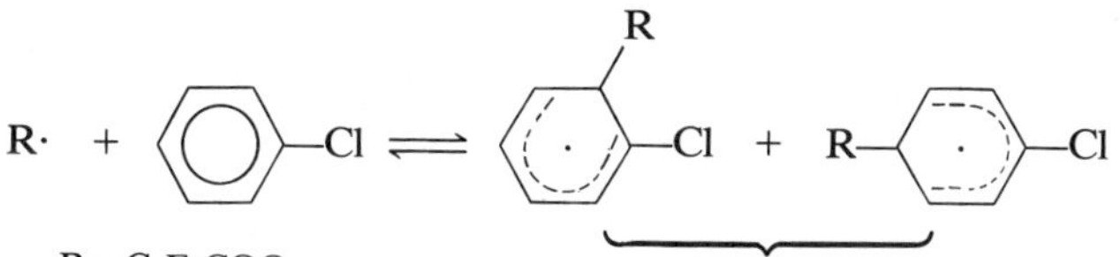

$$R = C_6F_5COO$$

29

The thermal decomposition of the *t*-butyl perester **30** in *o*-dichlorobenzene was studied by ^{13}C CIDNP.[22] The methylene carbon of escape product **33** was in strong emission, while the *S*-methyl carbon of disproportionation product **32** was in absorption. Other weak CIDNP signals from aromatic carbons were observed but not assigned. The authors contend that radical **31**, formed with anchimeric assistance by sulfur, rationalizes the observed polarization intensities. There must be large hyperfine interaction between the electron on sulfur and the adjacent methyl carbon in radical **31**. Open-shell CNDO/2 calculations support the bridged sulfuramyl radical structure **31**.

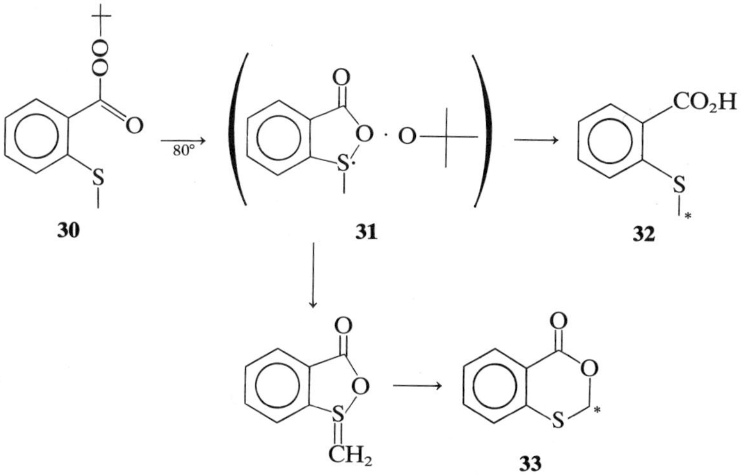

C. Aldehydes and Ketones

Benn and Dreeskamp studied the proton and ^{13}C CIDNP produced during the photolysis of benzylmethylketone, benzylphenylketone, phenylacetaldehyde, and acetaldehyde.[23] This work stressed the fact that the signs of electron—carbon hyperfine interaction constants could be obtained from interpretation of the observed CIDNP. This information would be difficult to obtain by esr spectroscopy in this case. They found a clear distinction between σ and π radicals. For σ radicals (e.g., acyl radicals), $A(^{13}C_\alpha) > 0$ and $A(^{13}C_\beta) > 0$ while for π radicals (e.g., benzyl) $A(^{13}C_\alpha) > 0$ and $A(^{13}C_\beta) > 0$. In addition, for the benzyl radical $A(^{13}C_{ortho}) > 0$, $A(^{13}C_{meta}) < 0$ and $A(^{13}C_{para}) > 0$. These findings provided experimental proof for alternating spin densities predicted by theory.

The photolysis of di-t-butylketone (**34**) in CCl_4 was found to lead to reaction from both singlet and triplet states.[24] Recombination of triplet pair **35** generates polarized starting ketone. This pathway was established by virtue of the emission signal from ketone **34** (Figure 6.5). Triplet pair **35** may disproportionate producing aldehyde **36** and olefin **37**. Escape of radicals from **35** leads to abstraction products **38** and **39**. Aldehyde **36** and acid chloride **38** are formed uniquely from triplet pair **35**. Olefin **37** and chloride **39** may have other origins. Triplet pair **35** may undergo pair substitution to give triplet pair **40**. A singlet exciplex may give singlet pair **40**. Production of olefin **37**, chloride **39**, trichloro compound **41**, and $CHCl_3$ will give the same polarization signs if they come from triplet pair **40** or singlet pair **40**. Thus, a definite mechanistic choice was not possible for these products (see below).

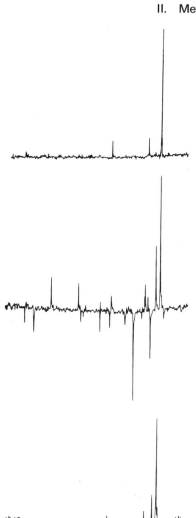

Figure 6.5
^{13}C FT spectra of a 25% solution of di-*t*-butyl ketone in CCl$_4$ obtained before (top), during (middle) and after (bottom) irradiation. From Ref. 24.

200   100   0

δ_c

^{13}C, ^{1}H, and ^{19}F CIDNP signals have been reported during irradiation of α,α,α-trifluoroacetophenone (**42**) in acetonitrile solvent with substituted benzenes as triplet quenchers[25] (see Figure 6.6). The CIDNP signals do not appear unless a trace amount of acid is present. In this photoreaction, starting material is not consumed and no products are formed. The acid-catalyzed CIDNP results from spin polarization in protonated triplet

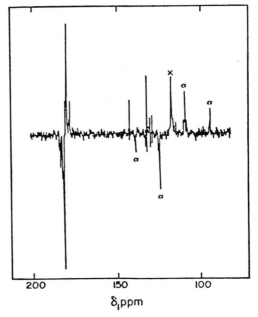

Figure 6.6
^{13}C FT spectrum observed after 5400 pulses during continuous irradiation of a degassed acetonitrile solution 0.1 M in p-chloro-α-trifluoroacetophenone, 0.065 M 1,4-dimethoxybenzene, and $\approx 0.3\ M$ in acetic acid. Peaks marked α represent CF$_3$ quartet; peak marked x is a solvent peak; absorbances between δ 130–150 are aromatic carbons. From Ref. 25.

ketone-aromatic exciplex **43**. Polarized ground state ketone is regenerated from **43** by in-cage electron transfer followed by rapid deprotonation. In the absence of acid, in-cage polarization is cancelled by rapid out-of-cage degenerate electron transfer between free polarized ketyls and ground-state ketone.

$$\text{ArCOCF}_3 + \text{Q} \longrightarrow {}^3(\text{ArCOCF}_3 \dots \text{Q}) \xrightarrow{\text{H}^+} (\text{Ar}\overset{\text{OH}}{\underset{\cdot}{\text{C}}}\text{—CF}_3 + \cdot\text{Q})^+$$

42   **43**

D. Rearrangements

The observation of ^{13}C CIDNP signals has elucidated the mechanism of thermal 1,3 rearrangements of oxime thionocarbamates.[26] Thermolysis of

44 at 85°C produced rearranged product **46** with the indicated polarized carbons. The polarization signs indicate a geminate recombination product-forming step directly from radical pair **45**. Escape product **47** does not exhibit CIDNP presumably because of the long lifetime of free iminyl radicals. (This is consistent with detection of iminyl radicals by esr). The authors contend that the polarization is lost by relaxation processes during this relatively long lifetime.

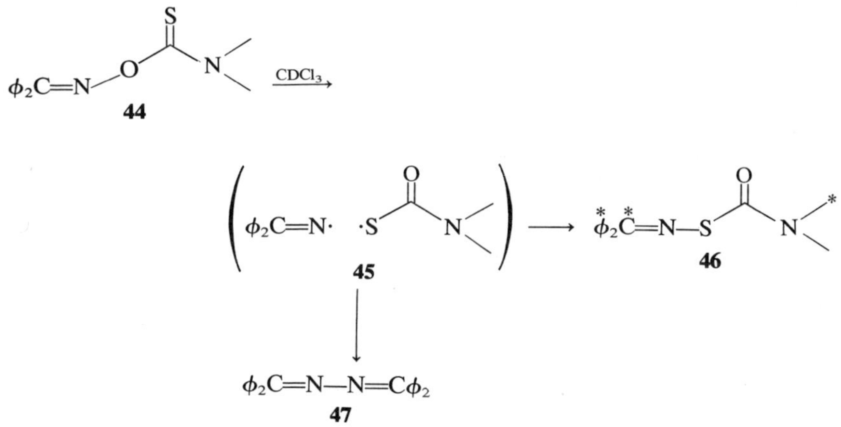

III. KINETIC ASPECTS OF ^{13}C CIDNP

A. Planning the Experiment

1. Quantitative ^{13}C NMR

CIDNP has been observed in ^{13}C CW experiments, but only in a few thermally induced reactions. The increased sensitivity of FT nmr offers the possibility of studying ^{13}C CIDNP reactions induced at low temperatures or those triggered photochemically. However, FT nmr, by its nature, requires extreme care in planning, conducting and analyzing CIDNP experiments. This is true of both mechanistic and kinetic CIDNP FT studies.

To extract kinetic data from a CIDNP experiment, quantitatively valid data must be obtained. For ^{13}C FT this is no simple matter. Only with careful attention to experimental and procedural details can conditions be worked out for good quantitative results.

The measured intensities of ^{13}C nmr signals are not always proportional to the number of carbon atoms contributing to them. Three factors bearing on this are (a) long ^{13}C spin-lattice relaxation times, (b) variable nuclear Overhauser effects (NOEs), and (c) nonuniform pulse power or instrument response across the spectral sweep width. The last factor can be overcome by proper design of the instrument. It is to be hoped that commercial instrumentation at least has such shortcomings minimized. The careful investigator must calibrate the spectrometer response and periodically check to make sure that instrument performance has not degraded.

Shoolery has discussed two methods commonly used for quantitating ^{13}C spectra.[27] These are (a) use of a paramagnetic relaxation agent to shorten T_1s and quench NOE, and (b) use of long pulse intervals (to assure complete relaxation of all ^{13}C nuclei) and gated decoupling (to minimize NOE effects).

Neither of these methods is suitable for a ^{13}C CIDNP kinetic experiment. Relaxation reagents distort or eliminate the intensity of the CIDNP signal. If signal averaging is required, long pulse intervals and the associated long collection times may mask short-term changes in CIDNP intensities.

A third approach for quantitative ^{13}C nmr is to ignore differences in T_1s and NOE factors; samples are run under the same conditions and include a known amount of a standard compound. Weighting factors are computed from standard solutions and applied to the solutions being analyzed.[28,29]

We have found this last method useful for monitoring concentrations of starting materials and products during the CIDNP reaction. Although the CIDNP signals are described by enhancement factors and not concentration factors, the internal standard provides a reference against which the relative intensity of a CIDNP peak can be measured during the course of the experiment. An ideal internal standard should be inert and have a minimum of ^{13}C signals, which are in the vicinity of the signals of interest.

There are two other questions that must be discussed before proceeding. The first is, "Are nuclear spin lattice relaxation times affected during a free radical reaction?" Christensen et al,[30] found no decreases in ^{13}C T_1s for various products during thermolysis of benzoyl peroxide. They estimated a radical flux of approximately $10^{-6} M$ in their experiments. They concluded that changes in T_1s may occur at higher radical concentrations, or for nuclei that are more sensitive to medium effects, such as protons or fluorine.

The second question concerns possible distortions of FT nmr spectra during CIDNP. Ernst and co-workers[31,32] have considered FT nmr of nonequilibrium systems. In systems with no homonuclear spin coupling, such as natural abundance ^{13}C, (the subject of this chapter), FT spectra are equivalent to those obtained in a CW experiment.

2. FT Parameters

Two factors present in FT nmr but not in CW nmr are the pulse interval (or pulse repetition time), τ and the flip angle (or pulse width), θ. These factors have been discussed fully elsewhere,[33] but it is important to review their effects in the context of the CIDNP experiment.

Pulse Interval (τ). The excess magnetization occurring via CIDNP is "unnatural," i.e., non-Boltzmann. The system will relax to equilibrium by spin-lattice processes characterized by the spin-lattice relaxation time, T_1. Thus, in the interval between pulses some CIDNP signal is lost due to relaxation. If τ is very short, significant new polarization may not have been generated. In a photolysis, where a steady-state concentration of the polarized species may eventually exist, setting τ too short may not allow the steady state to be reached.

Consider the case of two consecutive first-order reactions. For CIDNP these would be the formation of polarized intermediate, with rate constant k, and its relaxation, with rate constant $k_r (\equiv 1/T_1)$. The maximum concentration of polarized intermediate and hence the maximum CIDNP signal is obtained when $\tau = \tau_{max}$: $\tau_{max} = (k_r - k)^{-1} \, ln(k_r/k)$.[34] For ^{13}C T_1s of 10–100 sec and ks of 10^{-2}–10^{-4} sec^{-1}, τ_{max} ranges from tens of seconds to several hundred seconds. In our photolysis experiments, we have found that τ's of 10–20 sec give optimum results.

Flip Angle (θ). The flip angle determines how much signal will be measured at each pulse, i.e., the measurable signal is $M \sin \theta$. At low θ, $< 15°$, the system is not greatly perturbed, and the data may be approximated by a CW treatment. At $\theta = 90°$, all of the magnetization aligned along the Z axis is rotated into the XY plane. Thus, whatever excess magnetization due to CIDNP is present at the time of the 90° sampling pulse is irrevocably lost. Only the magnetization due to the "normal" product will be restored by natural T_1 processes during the pulse interval. The implications of θ in the kinetic analysis are covered in Section IIIB.

3. Kinetic Model

In order that rate constants may be extracted from CIDNP intensity-versus-time data, a rate expression must be set up from a kinetic model. From an analysis of the CIDNP data, i.e., the location and sense of the CIDNP peaks, one first proposes a reaction mechanism and hopes to generate a tractable set of rate expressions. Some reaction systems are so complex that the rate expressions written for various products may not be useful. They may contain combinations of the various rate constants in a variety of sums, products, and quotients that are impossible to separate. If this situation exists, the model may be simplified by eliminating "minor" reaction paths. This, of course, can be assisted by quantitative chemical assay of the products.

One must avoid drawing conclusions about concentrations from CIDNP intensities. CIDNP polarizations for ^{13}C may be so large that strong signals are observed from truly minor reactions products. On the other hand, CIDNP may be observed for a major product, but the route giving CIDNP may not be the only one, or even a primary one, for formation of that product.

An example of a system involving alternative routes to the same product is the photolysis of di-t-butyl ketone in CCl_4.[24] The ^{13}C CIDNP spectrum is quite rich and has been interpreted in terms of the possible reactions shown previously (see Section IIC). One of the products, t-butyl chloride can be formed by three paths. Analysis of these paths with the Kaptein equation shows all three to lead to the same expected polarization; enhanced absorption for carbon 1 and emission for carbon 2. Simplification by eliminating either the singlet or triplet path is not possible.

The rate constant expressions for the various polarized intermediates in the DTBK system, can be obtained in a complicated, but analytical, form. Kinetic analysis of the DTBK ^{13}C CIDNP results proved to be enigmatic, however. The CIDNP kinetic expressions (see below) for the 90° pulse case predict that the maximum CIDNP intensity will develop during the first collection period. This was not observed for DTBK photolysis.[35] A mechanism, perhaps sensitization brought about by a photoproduct, generates greater CIDNP at an intermediate time during the photolysis. The "late" maximum was confirmed in several replicate runs; however, the unknown mechanism is apparently extremely dependent on experimental conditions. Consequently, the scatter in measured polarizations was too large to allow a definitive kinetic analysis of this system to be made.

This example illustrates the complexity one may face trying to make kinetic measurements from CIDNP data, even in a CW mode. It seems clear that the first step should be to determine if the mechanism proposed to explain CIDNP results will allow meaningful kinetic expressions to be written before starting any kinetic studies.

B. Analysis of Data

This section will consider only the simple, first-order reaction $R \xrightarrow{k} P$. Expressions for other types of reactions have been given by Buchachenko.[36] The model is the polarized intermediate approach of Walling and Lepley[6], as shown in Eq. 1. Here, the reactant R forms the polarized product P^* with chemical rate constant k, and P^* forms equilibrium P with spin lattice relaxation rate constant, $k_r = 1/T_1$.

$$R \xrightarrow{k} P^* \xrightarrow{k_r} P \tag{1}$$

Associated with P^* and P are so-called nmr absorption coefficients, α and β, whose quotient, α/β, is called the "enhancement" factor.

There are three ways of measuring the CIDNP occurring in the reaction: the CW experiment and two FT approaches, $\theta = 90°$ and $\theta =$ intermediate angle. Remember that the equations presented here apply only to nuclei that are not involved in homonuclear spin coupling.

For the CW case the exact expression[37] for the measured nmr intensity due to P at any time, I_t, is given by Eq. 2 where R_0 is the initial concentration of R.

$$I_t = \beta R_0 + \frac{R_0}{k_r - k}(\alpha k - \beta k_r)e^{-kt} + \frac{kR_0}{k_r - k}(\beta - \alpha)e^{-k_r t} \tag{2}$$

or

$$I_t = A + Be^{-kt} + Ce^{-k_r t} \tag{3}$$

With the steady-state approximation for P^*, Eq. 4 results.

$$I_t^{ss} = \beta R_0 + \left(\alpha \frac{k}{k_r} R_0 - \beta R_0\right)e^{-kt} \tag{4}$$

$$= A + B'e^{-kt} \tag{5}$$

Standard curve fitting routines will yield the five constants in Eq. 3 if there are sufficient experimental data. Since k and k_r are explicitly determined, and R_0 is known, α and β are easily calculated. Although A, B', and k can be obtained from fitting Eq. 5, it is not possible to determine α or k_r from B'. A separate determination of T_1 is necessary to evaluate α.

$$B' = \alpha \frac{k}{k_r} R_0 - \beta R_0 \tag{6}$$

$$\frac{B' + \beta R_0}{kR_0} = \frac{\alpha}{k_r} = \alpha T_1 \tag{7}$$

But α and β are not necessarily constants. In the CW experiment β is certainly constant, since $I_p \propto [P]$ or $I_p = \beta[P]$. In the FT experiment $I_p = f(\theta, \tau, T_1)[P]$. Thus $\beta = f(\theta, \tau, T_1)$ can can be evaluated for a *given set* of FT conditions. The true significance of α is more difficult to see. It indicates the perturbation of the nuclear spin levels, which depends on more than reaction kinetics alone. In a reaction where a polarized product can form by more than one path will αs be the same for each path? It seems unlikely. The chief value of CIDNP kinetics is in measuring chemical-rate constants for reactions that would otherwise be difficult to study.

Just as in the CW case, in the FT experiment the intensity measured at each pulse consists of a contribution from relaxed and polarized product.

$$I_n = \alpha[P^*] + \beta[P] \tag{8}$$

For $\theta = 90°$, the amount of P^* is only that which has formed during τ since the preceding pulse, less that which has relaxed to P in the same interval. Remember that for $\theta = 90°$, all of M_Z is "destroyed" by the pulse, and the P^* present before the pulse can only be restored to a natural (Boltzmann) equilibrium. So for $\theta = 90°$, the intensity measured at pulse n is[37]

$$I_n = \frac{\alpha k}{k_r - k} R_0 e^{-k(n-1)\tau}(e^{-k\tau} - e^{-k_r\tau}) + \beta[R_0 - R_0 e^{-k(n-1)\tau}](1 - e^{-\tau/T_1}) \tag{9}$$

Equation 9 is a strict application of consecutive two-step first-order reaction kinetics to the events occurring during the pulse interval just before pulse n. During the pulse interval, τ, P^* has to be formed from scratch, from reactant that is present at the time of the previous pulse, $R_0 e^{-k(n-1)\tau}$ where $(n-1)\tau$ is the total time from the start of reaction. The contribution by P is only from P which has been formed up to pulse $n-1$ and has relaxed to equilibrium during τ. The contribution to P from relaxation of P^* during τ is neglected.

When θ is not 90° the bookkeeping becomes tricky. The concentration of P^* does not go to zero with each pulse, but a portion $(1 - \cos \theta)$ does "disappear." The amount remaining in the polarized state after the pulse is proportional to $\cos \theta$. This will relax during τ, contributing to the intensity of ground state product, P, at the next pulse.

The concentration of P^* at pulse n is the amount of P^* formed during $n\tau$, less the amount sampled or "destroyed" during the previous $n-1$ pulses, less the amount that has undergone T_1 relaxation during each of n pulse intervals. The concentrations of P^* just before pulse n is represented by the summation[37] in Eq. 10 where R' is a constant given in Eq. 11.

$$[P^*]_{n_-} = \sum_{j=0}^{j=n-1} R' e^{-k(n-1-j)\tau} e^{-k,j\tau}(\cos^j \theta) \tag{10}$$

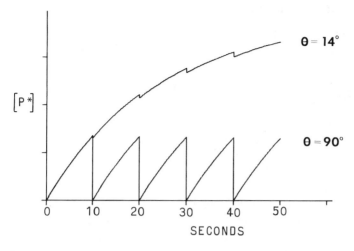

Figure 6.7.
Calculated time dependence of $[P^*]$ during early stages of a hypothetical reaction, $(R \xrightarrow{k} P^* \xrightarrow{k_r} P)$ under the influence of the FT nmr experiment. The curves are for two different values of the flip angle, θ, and the following other parameters: $\tau = 10$ sec, $k = 10^{-4}$ sec^{-1}, $k_r = 4 \times 10^{-2}$ sec^{-1} ($T_1 = 25$ sec), $R_0 = 1$ M. From Ref. 37.

$$R' = \frac{kR_0}{k_r - k}(e^{-k\tau} - e^{-k_r\tau}) \tag{11}$$

Figure 6.7 shows the behavior of $[P^*]$ during the early stages of a hypothetical reaction for the cases $\theta = 90°$ and $\theta = 14°$.

The expression for the concentration of P just before pulse n is given[37] in

$$[P]_{n_-} = \{R_0 - [R]_{n-1} - [P^*]_{(n-1)_-} + (1 - \cos \theta)[P^*]_{(n-1)_-}\}(1 - e^{-\tau/T_1})$$
$$+ (e^{-k_r\tau})[P^*]_{(n-1)_+} \tag{12}$$

Finally, the intensity measured at pulse n is

$$I_n = (\alpha[P^*]_{n_-} + \beta[P]_{n_-}) \sin \theta \tag{13}$$

Although they are fairly complex, Eqs. 10 and 12 can be evaluated if the kinetic parameters are known or assumed. However, the use of these expressions in Eq. 13 to fit experimental intensities in order to obtain the rate constant appears totally hopeless. If θ is small enough, $<15°$, then the results might be analyzed with the CW expression, Eq. 3.

C. Accumulation of Signals

A further complication arises when it is necessary to accumulate signals in order to obtain a useful spectrum. The accumulation can be carried out in two ways, in series or in parallel.

1. FT Parallel Method

In the parallel method, the FID obtained for each pulse during the experiment is individually stored in a memory device, such as a disk. The experiment is repeated N times and the FIDs of corresponding pulses are added. Data collected in this manner can be analyzed with the equations already described; Eq. 3 if $\theta < 15°$, Eq. 9 if $\theta = 90°$ and Eq. 13 if $15° < \theta < 90°$.

Parallel accumulation appears to be the ideal way to approach CIDNP kinetics. The two parameters that have to be controlled carefully are time and memory location. However, with the extremely versatile computer controlled systems now being implemented on FT nmr systems, errors in registration of these parameters should be minimal.

A recent reinvestigation of the thermal decomposition of benzoyl peroxide employed the parallel method, although an actual CIDNP kinetic analysis was not performed.[38] In this work the sample tube, preheated to 60°C, was placed in the probe which was set at 110°C. Thirty seconds later the first rf pulse was applied and the FID stored on a floppy disk. The pulse was applied every 60 sec thereafter, with each FID stored separately on the disk. The experiment was stopped after 25 min, a fresh sample was placed in the probe, and the cycle repeated. This time each FID was added to the previous FID obtained for the same time segment. This process was repeated 39 times. Thus, the final data were 25 composite FIDs, each representing a 1-min segment of the reaction. A 30° flip angle was used for these experiments.

Thermal lag is a problem with thermolysis experiments such as this. Significant polarization had already occurred and was waning during the five minutes it took the sample to warm from 60 to 110°C.[38] Similar warmup times were reported by Schulman et al.[39] in their CW study of the system.

2. FT Serial Method

Here the FID signals from N consecutive pulses during a time interval, T_N, are coadded in one block of memory, those from the next N pulses in a different block of memory, and so on.

Only the analysis for the FT case with $\theta = 90°$ will be discussed here. Clearly, the case for $15° < \theta < 90°$ is already so complex that the imposition of accumulation results in a useless exercise. No attempts have yet been made to describe the CW-CAT experiment.

For $\theta = 90°$ the signal accumulated during the ith collection period will be the sum of the NI_ns collected during that period.[37]

$$I_i^{total} = \left\{ \frac{\alpha k}{k_r - k} R_0(e^{-k\tau} - e^{-k_r\tau}) - \beta R_0(1 - e^{-\tau/T_1}) \right\} \sum_n^{n'} e^{-k(n-1)\tau}$$
$$+ N\beta R_0(1 - e^{-\tau/T_1}) \quad (14)$$

Here $n = (i-1)N + 1$ and $n' = iN$. Note that the time from start of the reaction, $(n-1)\tau$, appears in the summed exponential factor. This equation is not in a usable form because it contains N different times. So the summation is replaced with $Ne^{-kg(t)}$, where $g(t)$ is the time from the start of the reaction to the midpoint of the collection period, i. With this approximation Eq. 14 becomes

$$I_i^{total} = C_1 + C_2 e^{-C_3 g(t)} \quad (15)$$

where

$$C_1 = N\beta R_0(1 - e^{-\tau/T_1}) \quad (16a)$$

$$C_2 = NR_0 \left\{ \frac{\alpha k}{k_r - k} (e^{-k\tau} - e^{-k_r\tau}) - \beta(1 - e^{-\tau/T_1}) \right\} \quad (16b)$$

$$C_3 = k \quad (16c)$$

This approximation has been shown to be quite good for a particular set of trial parameters, but it has not been tested extensively and should be used with caution.

The earliest application of FT to ^{13}C CIDNP concerned the reaction of diazonium salts with base.[40] These reactions are complete in ~ 100 sec, and the spectra were accumulated during 30-sec segments. Another early report[41] used a pulse interval of 1.1 sec and a flip angle of 40°. Here, four 10-sec segments of 9 accumulations each were recorded during each run. The use of a 40° flip angle may invalidate, for kinetic analytical purposes, the intensity versus time curves obtained in that work. A similar approach was used in a study of the photolysis of benzoyl peroxide, with a 10-sec-pulse interval and a 90° flip angle.[19] This was the first effort to analyze FT CIDNP kinetics and will be discussed below.

3. CW Method

^{13}C CIDNP has been observed in the CW mode only for thermally induced reactions. For example, in the thermolysis of benzoyl peroxide in tetrachloroethane, Schulman et al.[39] could detect strong ^{13}C CIDNP peaks in a single scan. To observe weaker peaks, however, accumulations of 25–35 scans were required. No kinetic analysis was reported in this work. Lippmaa and coworkers[11b] studied the thermal decomposition of diazoaminobenzene as a function of concentration, temperature, and solvent. From the CW CIDNP intensity/time profile under normal spectrometer rf levels (-40 dB) they obtained ^{13}C T_1 values. From the CIDNP intensity-time profile run under saturating rf levels (0 dB) they measured the reaction ' rate constant. The changes in the CIDNP intensity/time profiles observed with changes in temperature and concentration were not discussed.

D. Example

Equation 15 was used to analyze CIDNP occurring during the photo-decomposition of benzoyl peroxide in chloroform.[19] A typical accumulated spectrum for an early stage of the reaction is given in Figure 6.8. The pulse angle was 90°, τ was 10 sec and 64 successive FIDs were collected during each interval. CIDNP was most prominent for benzene and for the C-1' carbon of phenyl benzoate. The fits of the data for these two cases are shown in Figures 6.9 and 6.10. The values for the rate constant for photolytic decomposition of benzoyl peroxide come immediately from the curve fitting ($k = C_3$ of Eq. 15). They agreed quite well with those obtained from a parallel product analysis of the same solutions. Because the kinetic scheme for the benzoyl peroxide was more complex than the one used to derive Eq. 15, the actual forms of C_1 and

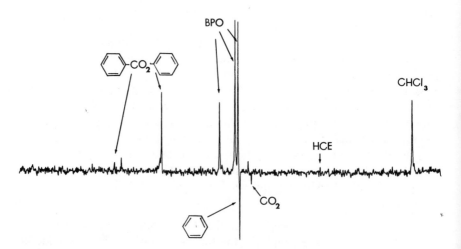

Figure 6.8
^{13}C CIDNP FT spectrum obtained during the photolysis of benzoyl peroxide in chloroform. From Ref. 19.

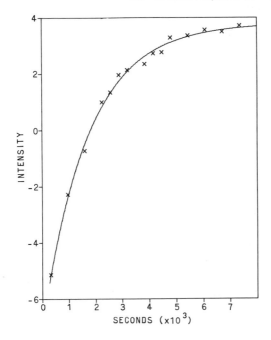

Figure 6.9
Relative measured intensity of the ^{13}C nmr peak of benzene obtained via accumulative FT nmr during the photodecomposition of benzoyl peroxide in chloroform. Each x represents the signal accumulated from 64 free induction decays ($\tau = 10$ sec) and is plotted at the midpoint of the collection period. The line represents the computer fit of these points to Eq. 15. From Ref. 19.

C_2 differ a bit from those given in Eq. 16 and 17, but they still allowed evaluation of α and β.

Figures 6.9 and 6.10 show maximum CIDNP intensity during the initial stages of the reaction. For a 90° flip angle this is predicted for primary products, as inspection of Figure 6.7 shows.[42] For the consecutive first-order reaction scheme $[P^*]$ is proportional to R_0 as shown in Eq. 17–19.

$$[P^*]_n = \frac{k}{k_r - k}[R]_{(n-1)}(e^{-k\tau} - e^{-k_r\tau}) \tag{17}$$

$$[R]_{(n-1)} = R_0 e^{-k(n-1)\tau} \tag{18}$$

$$[P^*]_1 \propto R_0 \tag{19a}$$

$$[P^*]_2 \propto R_0 e^{-k\tau} \tag{19b}$$

$$[P^*]_n \propto R_0 e^{-k(n-1)\tau} \tag{19c}$$

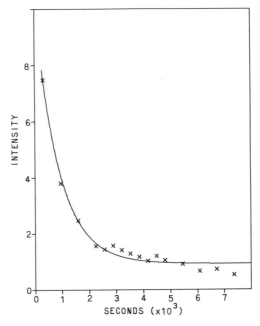

Figure 6.10
Relative measured intensity of the ^{13}C nmr peak of C-1′ carbon of phenyl benzoate obtained via accumulative FT nmr during the photodecomposition of benzoyl peroxide in chloroform. Each x represents the signal accumulated from 64 free induction decays ($\tau = 10$ sec) and is plotted at the midpoint of the collection period. The line represents the computer fit of these points to Eq. 15. From Ref. 19.

Thus $[P^*]$ is maximum at the first pulse and is smaller with each succeeding pulse. To get maximum intensity with each pulse, τ should equal τ_{max}, where $[P^*]$ is maximum under this kinetic scheme (see discussion in Section III.A.2).

IV. SUMMARY

During the past few years ^{13}C CIDNP has certainly been shown to be an excellent means for studying free radical reactions in solution. A clearer understanding of the requirements of the FT experiment is developing on both mechanistic and kinetic fronts. From a chemical viewpoint the kinetic problem has been solved analytically for the cases where the flip angle is small ($\leq 15°$) or $90°$. Either serial or parallel data accumulation may be utilized but the latter offers easier interpretation and less ambiguity. Questions on how best to run the experiment will depend on the reaction system and the equipment available. The thermal lag problem still has to be solved for kinetic studies of thermally induced reactions. In photochemical reactions the use of pulsed light sources[32] could open a whole new regime of ^{13}C mechanistic studies.

REFERENCES

1. (a) H. D. Roth, *Acc. Chem. Res.* **10,** 85 (1977); (b) J. K. S. Wan and A. J. Elliot, *Acc. Chem. Res.* **10,** 161 (1977); (c) P. G. Frith and K. A. McLauchlan, in *Nuclear Magnetic Resonance*, Vol. 3, Burlington House, London, 1974, Chap. 12; (d) A. R. Lepley and G. L. Closs, Ed., *Chemically Induced Magnetic Polarization*, Wiley, New York, 1973; (e) H. D. Roth, *Mol. Photochem.* **5,** 91 (1973); (f) H. R. Ward, *Acc. Chem. Res.* **5,** 18 (1972); (g) R. G. Lawler, *Acc. Chem. Res.* **5,** 25 (1972); (h) J. Potenza, *Adv. Mol. Relaxation Processes* **4,** 229 (1972); (i) S. H. Pine, *J. Chem. Educ.* **49,** 664 (1972); (j) G. L. Closs, *Spec. Lect. XXIIIrd Int. Congr. Pure Appl. Chem.* **4,** 19 (1971); (k) H. Fischer, *Fortschr. Chem. Fortsch.* **24,** 1 (1971); (l) H. Iwamura, *J. Syn. Org. Chem. Jap.* **29,** 15 (1971).

2. G. L. Closs and A. D. Trifunac, *J. Am. Chem. Soc.* **92,** 2183 (1970); G. L. Closs, *J. Am. Chem. Soc.* **91,** 4552 (1969); G. L. Closs and A. D. Trifunac, *J. Am. Chem. Soc.* **91,** 4554 (1969).

3. R. Kaptein, *J. Am. Chem. Soc.* **94,** 6251, 6262 (1972); R. Kaptein and J. A. denHollander, *J. Am. Chem. Soc.* **94,** 6269 (1972); R. Kaptein, J. Brokken-Zijp, and F. J. J. deKanter, *J. Am. Chem. Soc.* **94,** 6280 (1972); R. Kaptein and L. J. Oosterhoff, *Chem. Phys. Lett.* **4,** 195, 214 (1969).

4. F. J. Adrian, *J. Chem. Phys.* **54,** 3912, 3918 (1971); *Chem. Phys. Lett.* **10,** 70 (1971); *J. Chem. Phys.* **53,** 3374 (1970).

5. R. Kaptein, *Chem. Commun.* 732 (1971).

6. C. Walling and A. R. Lepley, *J. Am. Chem. Soc.* **94,** 2007 (1972).

7. W. B. Moniz and C. F. Poranski, Jr., *J. Am. Chem. Soc.* **97,** 3258 (1975).

8. A. D. Trifunac, K. W. Johnson, and R. H. Lowers, *J. Am. Chem. Soc.* **98,** 6067 (1976).

9. E. Lippmaa, T. Pehk, A. L. Buchachenko, and S. V. Rykov, *Chem. Phys. Lett.* **5**, 521 (1970).

10. For a review of the early ^{13}C CIDNP work see J. B. Stothers, in *Topics in Carbon-13 NMR Spectroscopy*, Vol. 1, G. C. Levy, Ed., Wiley-Interscience, New York, 1973, pp. 274ff.

11. (a) E. Lippmaa, T. Pehk, and T. Saluvere, *Ind. Chim. Belg.* **36**, 1070 (1971); (b) E. Lippmaa, T. Saluvere, T. Pehk, and A. Olivson, *Org. Mag. Res.* **5**, 429 (1973).

12. K. Albert, K. M. Dangel, A. Rieker, H. Iawmura, and Y. Imahashi, *Bull. Chem. Soc. Jap.* **49**, 2537 (1976).

13. (a) H. Iwamura and Y. Imahashi, *Tetrahedron Lett.* 1401 (1975); (b) H. Iwamura, Y. Imahashi, K. Kushida, K. Aoki, and S. Satoh, *Bull. Chem. Soc. Jap.* **49**, 1690 (1976).

14. H. Iwamura, Y. Imahashi, and K. Kushida, *J. Am. Chem. Soc.* **96**, 921 (1974),

15. H. Iwamura and Y. Imahashi, *Chem. Lett.* 357 (1976).

16. W. B. Moniz, S. A. Sojka, C. F. Poranski, Jr., and D. L. Birkle, *J. Am. Chem. Soc.*, in press.

17. For a preliminary report see S. A. Sojka, C. F. Poranski, Jr., and W. B. Moniz, *J. Am. Chem. Soc.* **97**, 5953 (1975).

18. S. A. Sojka, C. F. Poranski, Jr., and W. B. Moniz, *J. Mag. Res.* **23**, 417 (1976).

19. C. F. Poranski, Jr., W. B. Moniz, and S. A. Sojka, *J. Am. Chem. Soc.* **97**, 4275 (1975).

20. A. V. Kessenikh, P. V. Petrovskii, and S. V. Rykov, *Org. Mag. Res.* **5**, 227 (1973).

21. R. Kaptein, R. Freeman, H. D. W. Hill, and J. Bargon, *Chem. Commun.* 953 (1973).

22. W. Nakanishi, S. Koike, M. Inoue, and Y. Ikeda, *Tetrahedron Lett.* 81 (1977).

23. R. Benn and H. Dreeskamp, *Z. Phys. Chem. Neue Folge* **101**, 11 (1976).

24. W. B. Moniz, C. F. Poranski, Jr., and S. A. Sojka, *J. Org. Chem.* **40**, 2946 (1975).

25. M. J. Thomas and P. J. Wagner, *J. Am. Chem. Soc.* **99**, 3845 (1977).

26. C. Brown, R. F. Hudson, and A. J. Lawson, *J. Am. Chem. Soc.* **95**, 6500 (1973).

27. J. N. Shoolery, *Prog. NMR Spectrosc.* **11**, 79 (1977).

28. J. W. Blunt and M. H. G. Munro, *Aust. J. Chem.* **29**, 975 (1976).

29. M. L. Jozefowicz, I. K. O'Neill, and H. J. Prosser, *Anal. Chem.* **49,** 1140 (1977).

30. K. A. Christensen, D. M. Grant, E. M. Schulman, and C. Walling, *J. Phys. Chem.* **78,** 1971 (1974).

31. S. Schäublin, A. Höhener, and R. R. Ernst, *J. Mag. Res.* **13,** 196 (1974).

32. S. Schäublin, A. Wokaun, and R. R. Ernst, *J. Mag. Res.* **27,** 273 (1977).

33. P. Meakin and J. P. Jesson, *J. Mag. Res.* **13,** 354 (1974); B. W. Goodwin and R. Wallace, *J. Mag. Res.* **9,** 280 (1973); R. R. Ernst and R. E. Morgan, *Mol. Phys.* **26,** 49 (1973); D. E. Jones and H. Sternlicht, *J. Mag. Res.* **6,** 167 (1972); R. Freeman and H. D. W. Hill, *J. Mag. Res.* **4,** 366 (1971); P. Waldstein and W. E. Wallace, Jr., *Rev. Sci. Instrum.* **42,** 437 (1971).

34. K. B. Wiberg, *Physical Organic Chemistry*, Wiley, New York, 1964, p. 323.

35. W. B. Moniz, S. A. Sojka, and C. F. Poranski, Jr., unpublished results.

36. A. L. Buchachenko and Sh. A. Markarian, *Int. J. Chem. Kinet.* **4,** 513 (1972).

37. C. F. Poranski, Jr., S. A. Sojka, and W. B. Moniz, *J. Am. Chem. Soc.* **98,** 1337 (1976).

38. Y. Yokoyama, M. Arai, and A. Nishioka, *Nippon Kagaku Kaishi*, 238 (1977).

39. E. M. Schulman, R. D. Bertrand, D. M. Grant, A. R. Lepley, and C. Walling, *J. Am. Chem. Soc.* **94,** 5972 (1972).

40. S. Berger, S. Hauff, P. Niederer, and A. Rieker, *Tetrahedron Lett.* 2581 (1972).

41. H. Iwamura, M. Iwamura, M. Imanari, and M. Takeuchi, *Bull. Chem. Soc. Jap.* **46,** 3486 (1973).

42. This predicted behavior is not always found. See Kinetic Model Section.

INDEX

391